高等学校

# 研究生英语系列

# 科技英语阅读与翻译

## EST Reading and Translation

应用类

◎ 主编：李 健

编者：罗凌志 朱红梅 李 芝 王一娜

外语教学与研究出版社
FOREIGN LANGUAGE TEACHING AND RESEARCH PRESS
北京 BEIJING

**图书在版编目（CIP）数据**

科技英语阅读与翻译 = EST Reading and Translation / 李健主编 ；罗凌志等编．-- 北京 ：外语教学与研究出版社，2020.12（2022.9 重印）
高等学校研究生英语拓展系列．应用类
ISBN 978-7-5213-2360-3

Ⅰ．①科… Ⅱ．①李… ②罗… Ⅲ．①科学技术－英语－阅读教学－研究生－教材②科学技术－英语－翻译－研究生－教材 Ⅳ．①N43

中国版本图书馆 CIP 数据核字（2021）第 004571 号

出 版 人　王　芳
项目负责　张荣婕
责任编辑　张荣婕
责任校对　徐　洋
封面设计　牛茜茜
版式设计　平　原
出版发行　外语教学与研究出版社
社　　址　北京市西三环北路 19 号（100089）
网　　址　http://www.fltrp.com
印　　刷　北京盛通印刷股份有限公司
开　　本　787×1092　1/16
印　　张　21
版　　次　2021 年 3 月第 1 版　2022 年 9 月第 5 次印刷
书　　号　ISBN 978-7-5213-2360-3
定　　价　49.90 元

购书咨询：（010）88819926　电子邮箱：club@fltrp.com
外研书店：https://waiyants.tmall.com
凡印刷、装订质量问题，请联系我社印制部
联系电话：（010）61207896　电子邮箱：zhijian@fltrp.com

物料号：323600001

# Foreword

继2007年编写并出版了供大学生使用的《科技英语阅读》(外语教学与研究出版社，普通高等教育“十一五”国家级规划教材）后，今年我们又为研究生学习科技英语阅读与翻译编写了这本教材，这两本教材可以说是“姊妹篇”。

本教材的编写宗旨是满足研究生学习科技英语阅读和翻译的需要，通过各种练习的设计和安排，将科技英语阅读和翻译两项技能的学习融为一体，互相补充，互相促进。对于研究生来说，科技英语的学习已不仅仅是了解语言特点、熟悉科技词汇的使用和一般科技文章的翻译，而是要针对科研工作的需要，以语言的实际应用为目标，掌握更高层次的技能，包括灵活运用各种阅读技巧、准确深入理解文章内容、熟悉各种学术出版物的内容和体系、掌握科技英语翻译方法、了解英汉科技语言表达方面的差异、准确地进行翻译等。

本教材共12章，每章由“科技文章阅读与翻译”和“学术阅读”两大部分组成。科技文章节选自著名科技期刊《科学美国人》(*Scientific American*) 近期发表的论文，以及普林斯顿大学出版社近年出版的专著节选，内容涉及自然、物理、计算机、数学、气象、环境、医学、太空探索、农业技术、机器人等多个领域。

“科技文章阅读与翻译”的导读部分包括文章背景、主要内容以及作者的介绍，是阅读的“热身”准备阶段。为了帮助学生准确理解课文内容，每篇文章均附有词汇表和注释。阅读练习的设计依据实际应用的原则，依次分为对整个篇章的理解能力、对具体细节的查阅能力以及对语言内在含义深入了解的能力，目的是从不同层次和角度提高学生的阅读能力。翻译练习则是以阅读为基础展开，因为只有准确地理解文章内容才能谈得上准确、忠实、通顺地翻译，因此一部分翻译练习的内容来自阅读材料本身，另一部分练习的内容则与阅读材料的主题相关。此外，在翻译练习前，每章均设有“科技翻译技巧”板块，包括增词、减词、词性转换、从句转换、语序倒置、数字的翻译方法、前缀的翻译方法、科技文章题目的翻译方法等24个专题，分布于全书。每章的练习都有明确的目的，帮助学生逐步提高翻译技能。

“学术阅读”的目的是通过对各种学术资源的介绍，使学生熟悉获取科技信息和资料的途径和方式，其中既有传统意义上的学术专著、期刊、科技论文的篇章结构、百科全书、技术资料等，也有通过现代科技手段互联网提供的信息源，例如电子期刊、数字图书馆、网上信息源等。查阅资料是每个研究生必须具备的能力，因此这部分内容的设计也是与研究生学习紧密相关的。每篇文章后均有词汇表和相应的练习，以帮助阅读理解。教材提供全部练习的答案，作为学习参考。

在每章结尾处，我们还开辟了“著名科学家名言录”栏目，摘选了12位著名科学家有关科学研究和科学家素质的富有哲理的话语，并附上译文，这不仅仅是语言学习，更重要的是通过这些名言领会科学家的思想和精神，领悟科学研究的真谛。

本教材虽然是为研究生编写的，但也可作为英语专业高年级科技英语课程、具有一定英语水平的科技工作者以及英语爱好者学习和进修的教材。

我们衷心希望通过本教材的学习，同学们不仅能够提高科技英语阅读及翻译的能力和水平，而且能够触类旁通，认识和掌握科技英语阅读和翻译的规律和特点，逐步形成完整的知识体系，在今后的学习和科研工作中能够以此为基础，不断丰富和扩展这一体系，继续提高使用科技英语的能力。

本教材如有不当之处，诚请指正。

李　健

2009年5月

# Contents

# Chapter 1 Natural History Surveys and Biodiversity
# 自然史调查与生物多样性

## Section 1 Reading and Translation

### I What You Are Going to Read

We humans share Earth with 1.4 million known species and millions more species that are still unrecorded. Yet we know surprisingly little about the practical work that produced the vast inventory we have to date of our fellow creatures. How were these multitudinous creatures collected, recorded, and named? When, and by whom?

Here a distinguished historian of science tells the story of the modern discovery of biodiversity. Robert E. Kohler in his book *All Creatures: Naturalists, Collectors, and Biodiversity, 1850–1950* (Published by Princeton University Press, 2006) argues that the work began in the mid-eighteenth century and culminated around 1900, when collecting and inventory were organized on a grand scale in natural history surveys. Supported by governments, museums, and universities, biologists launched hundreds of collecting expeditions to every corner of the world.

What you are going to read in this section is an excerpt of the first chapter of the book.

### II About the Author

Robert E. Kohler is Professor Emeritus of History and Sociology of Science at the University of Pennsylvania. The recipient of an award for lifetime achievement in his field, he is the author of more than 30 papers and 5 books on experimental and field sciences.

### III Reading Passage

#### Naturalists, Collectors, and Biodiversity
#### 博物学家、标本采集者和生物多样性

By Robert E. Kohler

1 We humans are one in a million: to be exact, one species among 1,392,485, according to a recent **tally** by the **zoologist** Edward O. Wilson[1]. Those are the ones we know: estimates of the total number of

tally *n.* 纪录
zoologist *n.* 动物学家

pollinator *n.* 传授花粉的昆虫或动物
vertebrate *n.* 脊椎动物
cartilaginous *adj.* 软骨的
lamprey *n.* 七鳃鳗(又名八目鳗)
amphibia *n.* 两栖动物
invertebrate *n.* 无脊椎动物
tunicate *n.* 被囊类脊索动物
cephalochordate *n.* 头索动物亚门(脊索动物中最接近脊椎动物的类群)
mollusc (=mollusk) *n.* 软体动物(包括牡蛎、贻贝、蜗牛、蛞蝓等)
arthropod *n.* 节肢动物
protozoa *n.* [*pl* of protozoon] 原生动物
inventory *n.* 详细目录
earthling *n.* 居住在地球上的人；凡人
void *n.* 空间
biodiverse *adj.* 生物多样性的

living species range from five to thirty million and up, depending on how one reckons. A substantial majority of Earth's species are insects: something like 751,000 by Wilson's tally. Plants account for another 248,428, the vast majority being flowering plants (which coevolved with insect **pollinators**). Among the **vertebrates**, bony fishes are the largest group, with 18,150 species, leaving aside the 63 species of jawless fishes and the 843 **cartilaginous** fishes (**lampreys**, sharks). **Amphibia** and reptiles account for 4,184 and 6,300 more species; birds for 9,040, and mammals for 4,000, give or take. Not to mention **invertebrates** other than insects: **tunicates** and **cephalochordates** (1,273), **molluscs** (roughly 50,000), and **arthropods** (12,161). And single-cell organisms: algae (26,900), fungi (46,983), **protozoa** and microbes (36,560). Of our fellow vertebrates we have an **inventory** that is nearly complete—over 90 percent, it is estimated. On the plants and invertebrates, however, we may only have made a start. We **earthlings** sail through the **void** on an ark that is impressively **biodiverse**.[2]

conservationist *n.* 环境保护主义者

2 Biodiversity is a lively issue these days, mainly because of the number of species that are going extinct, either by natural causes, or because we space-hungry humans are destroying their habitats. Wilson estimates that perhaps 17,500 species (mostly insects) go extinct each year in tropical forests, and that we humans have accelerated the historical rate of extinction by a factor of one thousand to ten thousand. Biologists and **conservationists** are concerned that vast numbers of species may be forced into extinction ahead of schedule (extinction is the ultimate fate of all species) before they can be found and classified. There is concern, too, that in our ignorance we may be destroying species vital to the fabric of ecosystems on which we depend for our own survival.

taxonomy *n.* 生物分类学
humdrum *adj.* 平淡的；单调的；乏味的

3 Systematic biology, or **taxonomy**, is reputed to be a **humdrum**, cataloging science—a reputation entirely undeserved, let it be said. We depend on those few among us who collect, describe, name, and classify our fellow passengers on the global ark. But how exactly do we find, collect, identify, and order those millions of species? That is my subject here: not the biology or the ethics of biodiversity, but its practices and their history. Though people have always named plants and animals, the science of species inventory is relatively new,

beginning with the big bang of Carl von Linne's[3] invention of the (Linnaean) binomial system of naming[4] in the mid-eighteenth century. And though much has been written on theories of species, relatively little is known of the practical work that produced the empirical base for theorizing. When and how were those inventories created and made **robust**? Who organized and paid for collecting expeditions, collected and prepared specimens in the field, compiled lists, built museums and **herbaria**, and kept vast collections in good physical and conceptual order? Of these practical activities we do not as yet know much. This book is a step toward acquiring such knowledge.

robust *adj.* 有活力的；强健的
herbaria *n.* [*pl* of herbarium] 植物标本集；植物标本室；植物标本盒

4 The history of our knowledge of biodiversity is first and foremost a history of collecting and collections. Remarkably little has been written about the craft and social history of scientific collecting: it remains a "black box", as the historian Martin Rudwick[5] observed a few years ago, an activity that has "barely been described by historians, let alone analyzed adequately". There are now signs of a growing interest in the history of collecting science, but it is perhaps understandable why this black box is only now being opened. Although collecting is a widespread and varied **obsession**, modern scientific collecting is sober and businesslike, not irregular or **idiosyncratic**. It is done **en masse** and methodically, because modern taxonomy requires large and comprehensive collections. Scientific collecting is exacting and **quantitative** science, as methodical and organized as taking stock of galaxies, subatomic particles, or genes.[6] Modern specimen collections are quite unlike the romantic "cabinets of curiosities"[7] of earlier centuries. Modern herbaria consist of cases filled with hundreds of thousands of large **folios** of pressed plants in paper. Museum study collections are rooms of metal boxes, each with trays of animal skins and skulls in neat rows neatly labeled—all seemingly humdrum and unromantic.

obsession *n.* 牵挂；着迷；困扰
idiosyncratic *adj.* 特殊的，特质的
en masse [法] 全体，一起
quantitative *adj.* 定量的
folio *n.* 对折纸；对开纸

5 Yet the scientific visions that inspire collectors to go **afield**, and the varied activities that go into making large collections, are anything but humdrum. Collecting is an activity that has engaged diverse sorts of people—unlike laboratory science, which is restricted to a relatively few approved types. The **botanist** Edgar Anderson[8] once did an experiment, in which he took a manila folder at random from an herbarium case (a Southwestern grass, it turned out to be),

afield *adv.* 在野外
botanist *n.* 植物学家

flora *n.*（某地区或时期的）全部植物，植物区系
botany *n.* 植物学

to discover the kinds of people who had collected the specimens. It was an amazingly diverse lot: a botanist on the Mexican Boundary Survey of the early 1850s; an immigrant intellectual German who had come to America in 1848 to escape political persecution; the wife of a mining engineer stationed in a remote mountain range, who dealt with the isolation by studying the local **flora**; a Boston gentleman, who made collecting trips to New Mexico for thirty years; a Los Alamos scientist and amateur botanist; university professors of **botany**; and college students who bought a second auto and spent a summer holiday collecting. "Though they have sometimes been contemptuously referred to as 'taxonomic hay' by other biologists," Anderson concluded, "herbarium specimens can be quite romantic in their own dry way."

replicate *v.* 重复；复制
taxonomist *n.* 生物分类学家
naturalist *n.* 博物学家

6 Anderson's experiment is easily **replicated**: page through museums' accession lists, and you will see hundreds of names of people who contributed specimens to scientific collections, from a few odd skins to tens of thousands. Read **taxonomists**' checklists—which give for each species the name of the naturalist who first described it, and when—and you will glimpse a living community of collectors and **naturalists** stretching back 250 years, in which amateurs have the same honor and dignity as the most eminent professionals. Species collectors are as diverse as the species they collect, and no other community of scientists preserves such a deep sense of its collective identity and past. Taxonomists' elaborate system of keeping track of names, which anchors each species to the name historically first given to it and to the actual specimen first described—the "type" specimen—keeps the past forever present.[9] All sciences have their heroes and founding myths, but taxonomy is about the only one with a living memory of all past contributors, famous and obscure.

gene-sequencing 基因测序
logistics *n.* 后勤

7 Scientific collecting was (and is) also an unusually complex and varied kind of work. Collecting expeditions are more complex socially than anything one might find, say, in a biochemistry or **gene-sequencing** lab. They require a great deal of book knowledge, but also practical skills of woodcraft and **logistics**, as well as firsthand experience of animal habits and habitats. Modern natural history is an exacting science whose practitioners must also cope and improvise in difficult field conditions. Collecting expeditions afford an experience

of nature that mixes scientific and recreational culture in a way that lab sciences never do. Collecting parties usually travel light and depend on local inhabitants for information and support, making survey collecting a diversely social experience. And because of that diversity, the identity of scientific collectors has been less fixed than that of laboratory workers. In the black box of modern expeditionary collecting, there is much of interest.

8 We know nature through work, the environmental historian Richard White[10] has observed, whether it is poling canoes against the current of a great river (his particular case), or building dams across it to tap its energy, or hauling fish out of it, or diverting its waters for irrigated farming—or, historians may add, studying its **hydrology** and natural history. So too is our scientific knowledge of nature acquired through the work of mounting expeditions; observing plant and animal life; and collecting, preparing and sorting specimens. Historians have only recently begun to address the work of field science. And of all the field sciences, natural history survey is an exceptionally inviting subject—because the work of systematic, scientific collecting is so varied.

hydrology *n.* 水文学，水理学

9 One is also struck, paging through scientific inventories of species, by the **lumpiness** of the history of their discovery. Species have accumulated steadily, but more rapidly in certain periods than in others. The first such period of discovery was the Linnaean: roughly the second half of the eighteenth century. Then, after a pause of a few decades in the early nineteenth century, another period of rapid discovery set in from the 1830s to the 1850s, which I shall call "Humboldtian", after the **encyclopedic** author of *Cosmos*, Alexander von Humboldt. Following another pause, the pace of finding and naming again quickened from the 1880s into the 1920s, by which time a substantial proportion of vertebrate species had been found and named. Since the mid-twentieth century the pace of discovery of new vertebrate species has been a fitful trickle (though lists of invertebrates grow ever longer).

lumpiness *n.* 凹凸不平；不连贯
encyclopedic *adj.* 包含各种学科的；学识渊博的；广博的

10 These cycles of collecting and naming vary a good deal from one group of animals to another, depending on their accessibility and interest to us. Those that are large, fierce, **freakish**, beautiful, edible,

freakish *adj.* 不寻常的；反常的；奇怪的

carnivore *n.* 肉食动物
primate *n.* 灵长目动物
nocturnal *adj.*（指生物）夜间活动的
rodent *n.* 啮齿目动物（如鼠、松鼠或海狸）
insectivore *n.* 食虫动物

lovable, or dangerous were inventoried early on. These include birds, **carnivores**, **primates**, and large game. Inconspicuous or insignificant creatures, or those that do not appeal—because they are slimy, cold-blooded, annoying, **nocturnal**, or just very good at avoiding our notice—were not fully inventoried until the surveys of the late nineteenth and early twentieth century or even later. These groups include **rodents**, bats, **insectivores**, amphibians, and reptiles.

blip *n.*（雷达屏幕上的）光点
mammalian *adj.* 哺乳动物的
mammal *n.* 哺乳动物
prolific *adj.*（指动植物等）多产的，多育的
chiroptera *n.*（Greek for "hand-wings"）翼手目动物（如蝙蝠）
periodicity *n.* 周期性；定期性；间歇性
tabulate *v.* 将（事实或数字）列成表，列表显示

11 Birds—those visible, audible, and beloved objects of watchers and collectors—were so well inventoried in the Linnaean and Humboldtian periods that the discoveries of the later survey phase show up as mere **blips** on a declining curve of discovery. In contrast, discoveries of **mammalian** species display the most pronounced cyclic pattern, with marked activity in the first two phases, but the most productive collecting in the survey period. The pattern for North American **mammals** is even more pronounced, with discoveries more concentrated in the 1890s, and the earlier peak shifted from the 1830s and 1840s to the 1850s and 1860s. Different groups of mammals show some variation in this basic pattern. Most carnivore species were described in the eighteenth century, and most of the rest in the 1820s and 1830s—we humans have taken a keen interest in our closest competitors. Rodents, in contrast, were hardly known to Linnaean describers and not fully known to science until the age of survey, when it first became apparent just how **prolific** of species this group has been—it would appear that the Creator loves rodents as well as He does beetles. Insectivores display the same strikingly lumpy pattern of discovery; as do also **chiroptera** (bats), though with a stronger period of discovery in the mid-nineteenth century and a less striking peak in the early twentieth. Discoveries of North American reptiles and amphibians also display this **periodicity**, though less markedly: relatively few were described before 1800, most in the 1850s, with small peaks in the age of survey and after. (Data on world species of these groups is either absent or harder to **tabulate**.)

synchronize *v.*（使……）同步

12 These distinctive periods in the pace of collecting and describing suggest that the process of discovery was not random and individualistic, but that individual efforts were **synchronized** by larger cultural, economic, and social trends. This is not a novel thought. It is a commonplace (and doubtless true, as well) that early modern

naturalists were inspired by the flood of new knowledge that was a by-product of the expanding global reach of European trade and conquest. And we now also know that Linnaean taxonomy grew out of the widespread interest in Enlightenment Europe[11] in state-sponsored agricultural improvement, including schemes for acclimatizating exotic species to northern countries.

13 It is also clear that the early-nineteenth-century flowering of collecting and naming resulted from the greater affordability of transoceanic steam travel and from European imperial expansion and settlement, especially in the rich tropical environments of the southern hemisphere. In North America, naturalists like John James Audubon[12] followed the military frontier into the species-rich environments of the southeastern United States. And the western boundary and transport surveys of the 1850s took naturalists like Spencer Baird[13] into the faunally diverse and virtually unworked areas of the American West. No one has tried to map the historical geography of taxonomic knowledge onto that of imperial expansion and settlement, but I would expect a close correlation. If trade has followed flags, so also have naturalists and collectors. Access was crucial: wherever improved transportation technology and colonial infrastructure afforded ready access to places previously expensive or dangerous to reach, there the pace of discovery of new species will soon pick up.

14 The third of these cycles of collecting—I have without **fanfare** been calling it "survey" collecting—is the least well known and the most surprising. We do not think of the late nineteenth and early twentieth centuries as being a great age of discovery in natural history; but they were. One need only peruse the annual reports of national and civic museums to appreciate the enormous enthusiasm for expeditions and collecting. In the United States alone dozens or scores of collecting expeditions were dispatched each year to the far corners of the world between 1880 and 1930: hundreds in all, or thousands—perhaps as many as in the previous two hundred years of scientific expeditioning. They certainly produced as much knowledge of the world's biodiversity as any of the earlier episodes of organized collecting.

fanfare *n.* 夸耀，鼓吹

15 It was in the age of survey that scientists became fully aware of the world's biodiversity. In places that were explored but not

fauna *n.* （某地区或某时期的）全部动物

intensively worked, like the American West or much of South America, **faunas** and floras that had seemed closed books were reopened and vastly expanded. In its first two years of operation in the western states, the US Biological Survey turned up seventy-one new vertebrate species—an abundance that some zoologists found hard to credit. Inventories of vertebrate animals became so complete that subsequent discoveries of new species became media events.[14] Why, then, has this phase in the discovery of biodiversity remained the least well known?

unpretentious *adj.* 不炫耀的；不夸大的
exotica *n.* 奇特的东西
chronicler *n.* 编年史的编写者
polar *adj.* （南、北）极的；地极的；近地极的

16 One reason is that collecting expeditions were mostly small and **unpretentious**, unlike the grand voyages of imperial exploration. Scientific collecting in the age of survey was accomplished mostly by small parties (three to half a dozen) whose purpose was to send back not **exotica** and accounts of heroic adventure and discovery, but rather crates of specimens. It is the dramatic explorations of the earlier periods that have caught the eye, because they were designed to catch the eye—of investors, princes, publishers, readers, **chroniclers**. It is no accident that the heroic voyaging of eighteenth- and early-nineteenth-century explorers—Cook[15], Vancouver[16], La pérouse[17], Humboldt[18], Bougainville[19], Murchison[20]—is well documented and remembered. Or that historians have dwelt on the feats of American explorers from Lewis[21] and Clark[22] to later ventures like the Harriman Alaska Expedition[23], or the adventures of **polar** explorers, rather than on the more numerous but less flashy modern discoverers of biodiversity. Still, this imbalance needs to be set right, and I hope this book will help do that.

## Notes

1. **Edward O. Wilson:** 爱德华・威尔逊（1929– ），美国生物学家，研究蚁类及群居昆虫的权威。
2. **We earthlings sail through the void on an ark that is impressively biodiverse:** 我们这些生活在地球上的人们乘方舟漂流在天地之间，在这只方舟上有着各种各样的生物。本句运用了比喻的方法，句中的ark指《圣经》中诺亚及其家人和动物为躲避洪水而乘的大船。在洪水到来之前，上帝吩咐诺亚建造一只方舟，要他带上家人和地球上所有动物乘方舟躲避洪水之灾。洪水消退后，这些动物继续在地球上生息繁衍。
3. **Carl von Linne:** 卡尔・冯・林奈（1707–1778），瑞典博物学家，生物分类学奠基人。
4. **binomial system of naming:** 双名命名体系，双名法。由林奈提出的生物命名体系，动

植物和矿物均采用两个拉丁化的名字（拉丁双名）命名。例如植物命名，第一个名代表"属"(genus) 名，第二个名代表"种本名"(specific epithet)。由属名和种本名组合起来构成物种名（species name）。

5. **Martin Rudwick:** 马丁·罗德威克（1932– ），美国圣地亚哥加利福尼亚大学历史学荣誉教授。
6. **Scientific collecting is exacting and quantitative science, as methodical and organized as taking stock of galaxies, subatomic particles, or genes:** 以科学的方法进行物种收集是一门要求极高的定量学科，与观察星系、亚原子粒子或基因一样需要条理性。句中exacting是由动词exact变化而来的形容词，相当于making great demands, requiring great effort, 意思是"苛求的，严格的"。
7. **cabinets of curiosities:** (also known as *wonder-rooms*) 奇异珍品陈列室，用于珍藏自然史、地质、考古、宗教或历史等方面的遗物以及艺术品和古玩等。
8. **Edgar Anderson:** 埃德加·安德森（1897–1969），美国植物学家，他的*Introgressive Hybridization* (1949) 一书对植物遗传学做出了重要贡献。
9. **Taxonomists' elaborate system of keeping track of names, which anchors each species to the name historically first given to it and to the actual specimen first described—the "type" specimen—keeps the past forever present:** 生物分类学家建立了一套精确记录生物名称的体系，该体系将每一个物种与其在历史上的首次命名和最初描绘的标本（即"类型"标本）进行核对，从而将它的历史永久记录在案。句中anchor...to...原意为"把……固定在……上"，在此转义为"将……与……进行核实、查对（加以确认）"。
10. **Richard White:** 理查德·怀特（1947– ），美国历史学家。
11. **Enlightenment Europe:** 欧洲启蒙运动（17–18世纪）。指17–18世纪在欧洲思想界、科学界与哲学界兴起的运动，涉及宗教、哲学、经济、科学、史学、文学、美术等各个方面。启蒙思想家们用知识、科学启迪人们的心智，反对愚昧无知、传统偏见，打破旧的风俗习惯。
12. **John James Audubon:** 约翰·詹姆斯·奥杜邦（1785–1851），美国鸟类学家、博物学家。
13. **Spencer Baird:** 斯潘塞·贝尔德（1823–1887），美国动物学家。
14. **Inventories of vertebrate animals became so complete that subsequent discoveries of new species became media events:** 脊椎动物的目录已经非常完整，如果再发现新的物种就会在社会上引起轰动。本句中的became media event是比喻的用法，意为"成为媒体宣传的大事"。
15. **Cook:** 库克（James Cook, 1728–1779），英国探险家、航海家、制图员。
16. **Vancouver:** 温哥华（George Vancouver, 1758–1798），英国探险家、航海家。
17. **La Pérouse:** 拉普鲁斯（1741–1788?)，法国探险家。
18. **Humboldt:** 洪堡（1769–1859），德国博物学家、探险家。
19. **Bougainville:** 布干维尔（1729–1811），法国探险家。
20. **Murchison:** 穆奇森（1792–1871），英国地质学家。
21. **Lewis:** 路易斯（Meriwether Lewis, 1774–1809），美国探险家。
22. **Clark:** 克拉克（William Clark, 1770–1838），美国探险家。

23. **Harriman Alaska Expedition:** 哈里曼·阿拉斯加探险。1899年，由铁路大亨和金融家哈里曼在阿拉斯加组织的一次探险活动，此后发表了大量科研文章。

## Exercises

### I. Answer the following questions.

1. Why is biodiversity an issue that has attracted so much attention these days?
2. What is the relation between the study of biodiversity and species collection?
3. What is the attitude of the author toward the collectors of species throughout the history?
4. Why does the author call the history of collecting science a "black box"?
5. What questions does the author intend to answer in this book?

### II. The following statements are incomplete. Search the missing information in the passage and fill in the blanks.

1. The greatest number of living things on earth are ___________.
2. According to Wilson, perhaps ___________ species (mostly insects) go extinct each year in tropical forests.
3. The science of species inventory started in the ___________ century.
4. Species collectors usually travel light and depend on ___________ for information and support, making survey collecting a diversely social experience.
5. Species collecting requires not only book knowledge but also ___________ of woodcraft and logistics and ___________ of animal habits and habitats.
6. The cycles of species collecting and naming vary a good deal from one group of animals to another, depending on their ___________ and ___________ to us.
7. It seems that the development of ___________ steam travel promoted species collecting and naming in the early nineteenth century.
8. It was in the age of survey that scientists became fully aware of the world's ___________.

### III. Identify the implied meanings of the underlined parts of the following sentences according to the context of the passage, and translate the sentences into Chinese.

1. We depend on those few among us who collect, describe, name, and classify <u>our fellow passengers on the global ark</u>.
2. "Though they have sometimes been contemptuously referred to as 'taxonomic hay' by other biologists," Anderson concluded, "herbarium specimens can be quite <u>romantic</u> in their own dry way."
3. Remarkably little has been written about the craft and social history of scientific collecting: it remains a "<u>black box</u>".

4. Since the mid-twentieth century the pace of discovery of new vertebrate species has been a fitful trickle (though lists of invertebrates grow ever longer).
5. Rodents, in contrast, were hardly known to Linnaean describers and not fully known to science until the age of survey, when it first became apparent just how prolific of species this group has been—it would appear that the Creator loves rodents as well as He does beetles.
6. In places that were explored but not intensively worked, like the American West or much of South America, faunas and floras that had seemed closed books were reopened and vastly expanded.
7. It is a commonplace (and doubtless true, as well) that early modern naturalists were inspired by the flood of new knowledge that was a by-product of the expanding global reach of European trade and conquest.
8. If trade has followed flags, so also have naturalists and collectors.
9. Inventories of vertebrate animals became so complete that subsequent discoveries of new species became media events.
10. It is the dramatic explorations of the earlier periods that have caught the eye, because they were designed to catch the eye—of investors, princes, publishers, readers, chroniclers.

## Translation Techniques (1)

### Watch out the "Pitfalls" in Technical Translation
### （警惕科技翻译中的"陷阱"）

An English word may have different meanings when it is used in everyday conversations and in scientific papers. This kind of words are just like "pitfalls", which may mislead you and cause inaccuracy or even errors in your translation. For example, if you look up the word *crown* in a general-purpose or non-technical dictionary, it may only provide the meaning as "ornamental head-dress made of gold, jewels, etc. worn by a king or queen on official occasions" (王冠、皇冠). Obviously, this meaning is not appropriate for translating "the *crown* of the tree" because in botany *crown* often means "the upper part of a tree, which includes the branches and leaves" (树冠). Another example is the word *solution*, which means "answer to a problem or a question" (解决方法) when used in everyday life situations, but in chemistry it means "liquid in which something is dissolved" (溶液). Similarly, in steel industry, *pig* is no longer the animal that we are so familiar with but "ingot" (铸块) or "mould" (铸模).

If somebody says, "Through the keyhole, Tom saw the strange animal," you certainly know that keyhole in this sentence means "the hole in the door through which a key can be put for locking and unlocking the door" (锁孔); however, when the word is used in medicine and combined with the word *surgery*, the same translation will puzzle the readers—*keyhole surgery* should be translated as 针孔手术.

To make the matter more complicated, the same English word may be translated differently in different fields of science or professions. For example, the word *family* in zoology is usually translated as 科 (for example, *the cat family* 猫科动物), while in linguistics the same word is often translated as 语系 (for example, *the Indo-European family of languages* 印欧语系).

Similarly, one Chinese word can be translated into different English words in different situations. Take the word *tu* (图) for example. It can be translated into English as *figure, diagram, graph, plot, view, pattern, drawing, map, sketch, layout, line, scheme, draft, delineation, image, plan, detail, project, etc.*, depending on what kind of *tu* you are talking about.

## IV. Translate the following sentences into Chinese. Pay attention to the meanings of the italicized words and expressions, which may be misleading.

1. Among the vertebrates, *bony fishes* are the largest group, with 18,150 species, leaving aside the 63 species of jawless fishes and the 843 cartilaginous fishes (lampreys, sharks).
2. Wilson estimates that perhaps 17,500 species (mostly insects) go extinct each year in tropical forests, and that we humans have accelerated the historical rate of extinction *by a factor of one thousand to ten thousand*.
3. There is concern, too, that in our ignorance we may be destroying species vital to the *fabric of ecosystems* on which we depend for our own survival.
4. Though people have always named plants and animals, the science of species inventory is relatively new, beginning with the Big Bang of Carl von Linne's invention of the (Linnaean) binomial system of naming in the mid-eighteenth century.
5. The botanist Edgar Anderson once did an experiment, in which he took a *manila folder* at random from an herbarium case (a Southwestern grass, it turned out to be), to discover the kinds of people who had collected the specimens.
6. Modern natural history is an *exacting science* whose practitioners must also cope and improvise in difficult field conditions.
7. And because of that diversity, the identity of scientific collectors has been less fixed than that of *laboratory workers*.
8. These include birds, carnivores, primates, and large *game*.

## Translation Techniques (2)

### Amplification（增词法）

Although there should be no addition or subtraction in the meanings conveyed from one language to the other in translation, it is often necessary to reveal the meaning hidden in the original versions by adopting the technique of amplification—adding certain words or expressions—to make the translation more accurate and idiomatic. Amplification is also needed to make the logical relations explicit in the translated versions.

Amplification is adopted when some words are omitted in the original version, when a pronoun is used in the English sentence, or when the meaning is implied but can be understood in the original language. For example:

**原文** Biodiversity is a lively issue these days, mainly because of the number of species that are going extinct, either by natural causes, or because we space-hungry humans are destroying their habitats. (*Paragraph* 2)

Compare the following translations:

**译文1** 近来生物多样性是热点问题，主要是由于濒临灭绝的物种的数量，或者是自然造成的，或者是因为缺乏空间的人类将其生境破坏了。

**译文2** 近来生物多样性已成为热点问题，其主要原因是越来越多的物种濒临灭绝，这种情况与自然有关，也与人类因争夺空间而造成物种生境丧失有关。

It can be seen that the second translation is much better than the first one because such expressions as 已成为，越来越多的，这种情况与……有关，也与……有关 are used, and the pronoun "their" in "their habitats" is specified by 物种, so that the meaning is more accurate and clearer. For another example:

**原文** Collecting is an activity that has engaged diverse sorts of people—unlike laboratory science, which is restricted to a relatively few approved types. (*Paragraph* 5)

**译文** 物种收集与实验室科学不同，前者是各种不同人员参与的活动，而后者则只涉及很少几类经过挑选的人员。

The words added in the translation are 物种，前者是，而后者则是，and 人员 at the end of the sentence. It should also be noted that some changes have been made in the structure of the sentence as well.

**V. Discuss the translation technique and the ways of applying the technique to the translation of the following sentences. Complete each of the Chinese translations.**

1. Engineers predict that the new technique will make integrated optical circuits smaller, faster and cheaper.（工程师们预言，这项新技术将使集成光纤线路__________更小，__________更快，__________更低。）
2. Artificial intelligence is the key to a successful robot, but some of the simplest tasks for a human mind are difficult for a robot.（人工智能是保证机器人功能优良的关键，但是有些对人脑来说__________，放到机器人身上__________。）
3. New data show that the prevalence of advanced diseases in developed countries might be declining—an optimistic note.（一些新资料表明在发达国家，晚期疾病的发病率可能正在下降，__________乐观的现象。）
4. The prevalence of periodontitis increases with age.（牙周炎的患病率随年龄的__________而升高。）
5. Thanks to these new treatments, people with high blood pressure can live long and active lives.（由于有了这些新的治疗方法，高血压患者也能够长寿，__________积极的生活。）
6. The instruments contained in the probe will measure the temperature and pressure of the atmosphere, analyze the composition of atmospheric gases, and possibly even detect lightning.（探查器所携带的仪器和设备将测量大气的温度和压力，分析大气中所含气体的成分，如有可能甚至探测大气中的闪电__________。）
7. During the decade researchers will identify more and more human genes and the traits they govern.（在这10年中，研究人员将要鉴定越来越多的人类基因，并__________它们所影响的遗传特性。）
8. This theory is usually expressed in the famous formula: $E=mc^2$. E stands for energy, m for the mass, and c for the speed of light.（该理论通常用这个著名的公式$E=mc^2$来表示。__________，E代表能量，m__________质量，c__________光速。）

**VI. Translate the following sentences into Chinese. Pay attention to how the technique of amplification should be used.**

1. Collecting parties usually travel light and depend on local inhabitants for information and support, making survey collecting a diversely social experience.
2. Of all the field sciences, natural history survey is an exceptionally inviting subject—because the work of systematic, scientific collecting is so varied.
3. It is also clear that the early-nineteenth-century flowering of collecting and naming resulted from the greater affordability of transoceanic steam travel and from European imperial expansion and settlement, especially in the rich tropical environments of the southern hemisphere.

4. We do not think of the late nineteenth and early twentieth centuries as being a great age of discovery in natural history; but they were.
5. Scientific collecting in the age of survey was accomplished mostly by small parties (three to half a dozen) whose purpose was to send back not exotica and accounts of heroic adventure and discovery, but rather crates of specimens.

**VII. Translate the following passage into Chinese.**

Another limitation of this history is that it treats mainly vertebrate zoology and some botany, but insects and other invertebrates hardly at all. This is not an arbitrary limitation: survey collecting in my period, especially by museums, concentrated on vertebrate animals, because scientific fieldwork piggybacked on collecting for exhibits of vertebrate animals. (Insects, plants, and mollusks did not have quite the same potential for eye-catching displays.) In addition, invertebrates are discouragingly numerous for comprehensive survey inventories, and they remained the province of amateur specialists long after vertebrate animals became the objects of organized survey. Invertebrates have recently become the object of systematic inventory, but in ways quite different from earlier surveys.

Like any scientific (or any cultural) practice, natural history survey had its particular period and life cycle. It arose out of a particular set of environmental, cultural, and scientific circumstances; ran its course; then gave way to new and different ways of studying nature's diversity. It was especially well developed in the United States, though not exclusively there. My aim is to describe what natural history survey was in its heyday, the reasons it flourished where it did, and how it worked in practice.

**VIII. Translate the following passage into English.**

## 生物多样性

生物多样性指在某一特定的生态系统内生活的不同物种的数量。科学家对现存物种数量的估计各不相同，从300万种到3,000万种不等，其中250万种已进行了分类，包括90万种昆虫，41,000种脊椎动物，25万种植物，其余是无脊椎动物、真菌类、藻类和微生物。尽管还有尚未发现的物种，然而许多物种由于滥伐森林、污染以及建立人类居住区而濒临灭绝。

生物多样性主要集中在热带地区，特别是林区。处于平衡状态的生境意味着现有物种的数量与资源处于平衡状态。生物多样性受到资源、繁殖力与气候的影响。某一生物多样性的生境越原始，就越能更好地承受自然或人类带来的变迁或威胁，因为这种变迁可以通过生境内其他地方的调整加以平衡；遭到破坏的生境可能由于某一单一物种的灭绝，造成食物链的断裂，而最终消逝。由此可见，生物多样性有助于防止物种的灭绝，保持自然平衡。1992年，在联合国环境与发展大会上，150多个国家签署了保护地球生物多样性条约。

*In your research work, you may need to consult the studies of the others frequently or your supervisor may ask you to increase your knowledge and understanding of a subject by reading comprehensively. In the library you may find many of the scientific documents published in the form of monographs.*

*Then, what is a monograph?*

*According to* Oxford Advanced Learner's English-Chinese Dictionary, *a monograph is "detailed scholarly study of one subject". However, the above definition does not provide us with much idea about monograph. A more detailed description of monograph was made by National Research Council Canada (NRC) in the following passage with an extended definition.*

## Monograph

## 专著

### What Is a Monograph?

1 A **monograph** is a specialized scientific book. As **learned treatises** on clearly defined topics, which may be **intra-**, **inter-**, or **cross-disciplinary**, monographs generally are written by specialists for the benefit of other specialists. Although usually regarded as a component of the **review literature** of science, monographs are works that demand the highest standards of **scholarship**. Their preparation calls for exceptional breadth and depth of knowledge on the part of their authors, who, **inter alia**, must be able to collect, **collate**, analyze, integrate, and **synthesize** all relevant contributions to the **archival** literature of the scientific and engineering journals and to add original material as required. The value of monographs lies in the **coherence** and comprehensiveness of the information and knowledge they contain, which is important to the specialized researchers to whom they are directed and, therefore, to the **advancement** of science and engineering generally. Most monographic **manuscripts** are critically reviewed and tightly edited. The resulting books can be expected to have a reasonably long **shelf life**.

monograph *n.* 专著
learned *adj.* 学术性的
treatise *n.* (专题) 论文
intra- [构词成分] 在内，内部
inter- [构词成分] 表示"在……之间"
cross-disciplinary *adj.* 交叉学科的
review literature 评论性的著作或文献
scholarship *n.* 学问；学识；学术成就
inter alia [拉丁] 除了其他事物之外
collate *v.* 检点并整理（信息、书页等）
synthesize *v.* 合成
archival *adj.* 档案的，案卷的
coherence *n.* 连贯性；一致性
advancement *n.* 前进；进步
manuscript *n.* 原稿；草稿
shelf life *n.* 贮存寿命（物品的保存期限）

2 Monographs commonly are confused with other kinds of books; hence, some distinctions need to be drawn. Textbooks are **pedagogical** works which, even if written on fairly narrow subjects, are designed to serve broader and more junior readerships than specialized research communities. Textbooks are not monographs.

pedagogical *adj.* 教学法的

Neither are most books of **conference proceedings**, even though they may deal with specialized topics and be directed at specialized communities. Together with abstracts and the increasingly common "expanded abstracts", conference papers, valuable and necessary as they may be, commonly take the form of **premature** announcements of new scientific discoveries. Many are subsequently expanded and rendered in a form suitable for the scientific and engineering journals. Conference proceedings generally have a short shelf life. Certain books of scientific papers, which involve conference presentation in the course of their preparation, stand as notable exceptions, however, to the **foregoing** description of the conference literature. The papers in such books are designed from **inception** to review and **augment** existing knowledge of particular aspects of a specialized, **unitary** topic. The papers are prepared for inclusion more or less as "chapters" in a carefully planned and structured volume, and their conference presentation is intended primarily as a means of allowing invited contributors to the book to come together to discuss critically with one another the material they intend to include in their published "chapters". Many books produced in this way are indeed monographs, distinguished simply by having an unusually large number of authors.

conference proceedings 学术会议论文集
premature *adj.* 提前的；过早的；未到期的
foregoing *adj.* 在前的；上述的
inception *n.* 开始；开端
augment *v.* 增多；增大；增加
unitary *adj.* 单一的；一元的

3 In summary, therefore, monographs are generally regarded as scientific treatises of book length but otherwise variable format prepared by **acknowledged** experts on specialized topics for the benefit of others who have specialized in, or who wish to obtain a specialist's appreciation of, these topics. Monographs are externally reviewed and tightly edited. Textbooks and most volumes of conference proceedings are not monographs. As a component of the review literature in science and engineering, monographs **facilitate** the advancement of these fields of knowledge in a unique and important fashion.

acknowledged *adj.* 公认的，得到普通承认的
facilitate *v.* 促进

## The Main Components of a Monograph

4 In the era of "information explosion", you will find numerous books which relate to your research field directly or indirectly, and it is impossible, and often unnecessary, to read all of them from cover to cover. Therefore, previewing (going over the main parts of the book or browsing initially) is an important procedure before you study a certain book closely. Previewing is helpful in extracting information from books.

It saves time, provides you with important details, and gives you an overview. Each part of a book can yield useful information—as follows.

1. The author

This is a key item for identifying a book or recalling it later. It is important to know the author of a book when quoting from it. Always make a note of the author's full name.

2. Title and subtitle

The title should give you some idea of the book's content. If it doesn't, the subtitle might offer an explanation. Subtitles are particularly important if the main title is a quotation. Make a note of both for a full record in your notes.

3. Date of publication

This tells you when the book was first published. It may be important if you need information which is up-to-date. For full accuracy, make a note of the edition. The number of editions is an indication of the book's success.

4. Dust cover or blurb

On any serious book, this is more than just advertising. It gives you a rapid overview of the contents and approach. It might also say what the book contains and for whom it is written.

5. Contents page (or chapter headings)

This should be a list of the topics covered by the book. It might also have details of the sub-sections in each chapter. It's useful for knowing how useful the book will be for your needs.

6. Bibliography and index

These are usually included in any book intended for serious use. The bibliography is a list of books consulted by the author. It might also include suggestions for further reading. An index lists topics mentioned in the book with page references.

7. Illustrations

These might cover statistics, tables, graphs, diagrams, or pictures. This information should be clearly presented.

8. Preface or introduction

This provides an overview of the contents and the author's approach. At this point, authors say what their book is about, how the book came into being, or how the idea for the book was developed; this is often followed by thanks and acknowledgments to people who were helpful to the author during the time of writing.

Exercises

**I. Read the passage and decide whether the following statements are true or false. Write T for True and F for False in the brackets.**

1. Monographs are usually the scientific books that deal with the topics of certain areas or the areas which are related to one another. ( )
2. Most authors of the monographs are editors of the scientific and engineering journals. ( )
3. The authors of monographs should have rich knowledge and the ability to do research. ( )
4. Textbooks belong to the category of monographs. ( )
5. Textbooks are written on narrower subjects than those of monographs. ( )
6. Many books of conference proceedings are not monographs because they discuss new scientific discoveries. ( )
7. Only those conference proceedings which have a large number of authors are monographs. ( )
8. Different topics often lead to different format of monographs. ( )
9. Monographs are read by more readers because they are kept longer than other books in the library. ( )
10. The "acknowledged experts" in the passage means the experts who are rich in knowledge. ( )

**II. Read the passage again, and complete the following items.**

1. The general definition of a monograph: ___________
2. The value of monographs for scientific researches: ___________
3. The qualities of the authors of monographs: ___________
4. The differences between monographs and textbooks: ___________
5. The differences between monographs and books of conference proceedings: ___________
6. The main components of a monograph: ___________
7. An indication of the book's success: ___________
8. The function of the blurb: ___________

**Quotations from Great Scientists**

*Anyone who has never made a mistake has never tried anything new.*

*—Albert Einstein*

一个从未犯过错误的人也从未尝试过新的事物。

——阿尔伯特·爱因斯坦

# Chapter 2 Robots 机器人

## Section 1 Reading and Translation

### I What You Are Going to Read

Stories of artificial helpers and companions and attempts to create them have a long history, but fully autonomous machines—robots—only appeared in the 20th century. The first digitally operated and programmable robot was installed in 1961 to lift hot pieces of metal from a die casting machine and stack them. However, today, commercial and industrial robots are in widespread use performing jobs more cheaply or with greater accuracy and reliability than humans.

How will robots develop in the coming years? What will the future robots be like? What are the scientists doing in their research in this field? You may find the answers to these questions in the following essay *A Robot in Every Home* published in *Scientific American* (December, 2006).

### II About the Author

Bill Gates is co-founder and chairman of Microsoft, the world's largest software company. While attending Harvard University in the 1970s, Gates developed a version of the programming language BASIC for the first microcomputer, the MITS Altair. In his junior year, Gates left Harvard to devote his energies to Microsoft, the company he had begun in 1975 with his childhood friend Paul Allen. In 2000 Gates and his wife, Melinda, established the Bill & Melinda Gates Foundation, which focuses on improving health, reducing poverty and increasing access to technology around the world.

### III Reading Passage

**A Robot in Every Home**

**家家都有机器人**

By Bill Gates

*The leader of the PC revolution predicts that the next hot field will be robotics.*

1 Imagine being present at the birth of a new industry. It is an industry based on **groundbreaking** new technologies, **wherein** a handful of well-established corporations sell highly specialized devices for business use and a fast-growing number of **start-up** companies produce innovative toys, **gadgets** for **hobbyists** and other interesting **niche products**. But it is also a highly **fragmented** industry with few common standards or platforms. Projects are complex, progress is slow, and practical applications are relatively rare. In fact, for all the excitement and promise, no one can say with any certainty when—or even if—this industry will achieve critical mass. If it does, though, it may well change the world.

groundbreaking *adj.* 有创造力的；以创造力和创新为特征的
wherein *conj.* 在那里
start-up *n.* 启动，发动；创业（常用作定语名词）
gadget *n.* 小玩意儿
hobbyist *n.* 沉溺于某种癖好者
niche product 瞄准市场机会的产生；填补空缺的产品
fragmented *adj.* 分裂的，分化的

2 Of course, the paragraph above could be a description of the computer industry during the mid-1970s, around the time that Paul Allen[1] and I launched Microsoft. Back then, big, expensive computers ran the back-office operations for major companies, governmental departments and other institutions. Researchers at leading universities and industrial laboratories were creating the basic building blocks that would make the information age possible. Intel had just introduced the 8080 **microprocessor**, and Atari[2] was selling the popular electronic game Pong. At homegrown computer clubs, enthusiasts struggled to figure out exactly what this new technology was good for.

microprocessor *n.* 【计】微处理器

3 But what I really have in mind is something much more contemporary: the emergence of the robotics industry, which is developing in much the same way that the computer business did 30 years ago. Think of the manufacturing robots currently used on automobile **assembly** lines as the **equivalent** of yesterday's mainframes. The industry's niche products include robotic arms that perform surgery, **surveillance** robots **deployed** in Iraq and Afghanistan that dispose of roadside bombs, and domestic robots that vacuum the floor. Electronics companies have made robotic toys that can imitate people or dogs or **dinosaurs**, and hobbyists are anxious to get their hands on the latest version of the Lego robotics system.

assembly *n.* 装配，组装
equivalent *n.* 等价物，相等物
surveillance *n.* 监视，监督
deploy *v.* 部署，调度
dinosaur *n.* 恐龙

4 Meanwhile some of the world's best minds are trying to solve the toughest problems of robotics, such as visual recognition, navigation and machine learning. And they are succeeding. At the 2004 Defense Advanced Research Projects Agency (DARPA) Grand

autonomously *adv*. 自主地，自动地
rugged *adj*. 高低不平的，崎岖的
intriguing *adj*. 引起兴趣（或好奇心）的；有迷惑力的
precursor *n*. 前辈，先驱

Challenge, a competition to produce the first robotic vehicle capable of navigating **autonomously** over a **rugged** 142-mile course through the Mojave Desert, the top competitor managed to travel just 7.4 miles before breaking down.[3] In 2005, though, five vehicles covered the complete distance, and the race's winner did it at an average speed of 19.1 miles an hour. (In another **intriguing** parallel between the robotics and computer industries, DARPA also funded the work that led to the creation of Arpanet[4], the **precursor** to the Internet.)

5 What is more, the challenges facing the robotics industry are similar to those we tackled in computing three decades ago. Robotics companies have no standard operating software that could allow popular application programs to run in a variety of devices. The standardization of robotic processors and other hardware is limited, and very little of the programming code used in one machine can be applied to another. Whenever somebody wants to build a new robot, they usually have to start from square one.

entrepreneur *n*. 企业家，主办人
convergence *n*. 整合，集中
envision *v*. 想象，展望
ubiquitous *adj*. 到处存在的，（同时）普遍存在的
broadband *n*. 宽带

6 Despite these difficulties, when I talk to people involved in robotics—from university researchers to **entrepreneurs**, hobbyists and high school students—the level of excitement and expectation reminds me so much of that time when Paul Allen and I looked at the **convergence** of new technologies and dreamed of the day when a computer would be on every desk and in every home. And as I look at the trends that are now starting to converge, I can **envision** a future in which robotic devices will become a nearly **ubiquitous** part of our day-to-day lives. I believe that technologies such as distributed computing, voice and visual recognition, and wireless **broadband** connectivity will open the door to a new generation of autonomous devices that enable computers to perform tasks in the physical world on our behalf. We may be on the verge of a new era, when the PC will get up off the desktop and allow us to see, hear, touch and manipulate objects in places where we are not physically present.

## From Science Fiction to Reality

playwright *n*. 剧作家
mythology *n*. 神话
metalwork *n*. 金属制造，金属制品

7 The word "robot" was popularized in 1921 by Czech **playwright** Karel Capek[5], but people have envisioned creating robotlike devices for thousands of years. In Greek and Roman **mythology**, the gods of **metalwork** built mechanical servants made from gold. In the first

century A.D., Heron[6] of Alexandria—the great engineer credited with inventing the first steam engine—designed intriguing **automatons**, including one said to have the ability to talk. Leonardo da Vinci's[7] 1495 sketch of a mechanical knight, which could sit up and move its arms and legs, is considered to be the first plan for a **humanoid** robot.

automaton *n.* 自动装置
humanoid *adj.* 具有人形的，有人的特点的

8 Over the past century, **anthropomorphic** machines have become familiar figures in popular culture through books such as Isaac Asimov's[8] *I, Robot*, movies such as *Star Wars*[9] and television shows such as *Star Trek*[10]. The popularity of robots in fiction indicates that people are receptive to the idea that these machines will one day walk among us as helpers and even as companions. Nevertheless, although robots play a vital role in industries such as automobile manufacturing—where there is about one robot for every 10 workers—the fact is that we have a long way to go before real robots catch up with their science-fiction counterparts.

anthropomorphic *adj.* 被赋予人形（或人性）的

9 One reason for this gap is that it has been much harder than expected to enable computers and robots to sense their surrounding environment and to react quickly and accurately. It has proved extremely difficult to give robots the capabilities that humans take for granted—for example, the abilities to orient themselves with respect to the objects in a room, to respond to sounds and interpret speech, and to grasp objects of varying sizes, **textures** and **fragility**. Even something as simple as telling the difference between an open door and a window can be **devilishly** tricky for a robot.

texture *n.* 质地，质感
fragility *n.* 脆性，易脆性
devilishly *adv.* 过分地

10 But researchers are starting to find the answers. One trend that has helped them is the increasing availability of tremendous amounts of computer power. One **megahertz** of processing power, which cost more than $7,000 in 1970, can now be purchased for just pennies. The price of a **megabit** of storage has seen a similar decline. The access to cheap computing power has permitted scientists to work on many of the hard problems that are fundamental to making robots practical. Today, for example, voice-recognition programs can identify words quite well, but a far greater challenge will be building machines that can understand what those words mean in context. As computing capacity continues to expand, robot designers will have the processing power they need to tackle issues of ever greater complexity.

megahertz *n.* 【物】兆赫
megabit *n.* 百万位，兆位

sensor *n.* 传感器，感应器
servo *n.* 伺服器，伺服系统
manipulate *v.* 操作
ultrawideband *adj.* 超宽频的

11 Another barrier to the development of robots has been the high cost of hardware, such as **sensors** that enable a robot to determine the distance to an object as well as motors and **servos** that allow the robot to **manipulate** an object with both strength and delicacy. But prices are dropping fast. Laser range finders that are used in robotics to measure distance with precision cost about $10,000 a few years ago; today they can be purchased for about $2,000. And new, more accurate sensors based on **ultrawideband** radar are available for even less.

enhancement *n.* 增进，增强
defuse *v.* 卸除（爆炸物的）引信

12 Now robot builders can also add Global Positioning System chips, video cameras, array microphones (which are better than conventional microphones at distinguishing a voice from background noise) and a host of additional sensors for a reasonable expense.[11] The resulting **enhancement** of capabilities, combined with expanded processing power and storage, allows today's robots to do things such as vacuum a room or help to **defuse** a roadside bomb—tasks that would have been impossible for commercially produced machines just a few years ago.

## A BASIC Approach

13 In February 2004 I visited a number of leading universities, including Carnegie Mellon University, the Massachusetts Institute of Technology, Harvard University, Cornell University and the University of Illinois, to talk about the powerful role that computers can play in solving some of society's most pressing problems. My goal was to help students understand how exciting and important computer science can be, and I hoped to encourage a few of them to think about careers in technology. At each university, after delivering my speech, I had the opportunity to get a firsthand look at some of the most interesting research projects in the school's computer science department. Almost without exception, I was shown at least one project that involved robotics.

academia *n.* 学术界，学术环境
veteran *n.* 老手，有经验的人

14 At that time, my colleagues at Microsoft were also hearing from people in **academia** and at commercial robotics firms who wondered if our company was doing any work in robotics that might help them with their own development efforts. We were not, so we decided to take a closer look. I asked Tandy Trower[12], a member of my strategic staff and a 25-year Microsoft **veteran**, to go on an extended fact-finding mission

and to speak with people across the robotics community. What he found was universal enthusiasm for the potential of robotics, along with an industry-wide desire for tools that would make development easier. "Many see the robotics industry as a technological turning point where a move to PC architecture makes more and more sense," Tandy wrote in his report to me after his fact-finding mission. "As Red Whittaker[13], leader of [Carnegie Mellon's] entry in the DARPA Grand Challenge, recently indicated, the hardware capability is mostly there; now the issue is getting the software right."

15 Back in the early days of the personal computer, we realized that we needed an ingredient that would allow all of the pioneering work to achieve critical mass, to **coalesce** into a real industry capable of producing truly useful products on a commercial scale. What was needed, it turned out, was Microsoft BASIC. When we created this programming language in the 1970s, we provided the common foundation that enabled programs developed for one set of hardware to run on another. BASIC also made computer programming much easier, which brought more and more people into the industry. Although a great many individuals made essential contributions to the development of the personal computer, Microsoft BASIC was one of the key **catalysts** for the software and hardware innovations that made the PC revolution possible.

coalesce *v.* 合并，整合
catalyst *n.* 【化】催化剂；刺激（或促进）因素

16 After reading Tandy's report, it seemed clear to me that before the robotics industry could make the same kind of **quantum leap** that the PC industry made 30 years ago, it, too, needed to find that missing ingredient. So I asked him to assemble a small team that would work with people in the robotics field to create a set of programming tools that would provide the essential plumbing so that anybody interested in robots with even the most basic understanding of computer programming could easily write robotic applications that would work with different kinds of hardware.[14] The goal was to see if it was possible to provide the same kind of common, low-level foundation for integrating hardware and software into robot designs that Microsoft BASIC provided for computer programmers.

quantum leap 突飞猛进；巨大突破

17 Tandy's robotics group has been able to draw on a number of advanced technologies developed by a team working under the direction of Craig Mundie[15], Microsoft's chief research

loop *n.* 环，圈
precipice *n.*（危险）边缘，悬崖
trajectory *n.* 轨迹，轨道，轨线

and strategy officer. One such technology will help solve one of the most difficult problems facing robot designers: how to simultaneously handle all the data coming in from multiple sensors and send the appropriate commands to the robot's motors, a challenge known as concurrency. A conventional approach is to write a traditional, single-threaded program—a long **loop** that first reads all the data from the sensors, then processes this input and finally delivers output that determines the robot's behavior, before starting the loop all over again. The shortcomings are obvious: if your robot has fresh sensor data indicating that the machine is at the edge of a **precipice**, but the program is still at the bottom of the loop calculating **trajectory** and telling the wheels to turn faster based on previous sensor input, there is a good chance the robot will fall down the stairs before it can process the new information.[16]

orchestrate *v.* 组织，协调
multicore *adj.* 多芯的

18 Concurrency is a challenge that extends beyond robotics. Today as more and more applications are written for distributed networks of computers, programmers have struggled to figure out how to efficiently **orchestrate** code running on many different servers at the same time. And as computers with a single processor are replaced by machines with multiple processors and "**multicore**" processors—integrated circuits with two or more processors joined together for enhanced performance—software designers will need a new way to program desktop applications and operating systems. To fully exploit the power of processors working in parallel, the new software must deal with the problem of concurrency.

19 One approach to handling concurrency is to write multi-threaded programs that allow data to travel along many paths. But as any developer who has written multithreaded code can tell you, this is one of the hardest tasks in programming. The answer that Craig's team has devised to the concurrency problem is something called the concurrency and coordination runtime (CCR). The CCR is a library of functions—sequences of software code that perform specific tasks—that makes it easy to write multithreaded applications that can coordinate a number of simultaneous activities. Designed to help programmers take advantage of the power of multicore and multiprocessor systems, the CCR turns out to be ideal for robotics as well. By drawing on this library to write their programs, robot

designers can dramatically reduce the chances that one of their creations will run into a wall because its software is too busy sending output to its wheels to read input from its sensors.

20 In addition to tackling the problem of concurrency, the work that Craig's team has done will also simplify the writing of distributed robotic applications through a technology called **decentralized** software services (DSS). DSS enables developers to create applications in which the services—the parts of the program that read a sensor, say, or control a motor—operate as separate processes that can be orchestrated in much the same way that text, images and information from several servers are **aggregated** on a Web page. Because DSS allows software components to run in isolation from one another, if an individual component of a robot fails, it can be shut down and restarted—or even replaced—without having to **reboot** the machine. Combined with broadband wireless technology, this architecture makes it easy to monitor and adjust a robot from a remote location using a Web browser.

decentralize *v.* 分散，疏散
aggregate *v.* 聚集，集合
reboot *v.* 重新启动

21 What is more, a DSS application controlling a robotic device does not have to **reside** entirely on the robot itself but can be distributed across more than one computer. As a result, the robot can be a relatively inexpensive device that **delegates** complex processing tasks to the high-performance hardware found on today's home PCs. I believe this advance will pave the way for an entirely new class of robots that are essentially mobile, wireless **peripheral** devices that tap into the power of desktop PCs to handle processing-intensive tasks such as visual recognition and navigation. And because these devices can be networked together, we can expect to see the emergence of groups of robots that can work in concert to achieve goals such as mapping the seafloor or planting crops.

reside *n.* 存在，在于
delegate *v.* 分派，委派
peripheral *adj.* 边缘的；周围的；外表面的

22 These technologies are a key part of Microsoft Robotics Studio, a new software development kit built by Tandy's team. Microsoft Robotics Studio also includes tools that make it easier to create robotic applications using a wide range of programming languages. One example is a **simulation** tool that lets robot builders test their applications in a three-dimensional virtual environment before trying them out in the real world. Our goal for this release is to create an

simulation *n.* 模拟，模仿

affordable, open platform that allows robot developers to readily integrate hardware and software into their designs.

### Should We Call Them Robots?

23 How soon will robots become part of our day-to-day lives? According to the International Federation of Robotics, about two million personal robots were in use around the world in 2004, and another seven million will be installed by 2008. In South Korea the Ministry of Information and Communication hopes to put a robot in every home there by 2013. The Japanese Robot Association predicts that by 2025, the personal robot industry will be worth more than $50 billion a year worldwide, compared with about $5 billion today.

24 As with the PC industry in the 1970s, it is impossible to predict exactly what applications will drive this new industry. It seems quite likely, however, that robots will play an important role in providing physical assistance and even companionship for the elderly. Robotic devices will probably help people with disabilities get around and extend the strength and endurance of soldiers, construction workers and medical professionals. Robots will maintain dangerous industrial machines, handle hazardous materials and monitor remote oil pipelines. They will enable health care workers to diagnose and treat patients who may be thousands of miles away, and they will be a central feature of security systems and search-and-rescue operations.

25 Although a few of the robots of tomorrow may resemble the anthropomorphic devices seen in *Star Wars*, most will look nothing like the humanoid C-3PO[17]. In fact, as mobile peripheral devices become more and more common, it may be increasingly difficult to say exactly what a robot is. Because the new machines will be so specialized and ubiquitous—and look so little like the two-legged automatons of science fiction—we probably will not even call them robots. But as these devices become affordable to consumers, they could have just as profound an impact on the way we work, communicate, learn and entertain ourselves as the PC has had over the past 30 years.

Notes

1. **Paul Allen:** 保罗·艾伦（1953–2018），微软联合创始人，1975年与比尔·盖茨共同创立了微软公司。
2. **Atari:** 雅达利，一家美国电子游戏公司，1972年由Nolan Bushnell和Ded Dabney合作成立，开发了历史上第一台成功运行的电子游戏机"乒乓"（Ping Pong）。
3. **At the 2004 Defense Advanced Research Projects Agency (DARPA) Grand Challenge, a competition to produce the first robotic vehicle capable of navigating autonomously over a rugged 142-mile course through the Mojave Desert, the top competitor managed to travel just 7.4 miles before breaking down:** 在2004年美国国防部高级研究计划署（DARPA）举办的"大挑战"赛上，获胜者制造的车辆仅行驶了7.4英里就出了故障，尽管这次比赛的目的是要产生第一辆能够在美国加州莫哈韦沙漠142英里的崎岖赛道上自主操控行驶的机器人车。该句虽然较长，但仍是一个简单句，句子的主要部分是the top competitor managed to travel just 7.4 miles before breaking down，插入语a competition to produce the first robotic vehicle...是对the 2004 Defense Advanced Research Projects Agency (DARPA) Grand Challenge的说明。
4. **Arpanet:** 互联网的前身，由美国国防部高级研究计划署于1969年11月建立。只有四个结点，分布在洛杉矶的加利福尼亚州大学洛杉矶分校、加州大学圣巴巴拉分校、斯坦福大学、犹他州大学四所大学的四台大型计算机上，采用分组交换技术，通过专门的接口信号处理机（IMP）和专门的通信线路相互连接。
5. **Karel Capek:** 卡雷尔·恰佩克（1890–1938），奥地利著名的剧作家、科幻文学家和童话寓言家，是20世纪初期最重要的文学家之一。他在1920年出版的作品《罗素姆万能机器人》（*Rossum's Universal Robots*）中创造了"机器人"一词。
6. **Heron:** 希罗（10AD–70AD），古希腊数学家、力学家、机械学家和测量家。曾在罗马帝国的著名学术研究城市亚历山大（Alexandria）教授数学、物理学等。他发明了许多利用水力、蒸汽以及压缩空气作动力的机械，如喷泉及消防泵等。
7. **Leonardo da Vinci:** 列奥纳多·达·芬奇（1452–1519），意大利文艺复兴时期画家、科学家。
8. **Isaac Asimov:** 艾萨克·阿西莫夫（1920–1992），美国著名科幻作家，曾获代表科幻界最高荣誉的雨果奖和星云终身成就（大师）奖。他在作品《我，机器人》中提出了一套机器人和智能电脑的伦理定律。
9. ***Star Wars*:**《星球大战》，1977年由美国好莱坞拍摄的一部科幻影片。
10. ***Star Trek*:**《星际旅行》，科幻电视系列剧，在1966年9月8日由美国全国广播公司（NBC）首播。
11. **Now robot builders can also add Global Positioning System chips, video cameras, array microphones (which are better than conventional microphones at distinguishing a voice from background noise) and a host of additional sensors for a reasonable expense:** 现在，工程师可在合理的成本之下，为机器人加装全球定位系统晶片、摄影机、阵列传声器（比传统传声器更善于从背景噪音中分辨出特定声音），以及许多附加的感应器。句中a host of相当于a large number of, 意为"大量，许多"。
12. **Tandy Trower:** 唐迪·特劳尔，微软公司机器人业务集团总经理。

13. **Red Whittaker:** 莱德·惠塔克，卡内基·梅隆大学（Carnegie Mellon University）机器人工程学教授，世界上最为知名的机器人专家之一。
14. **So I asked him to assemble a small team that would work with people in the robotics field to create a set of programming tools that would provide the essential plumbing so that anybody interested in robots with even the most basic understanding of computer programming could easily write robotic applications that would work with different kinds of hardware:** 于是我让他召集人成立一个小组，与机器人领域的专家合作，研制出一套可提供基本组件的程序工具，这样任何对机器人感兴趣的人（即使只具备最基本计算机程序知识的人）都可以很容易地写出适用于不同硬件的应用程序。句中plumbing原意为"(建筑物的）管道装置，管件"，在本句中转义为"(计算机程序的）组件"。
15. **Craig Mundie:** 克雷格·蒙迪，微软首席研发与战略官。
16. **The shortcomings are obvious: if your robot has fresh sensor data indicating that the machine is at the edge of a precipice, but the program is still at the bottom of the loop calculating trajectory and telling the wheels to turn faster based on previous sensor input, there is a good chance the robot will fall down the stairs before it can process the new information:** 这个方式的缺点很明显：当机器人从感应器收到最新讯息时，机器已经临近一个危险的边缘，但由于程式还在回线后半部计算轨迹的部分，所以会根据先前输入的资料，命令轮子快点运转，机器人很可能根本没有机会处理新讯息，就跌下了楼梯。句中a good chance的意思是"很大可能"，相当于great possibility，chance后面省略了连接词that。又如：There's a faint chance that you'll find him at home.
17. **C-3PO:** 电影《星球大战》中的机器人形象。

## Exercises

**I. Answer the following questions.**

1. How has robot become the familiar figure in popular culture over the past century?
2. What is the reason for the gap between real robots and their science-fiction counterparts according to the passage?
3. What was the goal of Bill Gates' visiting several leading universities about computer talks in 2004?
4. What are the two greatest fruits that Craig Mundie's team has achieved?
5. In what fields will the future robots play their important part?

**II. The following statements are incomplete. Search the missing information in the passage and fill in the blanks.**

1. It was in the mid-1970s that big, expensive mainframe computers ran __________ for major companies, governmental departments and other institutions.

2. In spite of the difficulties in the robotics industry, Bill Gates actively communicates with people involved in robotics who are ___________ and he still holds an optimistic view on the future of robotics.
3. The word "robot" was first proposed by ___________.
4. Tandy Trower found that people across the robotics community showed great interest in ___________ and the whole robotics industry hoped to get tools for easier development.
5. Bill Gates believes that it was ___________ that made the PC revolution possible.
6. A conventional solution to concurrency problem is to ___________.
7. The purpose for Tandy's team releasing Microsoft Robotics Studio is to ___________.
8. Most of the robots tomorrow will ___________ the humanoid C-3PO.

**III. Identify the implied meanings of the underlined parts of the following sentences according to the context of the passage, and translate the sentences into Chinese.**

1. In fact, for all the excitement and promise, no one can say with any certainty when—or even if—this industry will achieve critical mass.
2. Electronics companies have made robotic toys that can imitate people or dogs or dinosaurs, and hobbyists are anxious to get their hands on the latest version of the Lego robotics system.
3. Meanwhile some of the world's best minds are trying to solve the toughest problems of robotics, such as visual recognition, navigation and machine learning.
4. In another intriguing parallel between the robotics and computer industries, DARPA also funded the work that led to the creation of Arpanet, the precursor to the Internet.
5. Whenever somebody wants to build a new robot, they usually have to start from square one.
6. I believe that technologies such as distributed computing, voice and visual recognition, and wireless broadband connectivity will open the door to a new generation of autonomous devices that enable computers to perform tasks in the physical world on our behalf.
7. Even something as simple as telling the difference between an open door and a window can be devilishly tricky for a robot.
8. After reading Tandy's report, it seemed clear to me that before the robotics industry could make the same kind of quantum leap that the PC industry made 30 years ago, it, too, needed to find that missing ingredient.
9. I believe this advance will pave the way for an entirely new class of robots that are essentially mobile, wireless peripheral devices that tap into the power of desktop PCs to handle processing-intensive tasks such as visual recognition and navigation.
10. And because these devices can be networked together, we can expect to see the

emergence of groups of robots that can work in concert to achieve goals such as mapping the seafloor or planting crops.

## Translation Techniques (1)

### Omission（减词法）

Just opposite to "amplification", "omission" means leaving some words out in the process of translating English into Chinese. This method is widely adopted because many words which are essential in English are not necessary in Chinese, such as the articles "a" and "the", and the pronouns like "his", "her", and "their", etc. For example, "*A* stroke occurs when *the* blood flow to a section of *the* brain stops; brain cells in the area lose *their* source of energy and within minutes begin to die."（血流在大脑的某个部位受阻时就会发生中风，这部分脑细胞失去了能量来源，几分钟内就开始坏死。）

Which words should be omitted depends on the meaning of the English original and the usage of the Chinese language. To translate the following sentence into Chinese, for example, it would make the meaning of the target language clear and concise to omit the words 如果 and 你: "*If you* melt two or more metals together, *you* can get a new metal."（将两种或多种金属熔合在一起可产生一种新金属。）

Observe the following English sentences and compare them with the Chinese translation, especially the italicized words in the original sentences.

1. Male cigarette smokers have *a* higher death rate from heart disease than non-smokers *males*.
   男性吸烟者的心脏病死亡率比不吸烟的高。
2. The software interprets these subtle numerical variations as different colors on *the* video terminal screen, *allowing the* physician to recognize change much more easily.
   该软件把这些细微数字变化用不同颜色在终端显示屏上显示出来，医生辨别这些变化就容易得多。
3. But for patients who have side effects, doctors often can adjust medication to relieve *them*.
   但是对有副作用的患者，医生常可调整用药来缓解。
4. *It is* generally thought that salt is *a* culprit in this disease, and many doctors prescribe *a* low-salt diet.
   人们普遍认为食盐是这种病的病因，于是许多医生建议低盐饮食。

Therefore, omission is needed to make your translation smooth and concise, and often sound more professional. Generally, amplification and omission should be integrated in translation.

**IV. Translate the following sentences into Chinese. Pay attention to how the technique of omission should be used.**

1. And as I look at the trends that are now starting to converge, I can envision a future in which robotic devices will become a nearly ubiquitous part of our day-to-day lives.
2. Leonardo da Vinci's 1495 sketch of a mechanical knight, which could sit up and move its arms and legs, is considered to be the first plan for a humanoid robot.
3. It has proved extremely difficult to give robots the capabilities that humans take for granted—for example, the abilities to orient themselves with respect to the objects in a room, to respond to sounds and interpret speech, and to grasp objects of varying sizes, textures and fragility.
4. Another barrier to the development of robots has been the high cost of hardware, such as sensors that enable a robot to determine the distance to an object as well as motors and servos that allow the robot to manipulate an object with both strength and delicacy.
5. At each university, after delivering my speech, I had the opportunity to get a firsthand look at some of the most interesting research projects in the school's computer science department.
6. Although a great many individuals made essential contributions to the development of the personal computer, Microsoft BASIC was one of the key catalysts for the software and hardware innovations that made the PC revolution possible.
7. Combined with broadband wireless technology, this architecture makes it easy to monitor and adjust a robot from a remote location using a Web browser.
8. It seems quite likely, however, that robots will play an important role in providing physical assistance and even companionship for the elderly.

## Translation Techniques (2)

### English Verbs and Chinese Translation（英语动词的翻译）

Verbs are the words indicating actions, events or states. English verbs are very active, and a considerable number of verbs have extended meanings, which may derive from their basic meanings. When these verbs are translated into Chinese, careful consideration should be given to the relations between the verbs and the contexts in which the verbs are used.

For example, the basic meaning of the verb *develop* is "(cause sb./sth.) to grow gradually（发展)", but when it is related to different contexts, additional meanings will be attached to it, and thus it is necessary to seek for suitable words to convey these meanings in Chinese translation. All the following sentences contain the word "develop", but different expressions are adopted in the target language.

1. When related to diseases:

   (1) Pneumonia *develops* from flu.
   肺炎是由流行性感冒引起的。

   (2) Some people *develop* skin rashes when they take sulfas.
   有些人在服用磺胺药物后会出现皮疹。

2. When related to animals and plants:

   (1) Land animals are believed to have *developed* from sea animals.
   陆地动物被认为是由海洋动物进化而来的。

   (2) After the chemical was applied, the bud *developed* into a blossom.
   在施用了这种化学药剂之后，花蕾开出了鲜花。

3. When related to new technology, inventions and discoveries:

   (1) Scientists are *developing* ways to make plastics as recyclable as metal or glass.
   科学家正在研究使塑料像金属或玻璃一样可以回收利用的方法。

   (2) The prototype of agricultural robot has been *developed* to pick melons or any other head crop.
   采摘甜瓜或其他球状农作物的农用机器人样机已经研制出来。

4. When related to photography:

   (1) When *developed*, this pattern will scatter light in exactly the same way as the original tissue, so it will look just like it.
   冲洗后，该图散射光线的状况恰好与原组织一样，所以看上去十分逼真。

   (2) The photographer *developed* the film by treating it with chemicals.
   摄影师用化学药剂使底片显影。

## V. Discuss the translation technique and the ways of applying the technique to the translation of the following sentences. Complete each of the Chinese translations.

1. It is reported that a number of children *developed* symptoms of consumption in this area.（据报道，该地区有些儿童__________肺结核的症状。）
2. Sure enough, 80% of the plants *developed* the disease.（果然80%的庄稼都__________这种病害。）
3. Girls who became overweight between 6 and 11 year of age were 7 times more likely to *develop* new asthma symptoms at age 11 or 13.（6至11岁超重女孩在11岁或13岁__________新的哮喘症状的可能性增加6倍。）
4. Sealed industrial cooling systems were originally *developed* to offer efficient and cost saving cooling alternatives to industry.（最初__________封闭式工业冷却系统的目的是给工业提供一种高效节约的冷却方法。）
5. In the late 1960s and early 1970s special nuclear warheads were *developed* that generated enhanced radiation in the X-ray portion of the electromagnetic spectrum.（20世纪60年代末和70年代初，特种核弹头__________，这种弹头在电磁光谱的X射线区产生加强辐射。）
6. The fetus *develop* lungs relatively late in the gestation period.（妊娠期胎儿的肺__________相对慢些。）
7. A biotech firm in California has *developed* a tomato that does not rot as fast as normal varieties.（加利福尼亚的一家生物技术公司已经__________一种不像普通品种烂得那样快的西红柿品种。）
8. If the offspring of patients *develop* the disease, they will tend to do so at an earlier age than their parents.（若患者的子女也__________此病，他们的发病年龄要比他们的父母早。）

## VI. Translate the following sentences into English. Each pair of the sentences contains the same verb (the italicized word). Pay attention to how the verbs should be translated properly.

1. a) Of course, the paragraph above could be a description of the computer industry during the mid-1970s, around the time that Paul Allen and I *launched* Microsoft.
   b) A spokesman for the dockyard said they hoped to *launch* the new submarine within two years.

2. a) The industry's niche products include robotic arms that *perform* surgery, surveillance robots deployed in Iraq and Afghanistan that dispose of roadside bombs, and domestic robots that vacuum the floor.

b) It has been proved that the new type of computers can *perform* many special tasks in the spacecraft.

3. a) Because DSS allows software components to run in isolation from one another, if an individual component of a robot *fails*, it can be shut down and restarted—or even replaced—without having to reboot the machine.

   b) The wheat *failed* last year because of the lack of rain.

4. a) Our goal for this release is to create an affordable, open platform that allows robot developers to readily *integrate* hardware and software into their designs.

   b) An *integrated* circuit is a very small electronic circuit which consists of a lot of small parts made on a piece of semiconducting material.

**VII. Translate the following passage into Chinese.**

*Robots in Home*: As their price falls, and their performance and computational ability rises, making them both affordable and sufficiently autonomous, robots are increasingly being seen in the home where they are taking on simple but unwanted jobs, such as vacuum cleaning, floor cleaning and lawn mowing. While they have been on the market for several years, 2006 saw a great increase in the number of domestic robots sold. By 2006, iRobot had sold more than two million vacuuming robots. They tend to be relatively autonomous, usually only requiring a command to begin their job. They then proceed to go about their business in their own way. At such, they display a good deal of agency, and are considered intelligent robots.

*Telerobots*: When a human cannot be present on site to perform a job because it is dangerous, far away, or inaccessible, teleoperated robots, or telerobots are used. Rather than following a predetermined sequence of movements a telerobot is controlled from a distance by a human operator. The robot may be in another room or another country, or may be on a very different scale to the operator. A laparoscopic surgery robot allows the surgeon to work inside a human patient on a relatively small scale compared to open surgery, significantly shortening recovery time. An interesting use of a telerobot is by the author Margaret Atwood, who has recently started using a robot pen to sign books remotely. At the other end of the spectrum, iRobot ConnectR robot is designed to be used by anyone to stay in touch with family or friends from far away. Still another robot is being used by doctors to communicate with patients, allowing the doctor to be anywhere in the world. This increases the number of patients a doctor can monitor.

## VIII. Translate the following passage into English.

### 机器人

究竟什么是机器人，其定义多种多样，因此往往很难比较不同国家所拥有的机器人的数量。为了提供一个普遍接受的定义，国际标准化组织在其ISO 8373中将机器人定义为："可自动控制、重复编程的、多种用途操作器，该操作器的编程可在三个或三个以上坐标轴上进行，既可以固定在某个地方，也可移动，用于工业自动化领域。"

除了国际标准化组织的定义以外，美国、日本等国家也有不同的定义。例如日本机器人的种类很多，因为在那里许多机器都被看作机器人。

由于没有哪一个定义能够使大家都满意，因此许多人都有他们自己的定义。例如，工业用机器人的开拓者Joseph Engelberger曾说："我虽然不会给机器人下定义，但我一眼就能看出哪一个是机器人。"

做那些需要速度、精确度、可靠性或忍耐性的工作，机器人远远超过人类，因此在工厂里许多曾经由人做的工作现在都已自动化了，从而使人们能够大批量生产价格低廉的产品，如汽车和电子产品等。自从人们最初使用Unimate机器人将高温金属自动从压铸机中取出以来，机器人在工厂中的使用至今已有50年，而且其数量的增长越来越快，目前全世界有100多万台机器人在运转。

## Section 2 Reading for Academic Purposes

*Besides monographs, academic journals consist of another important part in academic publishing, which are indispensable for your studies and research. Compared with monographs, academic journals are able to follow the latest development of the scientific and technological fields more closely owing to their regular publication at intervals, thus bringing fresh information and views to you timely to enrich your own research. Most established academic disciplines have their own journals, though many academic journals are somewhat interdisciplinary, and publish work from several distinct fields or subfields.*

*The following passage will tell you the main characteristics of the academic journals, including the types of the articles published by the journals, prestige of various academic journals, and their new developments.*

### Academic Journal
### 学术期刊

An academic journal is a **peer-reviewed periodical** in which scholarship relating to a particular academic **discipline** is published. Academic journals serve as forums for the introduction and presentation

peer-reviewed *adj.* 由同行评论的
periodical *n.* 期刊，杂志
discipline *n.* 学科

scrutiny *n.* 详细的检查，仔细的研究
critique *n.* 评论

for **scrutiny** of new research, and the **critique** of existing research. Content typically takes the form of articles presenting original research, review articles, and book reviews. Academic or professional publications that are not peer-reviewed are usually called *professional magazines*.

humanities *n.* 人文科学
qualitative *adj.* 定性的

The term "academic journal" applies to scholarly publications in all fields; this article discusses the aspects common to all academic field journals. Scientific journal and journals of the quantitive social sciences vary in form and function from journals of the **humanities** and **qualitative** social sciences; their specific aspects are separately discussed. The similar American and British journal publication systems are primarily discussed here; practices differ in other regions of the world.

## Scholarly articles

unsolicited *adj.* 未被恳求的，主动提供的
submission *n.* 提交，呈递
outright *adv.* 完全地，彻底地
anonymous *adj.* 匿名的
referee *n.* 审阅人，专家

In academia, professional scholars typically make **unsolicited submissions** of their articles to academic journals. Upon receipt of a submitted article manuscript, the journal editor (or editors) determines whether to reject the submission **outright** or begin the process of peer review. In the latter case, the submission becomes subject to **anonymous** peer-review by outside scholars of the editor's choosing. The number of these peer reviewers (or "**referees**") varies according to each journal's editorial practice—typically, no fewer than two, and usually at least three outside peers review the article. The editor(s) uses the reviewers' opinions in determining whether to publish the article, return it to the author(s) for revision, or to reject it. Even accepted articles are subjected to further (sometimes considerable) editing by journal editorial staff before they appear in print. Typically, because the process is lengthy, an accepted article will not be published until months after its initial submission, while publication after a period of several years is not unknown.

The peer-review process is considered critical to establishing a reliable body of research and knowledge. Scholars can be expert only in a limited area of their fields; they rely upon peer-reviewed journals to provide reliable, credible research upon which they can build subsequent, related research.

### Review articles

Review articles, also called "reviews of progress", are checks on the research published in journals. Some journals are devoted entirely to review articles, others contain a few in each issue, but most do not publish review articles. Such reviews often cover the research from the preceding year, some for longer or shorter terms; some are devoted to specific topics, some to general surveys. Some journals are **enumerative**, listing all significant articles in a given subject, others are selective, including only what they think worthwhile. Yet others are **evaluative**, judging the state of progress in the subject field. Some journals are published in series, each covering a complete subject field year, or covering specific fields through several years.

enumerative *adj.* 列举的
evaluative *adj.* 评价性的

Unlike original research articles, review articles tend to be **solicited** submissions, sometimes planned years in advance. The authors of review articles are paid a few hundred dollars for reviews, and because of this, the standard definitions of open access do not require review articles to be open access, though many are so. They are typically relied upon by students beginning a study in a given field, or for current awareness of those already in the field.

solicit *v.* 征求

### Book reviews

Reviews of scholarly books are checks upon the research books published by scholars; generally, book reviews are also solicited. Journals typically have a separate book review editor determining which new books to review and by whom. If an outside scholar accepts the book review editor's request for a book review, he or she generally receives a free copy of the book from the journal in exchange for a timely review. Publishers send books to book review editors in the hope that their books will be reviewed. The length and depth of research book reviews varies much from journal to journal, as does the extent of textbook and trade book review.

### Prestige

An academic journal's prestige is established over time, and can reflect many factors, some but not all of which are expressible quantitatively. In each academic discipline there are dominant journals that receive the largest number of submissions, and therefore can be selective in choosing their content. Yet, not only the largest journals are of excellent quality. For example, among United States academic historians, the two dominant journals are the *American Historical Review* and the *Journal of American History*, but there are dozens of other American peer-reviewed history journals specializing in specific historical periods, themes, or regions, and these may be considered of equally high quality in their specialties.

proxy *n.* 代理人；代表者
citation *n.* 引证，引用；引文

In the natural sciences and in the "hard" social sciences, impact factor is a convenient **proxy**, measuring the number of later articles citing articles already published in the journal. There are other, possible quantitative factors, such as the overall number of **citations**, how quickly articles are cited, and the average "half-life" of articles, i.e. when they are no longer cited. There also is the question of whether or not any quantitative factor can reflect true prestige; natural science journals are categorized and ranked in the *Science Citation Index*, and social science journals in the *Social Science Citation Index*.

Anglo-American *adj.* 英美的
preliminary *adj.* 初步的；预备的
validity *n.* 有效（性）；效力

In the **Anglo-American** humanities, there is no tradition (as there is in the sciences) of giving impact-factors that could be used—however incorrectly—in establishing a journal's prestige. Perhaps a key reason for this is the relative unimportance of academic journals in these subjects, in contrast with the importance of academic monographs. Very recently, there has been **preliminary** work done for determining such a measurement's **validity**.

The categorization of journal prestige in some subjects has been attempted, using letters to rank their importance in academic world. This journal-ranking is administered by the Vienna University of Economics and Business Administration.

## Publishing

subsidize *v.* 资助；给……津贴

Humanities and social science academic journals are usually **subsidized** by universities or professional organizations, and do not exist to make a profit. However, they often accept advertising to pay for production costs. Publishers charge libraries higher subscription prices than are charged to individual subscribers; institutional subscriptions range between several hundred to several thousand dollars. Journal editors tend to have other professional responsibilities, most often as teaching professors. In the case of the very largest journals, there is paid staff assisting in the editing. The production of the journals is almost always done by publisher-paid staff. Subject journal publishers often are the university presses, such as Oxford University Press, Cambridge University Press, Harvard University Press and University of California Press.

**New developments**

The Internet has revolutionized the production of, and access to, academic journals, with their contents available online via services subscribed to by academic libraries. Individual articles are subject-indexed in databases such as *Google Scholar*. Some of the smallest, most specialized journals are prepared **in-house**, by an academic department, and published only online—such form of publication has sometimes been in the **blog** format.

in-house *adv.* 在机构内部
blog *n.* 博客

Currently, there is a movement in higher education encouraging open access, either via self **archiving**, **whereby** the author **deposits** his paper in a **repository** where it can be searched for and read, or via publishing it in a free open access journal, which does not charge for subscriptions, being either subsidized or financed with author page charges. However, to date, open access has affected science journals more than humanities journals.

archive *v.* 存档
whereby *adv.* 靠那个；凭借那个，借以
deposit *v.* 存放，寄存
repository *n.* 知识库

**Exercises**

**I. Read the passage and decide whether the following statements are true or false. Write T for True and F for False in the brackets.**

1. From the viewpoint of content, articles in the academic journal can be classified into original research articles, review articles and book reviews. ( )
2. If a submitted article is quickly rejected, the decision must be made by the journal editor(s). ( )
3. Even accepted articles need further peer-review by outside scholars. ( )
4. Authors who write book reviews are usually paid a few thousand dollars. ( )
5. When students start a study in a certain field, they usually firstly refer to related review articles. ( )
6. Many journals appoint a specific editor to choose some new books to review and scholars to write the book reviews. ( )
7. The dominant journals in each academic discipline surely bear the highest quality in

their specialty. (  )

8. It is often teaching professors who take the editorial job for humanities and social science academic journals. (  )
9. With the aid of Internet, people can get access to the contents of academic journals free online. (  )
10. Even if the author publishes his paper in a free open access journal, he may still get money from subsidy or from author page charges. (  )

**II. Read the passage again, and complete the following items.**

1. The general definition of an academic journal: ___________
2. The significance of peer-review process: ___________
3. The definition of review articles: ___________
4. One difference between original research articles and review articles: ___________
5. The places where science journals are authoritatively ranked: ___________
6. The possible quantitative factors to reflect an academic journal's prestige: ___________
7. The financial resources of humanities and social science academic journals: ___________
8. The role of internet in the production of , and access to, academic journals: ___________

## Quotations from Great Scientists

*If I have seen further than others, it is by standing upon the shoulders of giants.*

*—Isaac Newton*

如果说我比其他人看得更远，那是因为我站在了巨人的肩膀上。

——艾萨克·牛顿

# Chapter 3 Dark Energy 暗能量

## Section 1 Reading and Translation

### I What You Are Going to Read

When you see the title of the reading passage of this chapter *The Universe's Invisible Hand* (*Scientific American*, January, 2007), you may feel surprised: What is this invisible hand in universe?

The author of the article will introduce you to an unknown form of energy that surrounds each of us—dark energy. According to the author, dark energy may be the key link among several aspects of galaxy formation that used to appear unrelated. This discovery is regarded as one of the most revolutionary discoveries in 20th-century cosmology, and may lead to the development of new theories of physics.

If you want to know more about dark energy and its role in universe, start your reading now.

### II About the Author

Christopher J. Conselice is an astronomer and lecturer at the University of Nottingham in England, where he recently moved from the California Institute of Technology. He specializes in the formation of galaxies and leads several observational programs in infrared and visible light with telescopes both on the ground and in space. A lover of both the heavens and the earth, he comes from a family of Pennsylvanian farmers and spends his free time boating, fishing, biking and caving.

### III Reading Passage

**The Universe's Invisible Hand**

**宇宙中的隐形手**

By Christopher J. Conselice

*Dark energy[1] does more than hurry along the expansion of the universe. It also has a stranglehold on the shape and spacing of galaxies.*

tug *v.* 用力拖，使劲拉
cosmos *n.* 宇宙
anticipate *v.* 预期，预见
detection *n.* 探测；察觉
cosmology *n.* 宇宙学，宇宙论
bulk *n.* 大部分，主体

1 What took us so long? Only in 1998 did astronomers discover we had been missing nearly three quarters of the contents of the universe, the so-called dark energy—an unknown form of energy that surrounds each of us, **tugging** at us ever so slightly, holding the fate of the **cosmos** in its grip, but to which we are almost totally blind. Some researchers, to be sure, had **anticipated** that such energy existed, but even they will tell you that its **detection** ranks among the most revolutionary discoveries in 20th-century **cosmology**. Not only does dark energy appear to make up the **bulk** of the universe, but its existence, if it stands the test of time, will probably require the development of new theories of physics.

implication *n.* 含义
inhabitant *n.* 居民，居住者
galaxy cluster 星系团

2 Scientists are just starting the long process of figuring out what dark energy is and what its **implications** are. One realization has already sunk in: although dark energy betrayed its existence through its effect on the universe as a whole, it may also shape the evolution of the universe's **inhabitants**—stars, galaxies, **galaxy clusters**. Astronomers may have been staring at its handiwork for decades without realizing it.

clump *v.* 形成一堆（或一团等）
intergalactic *adj.* 星系间的
asteroid *n.* 小行星

3 Ironically, the very pervasiveness of dark energy is what made it so hard to recognize. Dark energy, unlike matter, does not **clump** in some places more than others; by its very nature, it is spread smoothly everywhere. Whatever the location—be it in your kitchen or in **intergalactic** space—it has the same density, about $10^{-26}$ kilogram per cubic meter, equivalent to a handful of hydrogen atoms. All the dark energy in our solar system amounts to the mass of a small **asteroid**, making it an utterly inconsequential player in the dance of the planets. Its effects stand out only when viewed over vast distances and spans of time.

recession *n.* 衰退

4 Since the days of American astronomer Edwin Hubble[2], observers have known that all but the nearest galaxies are moving away from us at a rapid rate. This rate is proportional to distance: the more distant a galaxy is, the faster its **recession**. Such a pattern implied that galaxies are not moving through space in the conventional sense but are being carried along as the fabric of space itself stretches. For decades, astronomers struggled to answer the obvious follow-up question: How does the expansion rate change over time? They

reasoned that it should be slowing down, as the inward gravitational attraction exerted by galaxies on one another should have **counteracted** the outward expansion.

counteract *v.* 抵消，中和

5 The first clear observational evidence for changes in the expansion rate involved distant **supernovae**, massive exploding stars that can be used as markers of cosmic expansion, just as watching driftwood lets you measure the speed of a river. These observations made clear that the expansion was slower in the past than today and is therefore accelerating. More specifically, it had been slowing down but at some point underwent a transition and began speeding up. This striking result has since been cross-checked by independent studies of the cosmic microwave background radiation by, for example, the Wilkinson Microwave **Anisotropy** Probe (WMAP)[3].

supernovae *n.* 超新星
anisotropy *n.* 【物】各向异性

***Dark energy may be the key link among several aspects of galaxy formation that used to appear unrelated.***

6 One possible conclusion is that different laws of gravity apply on **supergalactic** scales than on lesser ones, so that galaxies' gravity does not, in fact, resist expansion. But the more generally accepted **hypothesis** is that the laws of gravity are universal and that some form of energy, previously unknown to science, opposes and overwhelms galaxies' mutual attraction, pushing them apart ever faster. Although dark energy is inconsequential within our galaxy (let alone your kitchen), it adds up to the most powerful force in the cosmos.

supergalactic *adj.* 超星系的
hypothesis *n.* 假设

## Cosmic Sculptor

7 As astronomers have explored this new phenomenon, they have found that, in addition to determining the overall expansion rate of the universe, dark energy has long-term consequences for smaller scales. As you zoom in from the entire observable universe, the first thing you notice is that matter on cosmic scales is distributed in a **cobweblike** pattern—a **filigree** of **filaments**, several tens of millions of light-years long, **interspersed** with voids of similar size. Simulations show that both matter and dark energy are needed to explain the pattern.

cobweblike *adj.* 类似蜘蛛网的
filigree *n.* 金银丝细工饰品
filament *n.* 细丝
intersperse *v.* 散布；点缀

8 That finding is not terribly surprising, though. The filaments

detach *v.* 分开，分离
equilibrium *n.* 平衡，均衡

and voids are not coherent bodies like, say, a planet. They have not **detached** from the overall cosmic expansion and established their own internal **equilibrium** of forces. Rather they are features shaped by the competition between cosmic expansion (and any phenomenon affecting it) and their own gravity. In our universe, neither player in this tug-of-war is overwhelmingly dominant. If dark energy were stronger, expansion would have won and matter would be spread out rather than concentrated in filaments. If dark energy were weaker, matter would be even more concentrated than it is.

accrete *v.* 使依附；使连接
assemblage *n.* 集合；集合物

9 The situation gets more complicated as you continue to zoom in and reach the scale of galaxies and galaxy clusters. Galaxies, including our own Milky Way, do not expand with time. Their size is controlled by an equilibrium between gravity and the angular momentum[4] of the stars, gas and other material that make them up; they grow only by **accreting** new material from intergalactic space or by merging with other galaxies. Cosmic expansion has an insignificant effect on them. Thus, it is not at all obvious that dark energy should have had any say whatsoever in how galaxies formed.[5] The same is true of galaxy clusters, the largest coherent bodies in the universe—**assemblages** of thousands of galaxies embedded in a vast cloud of hot gas and bound together by gravity.

10 Yet it now appears that dark energy may be the key link among several aspects of galaxy and cluster formation that not long ago appeared unrelated. The reason is that the formation and evolution of these systems is partially driven by interactions and mergers between galaxies, which in turn may have been driven strongly by dark energy.

baryonic *adj.* 【物】重子的
proton *n.* 【核】质子

11 To understand the influence of dark energy on the formation of galaxies, first consider how astronomers think galaxies form. Current theories are based on the idea that matter comes in two basic kinds. First, there is ordinary matter, whose particles readily interact with one another and, if electrically charged, with electromagnetic radiation. Astronomers call this type of matter "**baryonic**" in reference to its main constituent, baryons, such as **protons** and neutrons. Second, there is dark matter (which is distinct from dark energy), which makes up 85 percent of all matter and whose salient property is that it comprises particles that do not react with radiation. Gravitationally, dark matter behaves just like ordinary matter.

12 According to models, dark matter began to clump immediately after the big bang, forming spherical **blobs** that astronomers refer to as "**halos**". The baryons, in contrast, were initially kept from clumping by their interactions with one another and with radiation. They remained in a hot, gaseous phase. As the universe expanded, this gas cooled and the baryons were able to pack themselves together. The first stars and galaxies **coalesced** out of this cooled gas a few hundred million years after the big bang. They did not materialize in random locations but in the centers of the dark matter halos that had already taken shape.

blob *n.* 斑点
halo *n.*（日月周围的）晕轮
coalesce *v.* 结合；合并

13 Since the 1980s a number of theorists have done detailed computer simulations of this process, including groups led by Simon D. M. White of the Max Planck Institute for Astrophysics[6] in Garching, Germany, and Carlos S. Frenk of Durham University in England. They have shown that most of the first structures were small, low-mass dark matter halos. Because the early universe was so dense, these low-mass halos (and the galaxies they contained) merged with one another to form larger-mass systems. In this way, galaxy construction was a bottom-up process, like building a dollhouse out of Lego bricks[7]. (The alternative would have been a top-down process, in which you start with the dollhouse and smash it to make bricks.) My colleagues and I have sought to test these models by looking at distant galaxies and how they have merged over cosmic time.

## Galaxy Formation Peters Out

14 Detailed studies indicate that a galaxy gets bent out of shape when it merges with another galaxy. The earliest galaxies we can see existed when the universe was about a billion years old, and many of these indeed appear to be merging. As time went on, though, the **fusion** of massive galaxies became less common. Between two billion and six billion years after the big bang—that is, over the first half of cosmic history—the fraction of massive galaxies undergoing a merger dropped from half to nearly nothing at all. Since then, the distribution of galaxy shapes has been frozen, an indication that smashups and mergers have become relatively uncommon.

fusion *n.* 熔化，熔解

15 In fact, fully 98 percent of massive galaxies in today's universe are either **elliptical** or spiral, with shapes that would be disrupted by a merger. These galaxies are stable and comprise mostly old stars,

elliptical *adj.* 椭圆的

morphological *adj.* 形态学的

which tells us that they must have formed early and have remained in a regular **morphological** form for quite some time. A few galaxies are merging in the present day, but they are typically of low mass.

cessation *n.* 停止
venue *n.* 发生地；集合地点
paradoxical *adj.* 自相矛盾的

16 The virtual **cessation** of mergers is not the only way the universe has run out of steam since it was half its current age.[8] Star formation, too, has been waning. Most of the stars that exist today were born in the first half of cosmic history, as first convincingly shown by several teams in the 1990s, including ones led by Simon J. Lilly, then at the University of Toronto, Piero Madau, then at the Space Telescope Science Institute, and Charles C. Steidel of the California Institute of Technology. More recently, researchers have learned how this trend occurred. It turns out that star formation in massive galaxies shut down early. Since the universe was half its current age, only lightweight systems have continued to create stars at a significant rate. This shift in the **venue** of star formation is called galaxy downsizing. It seems **paradoxical**. Galaxy formation theory predicts that small galaxies take shape first and, as they amalgamate, massive ones arise. Yet the history of star formation shows the reverse: massive galaxies are initially the main stellar birthing grounds, then smaller ones take over.

***The universe has run out of steam since it was half its current age. Mergers have ceased, and black holes are quiescent.***

quasar *n.* 类星体

17 Another oddity is that the buildup of supermassive black holes, found at the centers of galaxies, seems to have slowed down considerably. Such holes power **quasars** and other types of active galaxies, which are rare in the modern universe; the black holes in our galaxy and others are quiescent. Are any of these trends in galaxy evolution related? Is it really possible that dark energy is the root cause?

## The Steady Grip of Dark Energy

culprit *n.* 问题的起因

18 Some astronomers have proposed that internal processes in galaxies, such as energy released by black holes and supernovae, turned off galaxy and star formation. But dark energy has emerged as possibly a more fundamental **culprit**, the one that can link everything

together. The central piece of evidence is the rough coincidence in timing between the end of most galaxy and cluster formation and the onset of the domination of dark energy. Both happened when the universe was about half its present age.

19 The idea is that up to that point in cosmic history, the density of matter was so high that gravitational forces among galaxies dominated over the effects of dark energy. Galaxies rubbed shoulders, interacted with one another, and frequently merged. New stars formed as gas clouds within galaxies collided, and black holes grew when gas was driven toward the centers of these systems. As time progressed and space expanded, matter thinned out and its gravity weakened, whereas the strength of dark energy remained constant (or nearly so). The **inexorable** shift in the balance between the two eventually caused the expansion rate to switch from deceleration to acceleration. The structures in which galaxies reside were then pulled apart, with a gradual decrease in the galaxy merger rate as a result. Likewise, intergalactic gas was less able to fall into galaxies. Deprived of fuel, black holes became more quiescent.

inexorable *adj.* 不可阻挡的

20 This sequence could perhaps account for the downsizing of the galaxy population. The most massive dark matter halos, as well as their embedded galaxies, are also the most clustered; they reside in close proximity to other massive halos. Thus, they are likely to knock into their neighbors earlier than are lower-mass systems. When they do, they experience a burst of star formation. The newly formed stars light up and then blow up, heating the gas and preventing it from collapsing into new stars. In this way, star formation chokes itself off: stars heat the gas from which they emerged, preventing new ones from forming. The black hole at the center of such a galaxy acts as another damper on star formation. A galaxy merger feeds gas into the black hole, causing it to fire out jets that heat up gas in the system and prevent it from cooling to form new stars.

21 Apparently, once star formation in massive galaxies shuts down, it does not start up again—most likely because the gas in these systems becomes **depleted** or becomes so hot that it cannot cool down quickly enough. These massive galaxies can still merge with one another, but few new stars emerge for want of cold gas. As the massive galaxies

deplete *v.* 耗尽；大大减少

stagnate v. 停滞；不流动
modulate v. 调整，调节

**stagnate**, smaller galaxies continue to merge and form stars. The result is that massive galaxies take shape before smaller ones, as is observed. Dark energy perhaps **modulated** this process by determining the degree of galaxy clustering and the rate of merging.

22 Dark energy would also explain the evolution of galaxy clusters. Ancient clusters, found when the universe was less than half its present age, were already as massive as today's clusters. That is, galaxy clusters have not grown by a significant amount in the past six billion to eight billion years. This lack of growth is an indication that the infall of galaxies into clusters has been curtailed since the universe was about half its current age—a direct sign that dark energy is influencing the way galaxies are interacting on large scales. Astronomers knew as early as the mid-1990s that galaxy clusters had not grown much in the past eight billion years, and they attributed this to a lower matter density than theoretical arguments had predicted. The discovery of dark energy resolved the tension between observation and theory.

retinue n.（要人的）一批随员

23 An example of how dark energy alters the history of galaxy clusters is the fate of the galaxies in our immediate vicinity, known as the Local Group[9]. Just a few years ago astronomers thought that the Milky Way and Andromeda[10], its closest large neighbor, along with their **retinue** of satellites, would fall into the nearby Virgo cluster[11]. But it now appears that we shall escape that fate and never become part of a large cluster of galaxies. Dark energy will cause the distance between us and Virgo to expand faster than the Local Group can cross it.

24 By throttling cluster development, dark energy also controls the makeup of galaxies within clusters. The cluster environment facilitates the formation of a zoo of galaxies such as the so-called lenticulars[12], giant ellipticals and dwarf ellipticals[13]. By regulating the ability of galaxies to join clusters, dark energy dictates the relative abundance of these galaxy types.

***Space is emptying out, leaving our Milky Way galaxy and its neighbors an increasingly isolated island.***

25 This is a good story, but is it true? Galaxy mergers, black hole

activity and star formation all decline with time, and very likely they are related in some way. But astronomers have yet to follow the full sequence of events. Ongoing surveys with the Hubble Space Telescope[14], the Chandra X-ray Observatory[15] and sensitive ground-based imaging and spectroscopy will scrutinize these links in coming years. One way to do this is to obtain a good census of distant active galaxies and to determine the time when those galaxies last underwent a merger. The analysis will require the development of new theoretical tools but should be within our grasp in the next few years.

## Notes

1. **dark energy:** 暗能量，暗能。20世纪90年代宇宙学家发现，宇宙是由普通物质和看不见的暗物质（dark matter）组成，宇宙深处的星体爆炸提示，一种宇宙力在导致宇宙分离，天文学家把这种力称为“暗能”。据研究，宇宙中70%是暗能，15%是传统的暗物质，仅仅5%是构成诸如地球和星体这样的普通物质。
2. **Edwin Hubble:** 爱德温·哈勃（1889–1953），美国天文学家，研究现代宇宙理论最著名的科学家之一，建立了哈勃定律，银河外天文学的奠基人和提供宇宙膨胀实证的第一人。
3. **the Wilkinson Microwave Anisotropy Probe (WMAP):** 威尔金森宇宙微波各向异性探测卫星，是美国航空航天局于2001年6月发射的一颗人造卫星，目的是探测宇宙中大爆炸残留的辐射热，以帮助测试有关宇宙产生的各种理论。
4. **angular momentum:** 角动量。描述物体转动状态的量，又称动量矩。数学上，角动量为质量、速度和旋转半径的乘积。
5. **Thus, it is not at all obvious that dark energy should have had any say whatsoever in how galaxies formed:** 因此，在银河系的形成过程中，暗能所起的作用并不明显。句中的say意为“发言机会，发言权”，该句采用了拟人的修辞方法。
6. **the Max Planck Institute for Astrophysics:**（德国）马克士普朗克天体物理研究所。
7. **Lego bricks:** 乐高积木。1932年，丹麦木匠奥尔·科克·克里斯蒂安森发明了一种可以互相拼插的塑料玩具，并将Leg和Godt（丹麦语“玩得好”）合在一起，创造了LEGO（乐高）这一品牌。巧合的是lego在拉丁语中的意思是“拼在一起”。此后还出现了乐高砖块手机游戏。
8. **The virtual cessation of mergers is not the only way the universe has run out of steam since it was half its current age:** 在宇宙形成到目前的后半段时间里，星系之间的融和实际上已经停止了，而这一现象并不是宇宙失去动力的唯一方式。句中短语run out of steam常用于口语，意为“失去动力，筋疲力尽”，相当于become exhausted。
9. **Local Group:** 本星系群，为典型的疏散星系团，没有明显的向心趋势。成员星系约40个。银河系和仙女星系是本星系群中最大的两个星系。
10. **Andromeda:** 仙女座，距地球约250万光年，为本星系群的主要成员。它的直径达12万光年，含有10,000亿颗恒星。

11. **Virgo cluster:** 室女座星系团，是距离银河系最近的星系团。
12. **lenticulars:** 透镜状星系（lenticular galaxies)，为介于椭圆星系和螺旋星系之间的星系，呈扁平盘状，类似椭圆星系；但同时旋臂结构不明显，类似螺旋星系。侧对地球的透镜状星系往往难于和椭圆星系相区分。
13. **giant ellipticals and dwarf ellipticals:** 巨椭圆星系和矮椭圆星系。椭圆星系是河外星系的一种，呈圆球型或椭球型。其中质量最小的是矮椭圆星系，和球状星团相当；而质量最大的超巨型椭圆星系可能是宇宙中最大的恒星系统，质量约为太阳的千万倍到百万亿倍。
14. **the Hubble Space Telescope:** 哈勃太空望远镜，1990年4月由美国航空航天局利用航天飞机发射的一台巨大的太空望远镜。该望远镜镜面直径达240厘米，在离地球表面580千米高空的轨道上运行，已向地面发回大量极有价值的太空图片。
15. **the Chandra X-ray Observatory:** 钱德拉X射线天文台（缩写为CXO)，是美国航空航天局于1999年7月发射的一颗X射线天文卫星，为大型轨道天文台计划的第三颗卫星，目的是观测天体的X射线辐射。

## Exercises

### I. Answer the following questions.

1. What is the importance of dark energy?
2. What makes dark energy so hard to be recognized?
3. Why would scientists think that the expansion rate of galaxies should be slowing down?
4. Why does the author say dark energy may be the key link among several aspects of galaxy and cluster formation that not long ago appeared unrelated?
5. According to the author, what is the possible proof that it was dark energy that turned off galaxy and star formation?

### II. The following statements are incomplete. Search the missing information in the passage and fill in the blanks.

1. Scientists are just starting the long process of figuring out what dark energy is and what its __________ are.
2. All the dark energy in our solar system amounts to the mass of __________.
3. The rate at which galaxies are moving away from us is proportional to __________: the more __________ a galaxy is, the __________ its recession.
4. The observations made clear that the expansion of universe was __________ in the past than today and is therefore __________.
5. As you zoom in from the entire observable universe, the first thing you notice is that matter on cosmic scales is distributed in a __________ pattern—a filigree of filaments,

several tens of millions of __________ long, interspersed with voids of similar size.

6. The earliest galaxies we can see existed when the universe was about __________ years old, and many of these indeed appear to be __________.
7. Detailed studies indicate that a galaxy gets __________ when it merges with another galaxy.
8. New stars formed as __________ within galaxies collided, and __________ grew when gas was driven toward the centers of these systems.

## III. Identify the implied meanings of the underlined parts of the following sentences according to the context of the passage, and translate the sentences into Chinese.

1. Dark energy does more than hurry along the expansion of the universe. It also has a stranglehold on the shape and spacing of galaxies.
2. All the dark energy in our solar system amounts to the mass of a small asteroid, making it an utterly inconsequential player in the dance of the planets.
3. The situation gets more complicated as you continue to zoom in and reach the scale of galaxies and galaxy clusters.
4. In our universe, neither player in this tug-of-war is overwhelmingly dominant.
5. According to models, dark matter began to clump immediately after the Big Bang, forming spherical blobs that astronomers refer to as "halos".
6. Galaxies rubbed shoulders, interacted with one another, and frequently merged.
7. This sequence could perhaps account for the downsizing of the galaxy population.
8. Thus, they are likely to knock into their neighbors earlier than are lower-mass systems.
9. In this way, star formation chokes itself off: stars heat the gas from which they emerged, preventing new ones from forming.
10. The cluster environment facilitates the formation of a zoo of galaxies.

## Translation Techniques (1)

### Conversion Between English Nouns and Chinese Verbs (英语名词与汉语动词的转换)

The fact that English and Chinese belong to two different language families determines that it is often necessary to change the "parts of speech" when we translate English into Chinese or vice versa. Generally speaking, technical English uses nouns more widely in order to achieve formality in style while Chinese tends to use verbs more often. Therefore, the same concept or idea may be expressed in the form of nouns in English and in the form of verbs in Chinese. Consequently, the translator needs to convert nouns into verbs when translating the English versions into Chinese, and convert verbs when translating the Chinese versions into English.

For example, in the translation of the sentence "A team of British researchers believe that if the comet contained flammable gases, a collision with earth would liberate the gases into the earth's atmosphere where it would be ignited by lightning bolts", it is necessary to change the noun *collision* into the verb: 英国的一组研究人员认为，如果彗星含有可燃气体，在与地球相撞时，将会把这些气体排放到地球的大气并被闪电引燃。

Here are a few more examples:

1. Basically applied science starts with the clear understanding of a human need and then uses all the available scientific knowledge to assist in the *satisfaction* of this need.
   归根到底，应用科学的出发点是清楚地了解人类的某一项需要，然后利用一切现有的科学知识去帮助满足这种需要。
2. Pure science is primarily concerned with the *development* of theories and the *establishment* of the relations between the phenomena of the universe.
   理论科学的主要任务是发展理论，建立宇宙间各种现象之间的关系。
3. Over the years, NASA's computer aces had radioed up so many *improvements* to the on-board computer that the craft had essentially received a long-distance brain transplant.
   在过去的几年中，（美国）国家航空航天局的计算机专家通过发射无线电信号，多次改善飞船上的计算机，飞船实质上接受了长距离的大脑移植。
4. Increasingly, there is *talk* of viewer *participation* in television; two-way TV programs will make it possible to take part in a soap opera or football game as if the viewer were there.
   人们越来越多地谈论着电视观众参与电视节目的问题，双向电视节目将使电视观众有可能参加电视连续剧的演出或足球比赛，犹如身临其境。
5. In general, drying a solid means the *removal* of relatively small amounts of water or other liquid from the solid material to reduce the content of residual liquid to an acceptable low value.
   一般情况下，固体干燥指的是从固体材料中除去相对少量的水或其他液体，从而使残留液体的含量减少到可接受的低水平。

However, it is also possible that the verb in the English sentence is changed into the noun in Chinese translation. For example, "Some researchers, to be sure, had anticipated that such energy existed, but even they will tell you that its detection ranks among the most revolutionary discoveries in 20th-century cosmology."（可以肯定地说，有些研究者在此之前就已经预见到这种能量的存在，但是即使是他们也会告诉你，这一发现是20世纪宇宙学最重大的发现之一。）

## IV. Translate the following sentences into Chinese. Pay attention to the conversion between the English nouns and the Chinese verbs.

1. Not only does dark energy appear to make up the bulk of the universe, but its existence, if it stands the test of time, will probably require the development of new theories of physics.
2. The same is true of galaxy clusters, the largest coherent bodies in the universe—assemblages of thousands of galaxies embedded in a vast cloud of hot gas and bound together by gravity.
3. Since then, the distribution of galaxy shapes has been frozen, an indication that smashups and mergers have become relatively uncommon.
4. The structures in which galaxies reside were then pulled apart, with a gradual decrease in the galaxy merger rate as a result.
5. These massive galaxies can still merge with one another, but few new stars emerge for want of cold gas.
6. The analysis will require the development of new theoretical tools but should be within our grasp in the next few years.
7. Scientists say that government construction of a multi-billion-dollar, permanent space station could aid in detecting natural disasters on earth in advance, conducting medical research and collecting solar energy to transmit back.
8. So the search for cosmic gamma rays is also the search for the origin of cosmic rays.

## Translation Techniques (2)

### Inversion（语序倒置）

As a translation technique, "inversion" means the change of the word order in a sentence according to the meaning and the usage of the language to be translated into. The change of the word-order is often necessary or even inevitable in translation because each language has its own "natural word order" which must be followed.

For example, when translating the sentence "He looked at radio signals from two nearby stars while studying interstellar gas with a radio telescope at Greenbank, West Virginia", we need to change the positions of "at Greenbank", "Virginia", "with a radio telescope", "radio signals", and "from two nearby stars", etc. So the Chinese sentence is completely different from the original English sentence in word order and structure: 他在西弗吉尼亚州的格林班克用射电望远镜研究星际气体时，看到了来自附近两颗恒星的无线电信号。

A careful comparison between English and Chinese reveals to us that there are many situations in which inversion is needed. The following examples are typical in technical translation.

1. Physicist Edward Teller, *the father of the hydrogen bomb*, argues that the nuclear weapons are "uniquely designed for defensive purposes" and that "we need to know what the other side is doing and how to defend against it".
   氢弹之父、物理学家爱德华·泰勒争辩说：核武器是"专门为防御目的而设计的"，而且"我们需要知道对方在干什么，以及如何防备。"
2. U.S. and Japanese firms have made prototype ceramic diesel engines that run without a cooling system, allowing *higher efficiency* and *lower weight*.
   美国和日本的公司已经制造出不需要冷却系统，效率高而重量轻的陶瓷柴油发动机样机。
3. Scientists believe that the nucleus of an average comet is *only a mile or two in diameter*.
   科学家认为普通彗星的慧核，其直径只不过一两英里。
4. *The two phenomena*, *inertia and gravitation*, seem so different from each other that one can't help but wonder why these two different ways of measuring mass always give the same answer.
   惯性与重力这两种现象看起来如此不同，人们禁不住要问为什么两种不同方法所测得的质量总是相同的。
5. It is only recently that astronomers have begun *specific research into black holes*.
   最近天文学家才开始对黑洞进行具体的研究。

## V. Discuss the translation technique and the ways of applying the technique to the translation of the following sentences. Complete each of the Chinese translations.

1. The top basaltic layer of the moon has been pounded by a steady stream of small meteorites, including many less than a millimeter in width. (__________不断受到微小的陨石流的冲击，其中包括__________。)
2. Franz Halberg, a young European scientist working in the United States, noticed that the number of white blood cells in laboratory mice was dramatically higher and lower at different times of day. (__________，曾注意到__________的多寡在每天不同的时间里有显著的差别。)
3. The rock sample brought back from the moon varied in age between 3 billion years in the maria to 4.6 billion years in the highlands. (__________，既有30亿年前的“海”中岩石，又有__________。)
4. All signs point to an easier job for application software vendors and, ultimately, better application software for end-users. (所有的迹象都表明，__________，而且最终__________。)
5. The holoendoscope is a stainless steel tube, 10 millimeters wide and 86 millimeters long. (全息内窥镜是__________。)
6. Frank Drake, an American radio astronomer, made the first SETI observations in 1960. (__________于1960年率先进行了搜寻外星人的天文观察。)
7. Wings vibrating, hindquarters waggling, a bee dancing on the side of the honeycomb, directs followers to nectar or pollen found on a recent journey. (__________，从而告诉同伴__________。)
8. A tremendous storm system, a countercyclone as big across as Earth and christened the Great Dark Spot, marks the southern hemisphere of Neptune.
   巨大的风暴体系（即__________）标志着海王星的南半球。

## VI. Translate the following sentences into Chinese. Pay attention to how the change of the word order should be made.

1. Since the days of American astronomer Edwin Hubble, observers have known that all but the nearest galaxies are moving away from us at a rapid rate.
2. My colleagues and I have sought to test these models by looking at distant galaxies and how they have merged over cosmic time.
3. Most of the stars that exist today were born in the first half of cosmic history, as first convincingly shown by several teams in the 1990s, including ones led by Simon J. Lilly,

then at the University of Toronto, Piero Madau, then at the Space Telescope Science Institute, and Charles C. Steidel of the California Institute of Technology.

4. The result is that massive galaxies take shape before smaller ones, as is observed.
5. Galaxy mergers, black hole activity and star formation all decline with time, and very likely they are related in some way.

## VII. Translate the following passage into Chinese.

Astronomers know dark matter is there by its gravitational effect on the matter that we see and there are ideas about the kinds of particles it must be made of. By contrast, dark energy remains a complete mystery. The name "dark energy" refers to the fact that some kind of "stuff" must fill the vast reaches of mostly empty space in the Universe in order to be able to make space accelerate in its expansion. In this sense, it is a "field" just like an electric field or a magnetic field, both of which are produced by electromagnetic energy. But this analogy can only be taken so far because we can readily observe electromagnetic energy via the particle that carries it, the photon.

Some astronomers identify dark energy with Einstein's Cosmological Constant. Einstein introduced this constant into his general relativity when he saw that his theory was predicting an expanding universe, which was contrary to the evidence for a static universe that he and other physicists had in the early 20th century. This constant balanced the expansion and made the universe static. With Edwin Hubble's discovery of the expansion of the Universe, Einstein dismissed his constant. It later became identified with what quantum theory calls the energy of the vacuum.

In the context of dark energy, the cosmological constant is a reservoir which stores energy. Its energy scales as the universe expands. Applied to the supernova data, it would distinguish effects due to the matter in the universe from those due to the dark energy. Unfortunately, the amount of this stored energy required is far more than observed, and would result in very rapid acceleration (so much so that the stars and galaxies would not form). Physicists have suggested a new type of matter, "quintessence", which would fill the universe like a fluid which has a negative gravitational mass. However, new constraints imposed on cosmological parameters by Hubble Space Telescope data rule out at least simple models of quintessence.

## VIII. Translate the following passage into English.

### 暗物质

在物理学和宇宙学中，暗物质指的是一种假想的物质，它与电磁力不发生相互作用，但从万有引力对可见物质的影响可以推断它的存在。根据大爆炸宇宙学，以及目前对那些大于星系的结构体的观测，在可观察的宇宙空间，绝大部分物质都是由暗物质和暗能量组成的。迄今人们观察到了星系的旋转速度，星系在星系团中的轨道运行速度，星系团（如“子弹星系团”）附近背景物体的引力透镜效应，以及星系和星系团中热气体的温度分布等现象，这些现象意味着暗物质的存在。暗物质不仅在结构形成和星系演化过程中起着关键作用，对宇宙微波背景辐射的各向异性也有着显著影响。所有这些观察结果都表明，星系、星系团乃至整个宇宙所拥有的物质远远超过与电磁辐射相互作用的物质：其余这部分就叫做“暗物质成分”。

暗物质成分的质量远远大于宇宙的“可见”成分。据估计，目前宇宙中普通重子和辐射的密度大约相当于每立方米一个氢原子，可以直接观察到的能量密度仅占总量的4%，人们认为大约22%的能量密度是由暗物质组成，而其余74%由暗能量构成，后者是一种更加奇特的成分，稀薄地分布在宇宙中。据称，某种很难观测到的重子物质是暗物质的构成部分，但也只是一小部分而已。确定暗物质这种不可见物质的性质是现代宇宙学和粒子物理学最重要的课题之一。

## Section 2 Reading for Academic Purposes

*Reading for the purpose of academic researches requires you to read actively, which actually means a series of things. Perhaps most importantly active reading means the awareness of the purpose for reading. Far too often students read aimlessly, hoping that the key ideas will somehow "sink in" and then eventually "surface" when they need to. Having a purpose is another way of saying that you have set goals for your readings.*

*In addition to setting goals and purposes for reading, active reading may also involve using the structure of your reading to construct an overview for your reading, which you use to select a focus. The structures of the reading materials vary almost as much as the readings themselves, but there are some common features associated with various kinds of readings that readers can make effective use of.*

*All of these structures assist the reader in developing an overview of what is about to be read and this allows readers to guide themselves through the text with a focus in mind. The following article will familiarize you with the structure of the scientific papers, and explain the characteristics of each section.*

### Organization of a Scientific Paper
### 科技论文的篇章结构

The main purpose of a scientific paper is to report new results, usually experimental, and to relate these results to previous knowledge in the field. Papers are one of the most important ways that we communicate with one another.

In most scientific journals, scientific papers follow a standard format *n.* 版本，格式

**format**. They are divided into several sections, and each section serves a specific purpose in the paper. We first describe the standard format, then some variations on that format.

A paper begins with a short Summary or Abstract. Generally, it gives a brief background to the topic; describes concisely the major findings of the paper; and relates these findings to the field of study. As will be seen, this logical order is also that of the paper as a whole.

consistent *adj.* 一致的

The next section of the paper is the Introduction. In many journals this section is not given a title. As its name implies, this section presents the background knowledge necessary for the reader to understand why the findings of the paper are an advance on the knowledge in the field. Typically, the Introduction describes first the accepted state of knowledge in a specialized field; then it focuses more specifically on a particular aspect, usually describing a finding or set of findings that led directly to the work described in the paper. If the authors are testing a hypothesis, the source of that hypothesis is spelled out, findings are given with which it is **consistent**, and one or more predictions are given. In many papers, one or several major conclusions of the paper are presented at the end of this section, so that the reader knows the major answers to the questions just posed. Papers more descriptive or comparative in nature may begin with an introduction to an area which interests the authors, or the need for a broader database.

replicate *v.* 复制；重复

The next section of most papers is the Materials and Methods. In some journals this section is the last one. Its purpose is to describe the materials used in the experiments and the methods by which the experiments were carried out. In principle, this description should be detailed enough to allow other researchers to **replicate** the work. In practice, these descriptions are often highly compressed, and they often refer back to previous papers by the authors.

The third section is usually Results. This section describes the experiments and the reasons they were done. Generally, the logic of the Results section follows directly from that of the Introduction. That is, the Introduction poses the questions addressed in the early part of Results. Beyond this point, the organization of Results differs from one paper to another. In some papers, the results are presented without extensive discussion, which is reserved for the following section. This is appropriate when the data in the early parts do not need to be interpreted extensively to understand why the later experiments were done. In other papers, results are given, and then they are **interpreted**, perhaps taken together with other findings not in the paper, so as to give the logical basis for later experiments.

interpret *v.* 解释；说明

The fourth section is the Discussion. This section serves several purposes. First, the data in the paper are interpreted; that is, they are analyzed to show what the authors believe the data show. Any limitations to the interpretations should be **acknowledged**, and fact should clearly be separated from speculation. Second, the findings of the paper are related to other findings in the field. This serves to show how the findings contribute to knowledge, or correct the errors of previous work. As stated, some of these logical arguments are often found in the Results when it is necessary to clarify why later experiments were carried out. Although you might argue that in this case the discussion material should be presented in the Introduction, more often you cannot grasp its significance until the first part of Results is given.

acknowledge *v.* 承认

Finally, papers usually have a short Acknowledgments section, in which various contributions of other workers are recognized, followed by a Reference list giving references to papers and other works cited in the text.

Papers also contain several Figures and Tables. These contain data described in the paper. The figures and tables also have **legends**, whose purpose is to give details of the particular experiment or experiments shown there. Typically, if a procedure is used only once in a paper, these details are described in Materials and Methods, and the Figure or Table legend refers back to that description. If a procedure is used repeatedly, however, a general description is given in Materials and Methods, and the details for a particular experiment are given in the Table or Figure legend.

legend *n.* 图例；插图的说明

## Variations on the organization of a paper

In most scientific journals, the above format is followed. Occasionally, the Results and Discussion are combined, in cases in which the data need extensive discussion to allow the reader to follow the train of logic developed in the course of the research. As stated, in some journals, Materials and Methods follows the Discussion. In certain older papers, the Summary was given at the end of the paper.

discrete *adj.* 分离的，互不连接的
self-contained *adj.* 独立的
circumvent *v.* 设法克服或避免；回避

The formats for two widely-read journals, *Science* and *Nature*, differ markedly from the above outline. These journals reach a wide audience, and many authors wish to publish in them; accordingly, the space limitations on the papers are severe, and the prose is usually highly compressed. In both journals, there are no **discrete** sections, except for a short abstract and a reference list. In *Science*, the abstract is **self-contained**; in *Nature*, the abstract also serves as a brief introduction to the paper. Experimental details are usually given either in endnotes (for *Science*) or Figure and Table legends and a short Methods section (in *Nature*). Authors often try to **circumvent** length limitations by putting as much material as possible in these places.

Many other journals also have length limitations, which similarly lead to a need for conciseness. For example, the *Proceedings of the National Academy of Sciences* (*PNAS*) has a six-page limit; *Cell* severely edits many papers to shorten them, and has a short word limit in the abstract; and so on.

condense *v.* 压缩，精简
accessible *adj.* 易懂的；可以理解的

In response to the pressure to edit and make the paper concise, many authors choose to **condense** or, more typically, omit the logical connections that would make the flow of the paper easy. In addition, much of the background that would make the paper **accessible** to a wider audience is condensed or omitted, so that the less-informed reader has to consult a review article or previous papers to make sense of what the issues are and why they are important. Finally, again, authors often circumvent page limitations by putting crucial details into the Figure and Table legends, especially when (as in *PNAS*) these are set in smaller type.

## Exercises

**I. Read the passage and decide whether the following statements are true or false. Write T for True and F for False in the brackets.**

1. The main purpose of a scientific paper is to report new results, usually experimental, and to relate these results to previous knowledge in the field. (    )

2. A paper begins with a short Summary or Abstract. (  )
3. In many journals the section of introduction is given a title. (  )
4. The next section of most papers is the Materials and Methods. In some journals this section is the last one. (  )
5. Generally, the logic of the Results section follows indirectly from that of the Introduction. (  )
6. Finally, papers usually have a short Acknowledgements section, followed by a Reference list. (  )
7. The figures and tables also have legends, whose purpose is to give details of the particular experiment or experiments shown there. (  )
8. As stated, in all journals, Materials and Methods follows the Discussion. (  )
9. In response to the pressure to edit and make the paper concise, many authors choose to condense. (  )
10. Authors often handle page limitations by putting unimportant details into the Figure and Table legends. (  )

**II. Read the passage again, and complete the following items.**

1. In most scientific journals, scientific papers include the following sections: ___________
2. The content of Summary or Abstract: ___________
3. The Introduction section deals with the following two points: ___________
4. The purpose of Materials and Methods: ___________
5. The two ways of organizing Results: ___________
6. The purposes of the Discussion section: ___________
7. The reason for combining the Results and Discussion: ___________
8. The difference between the abstracts in *Science* and those in *Nature*: ___________

## Quotations from Great Scientists

*A scientist in his laboratory is not a mere technician: he is also a child confronting natural phenomena that impress him as though they were fairy tales.*

*—Marie Curie*

科学家在他的实验室里不仅是个从事技术研究的人员：他也是个孩子，他所面对的是令他神往的童话般的自然现象。

——玛丽亚·居里

# Chapter 4

## Space and Time
## 空间与时间

## Section 1 Reading and Translation

### I What You Are Going to Read

Nothing is more mystically alluring to human beings than the far-reaching speculation on space and time. From Newton to Einstein, the basic laws of physics work equally well forward or backward in time, yet we perceive time to move in one direction only—toward the future. Why?

To account for it, we have to delve into the prehistory of the universe, to a time before the Big Bang. Our universe may be part of a much larger multiverse, which as a whole is time-symmetric. Time may run backward in other universes.

The following is an excerpt of an article published in the June 2008 issue of *Scientific American*, which will reveal this mysterious world to you.

### II About the Author

Sean M. Carroll is a senior research associate in physics at the California Institute of Technology. His research ranges over cosmology, particle physics and Einstein's general theory of relativity, with a particular expertise in dark energy. He has been awarded fellowships from the Sloan and Packard foundations, as well as the M.I.T. Graduate Student Council Teaching Award and the Villanova University Arts and Sciences Alumni Medallion. Outside of academia, Carroll is best known as a contributor to the blog *Cosmic Variance*, which is not only one of the most thoughtful science blogs but also the way he met his wife, science writer Jennifer Ouellette.

### III Reading Passage

**Does Time Run Backward in Other Universes?**

**时间在其他宇宙中是倒转的吗？**

By Sean M. Carroll

*One of the most basic facts of life is that the future looks different from the past. But on a grand cosmological scale, they may look the same.*

1 The universe does not look right. That may seem like a strange thing to say, given that **cosmologists** have very little standard for comparison.[1] How do we know what the universe is supposed to look like? Nevertheless, over the years we have developed a strong **intuition** for what counts as "natural"—and the universe we see does not qualify.

cosmologist *n.* 宇宙学家
intuition *n.* 直觉；凭直觉感知的事

2 Make no mistake: cosmologists have put together an incredibly successful picture of what the universe is made of and how it has evolved. Some 14 billion years ago the cosmos was hotter and denser than the interior of a star, and since then it has been cooling off and thinning out as the fabric of space expands. This picture accounts for just about every observation we have made, but a number of unusual features, especially in the early universe, suggest that there is more to the story than we understand.

3 Among the unnatural aspects of the universe, one stands out: time **asymmetry**[2]. The **microscopic** laws of physics that underlie the behavior of the universe do not distinguish between past and future, yet the early universe—hot, dense, **homogeneous**—is completely different from today's—cool, **dilute**, **lumpy**. The universe started off orderly and has been getting increasingly disorderly ever since. The asymmetry of time, the arrow that points from past to future, plays an unmistakable role in our everyday lives: it accounts for why we cannot turn an **omelet** into an egg, why ice cubes never **spontaneously** unmelt in a glass of water, and why we remember the past but not the future. And the origin of the asymmetry we experience can be traced all the way back to the orderliness of the universe near the Big Bang[3]. Every time you break an egg, you are doing observational cosmology.

asymmetry *n.* 不对称
microscopic *adj.* 微观的
homogeneous *adj.* 均质的；均匀的
dilute *adj.* 稀释的，冲淡的
lumpy *adj.* 多块的
omelet *n.* 煎蛋卷，炒鸡蛋
spontaneously *adv.* 自然地，自发地

4 The arrow of time[4] is arguably the most **blatant** feature of the universe that cosmologists are currently at an utter loss to explain. Increasingly, however, this puzzle about the universe we observe hints at the existence of a much larger spacetime[5] we do not observe. It adds support to the notion that we are part of a **multiverse** whose dynamics help to explain the seemingly unnatural features of our local **vicinity**.

blatant *adj.* 极为明显的
multiverse *n.* 多元宇宙
vicinity *n.* 附近，邻近

## The Puzzle of Entropy[6]

5 Physicists **encapsulate** the concept of time asymmetry in the celebrated second law of **thermodynamics**[7]: entropy in a closed system

encapsulate *v.* 总结，扼要概括
thermodynamics *n.* 热力学

microstate *n.* 微观状态
velocity *n.* 速度

never decreases. Roughly, entropy is a measure of the disorder of a system. In the 19th century, Austrian physicist Ludwig Boltzmann[8] explained entropy in terms of the distinction between the **microstate** of an object and its macrostate. If you were asked to describe a cup of coffee, you would most likely refer to its macrostate—its temperature, pressure and other overall features. The microstate, on the other hand, specifies the precise position and **velocity** of every single atom in the liquid. Many different microstates correspond to any one particular macrostate: we could move an atom here and there, and nobody looking at macroscopic scales would notice.

logarithm *n.* 对数
configuration *n.* 构造，结构
segregate *v.* 分开，隔离

6 Entropy is the number of different microstates that correspond to the same macrostate. (Technically, it is the number of digits, or **logarithm**, of that number.) Thus, there are more ways to arrange a given number of atoms into a high-entropy **configuration** than into a low-entropy one. Imagine that you pour milk into your coffee. There are a great many ways to distribute the molecules so that the milk and coffee are completely mixed together but relatively few ways to arrange them so that the milk is **segregated** from the surrounding coffee. So the mixture has a higher entropy.

conspire *v.* 共同策划
of its own accord 自愿地，主动地
reshuffle *v.* 重新组合，改组

7 From this point of view, it is not surprising that entropy tends to increase with time. High-entropy states greatly outnumber low-entropy ones; almost any change to the system will land it in a higher-entropy state, simply by the luck of the draw[9]. That is why milk mixes with coffee but never unmixes. Although it is physically possible for all the milk molecules to spontaneously **conspire** to arrange themselves next to one another, it is statistically very unlikely. If you waited for it to happen **of its own accord** as molecules randomly **reshuffled**, you would typically have to wait much longer than the current age of the observable universe. The arrow of time is simply the tendency of systems to evolve toward one of the numerous, natural, high-entropy states.

8 But explaining why low-entropy states evolve into high-entropy states is different from explaining why entropy is increasing in our universe. The question remains: Why was the entropy low to start with? It seems very unnatural, given that low-entropy states are so rare. Even granting that our universe today has medium entropy, that

does not explain why the entropy used to be even lower. Of all the possible initial conditions that could have evolved into a universe like ours, the overwhelming majority have much higher entropy, not lower.

9 In other words, the real challenge is not to explain why the entropy of the universe will be higher tomorrow than it is today but to explain why the entropy was lower yesterday and even lower the day before that. We can trace this logic all the way back to the beginning of time in our observable universe. Ultimately, time asymmetry is a question for cosmology to answer.

## The Disorder of Emptiness

10 The early universe was a remarkable place. All the particles that make up the universe we currently observe were squeezed into an extraordinarily hot, dense volume. Most important, they were distributed nearly uniformly throughout that tiny volume. On average, the density differed from place to place by only about one part in 100,000.[10] Gradually, as the universe expanded and cooled, the pull of gravity enhanced those differences. Regions with slightly more particles formed stars and galaxies, and regions with slightly fewer particles emptied out to form voids.

11 Clearly, gravity has been crucial to the evolution of the universe. Unfortunately, we do not fully understand entropy when gravity is involved. Gravity arises from the shape of spacetime, but we do not have a comprehensive theory of spacetime; that is the goal of a quantum theory of gravity. Whereas we can relate the entropy of a fluid to the behavior of the molecules that constitute it, we do not know what constitutes space, so we do not know what gravitational microstates correspond to any particular macrostate.

12 Nevertheless, we have a rough idea of how entropy evolves. In situations where gravity is **negligible**, such as a cup of coffee, a uniform distribution of particles has a high entropy. This condition is a state of **equilibrium**. Even when particles reshuffle themselves, they are already so thoroughly mixed that nothing much seems to happen **macroscopically**. But if gravity is important and the volume is fixed, a smooth distribution has relatively low entropy. In this case, the system is very far from equilibrium. Gravity causes particles to **clump** into stars and galaxies, and entropy increases noticeably—consistent with

negligible *adj.* 可忽略的，无足轻重的
equilibrium *n.* 平衡，均势
macroscopically *adv.* 宏观地，肉眼可见地
clump *v.* 簇拥，集结

the second law.

provocative *adj.* 煽动性的，具有挑衅意味的

13 Indeed, if we want to maximize the entropy of a volume when gravity is active, we know what we will get: a black hole. In the 1970s Stephen Hawking[11] of the University of Cambridge confirmed a **provocative** suggestion of Jacob Bekenstein[12], now at the Hebrew University of Jerusalem, that black holes fit neatly into the second law. Like the hot objects that the second law was originally formulated to describe, black holes emit radiation and have entropy—a lot of it. A single million-solar-mass[13] black hole, such as the one that lives at the center of our galaxy, has 100 times the entropy of all the ordinary particles in the observable universe.

evaporate *v.* 蒸发，消失

14 Eventually even black holes **evaporate** by emitting Hawking radiation[14]. A black hole does not have the highest possible entropy—but just the highest entropy that can be packed into a certain volume. The volume of space in the universe, however, appears to be growing without limit. In 1998 astronomers discovered that cosmic expansion is accelerating. The most straightforward explanation is the existence of dark energy, a form of energy that exists even in empty space and does not appear to dilute away as the universe expands. It is not the only explanation for cosmic acceleration, but attempts to come up with a better idea have so far fallen short.[15]

for all intents and purposes (=to all intents and purposes) 实际上，实质上
max out 达到最高状态

15 If dark energy does not dilute away, the universe will expand forever. Distant galaxies will disappear from view. Those that do not will collapse into black holes, which in turn will evaporate into the surrounding gloom as surely as a puddle dries up on a hot day. What will be left is a universe that is, **for all intents and purposes**, empty. Then and only then will the universe truly have **maxed out** its entropy. The universe will be in equilibrium, and nothing much will ever happen.

16 It may seem strange that empty space has such a huge entropy. It sounds like saying that the most disorganized desk in the world is a completely empty desk. Entropy requires microstates, and at first glance empty space does not have any. In actuality, though, empty space has plenty of microstates—the quantum-gravitational microstates built into the fabric of space. We do not yet know what exactly these states are, any more than we know what microstates account for the entropy

of a black hole, but we do know that in an accelerating universe the entropy within the observable volume approaches a constant value **proportional** to the area of its boundary. It is a truly enormous amount of entropy, far greater than that of the matter within that volume.

proportional *adj.* 比例的，成比例的

## Past vs. Future

17 The striking feature of this story is the pronounced difference between the past and the future. The universe starts in a state of very low entropy: particles packed together smoothly. It evolves through a state of medium entropy: the lumpy distribution of stars and galaxies we see around us today. It ultimately reaches a state of high entropy: nearly empty space, featuring only the occasional stray low-energy particle.

18 Why are the past and future so different? It is not enough to simply **posit** a theory of initial conditions—a reason why the universe started with low entropy. As philosopher Huw Price[16] of the University of Sydney has pointed out, any reasoning that applies to the initial conditions should also apply to the final conditions, or else we will be guilty of assuming the very thing we were trying to prove—that the past was special[17]. Either we have to take the profound asymmetry of time as a **blunt** feature of the universe that escapes explanation, or we have to dig deeper into the workings of space and time.

posit *v.* 假定；假设
blunt *adj.* 直言不讳的

19 Many cosmologists have tried to attribute the time asymmetry to the process of cosmological inflation. Inflation is an attractive explanation for many basic features of the universe. According to this idea, the very early universe (or at least some part of it) was filled not with particles but rather with a temporary form of dark energy, whose density was enormously higher than the dark energy we observe today. This energy caused the expansion of the universe to accelerate at a fantastic rate, after which it decayed into matter and radiation, leaving behind a tiny **wisp** of dark energy that is becoming relevant again today. The rest of the story of the Big Bang, from the smooth **primordial** gas to galaxies and beyond, simply follows.

wisp *n.* 一绺，一缕
primordial *adj.* 最早的；原始的

20 The original motivation for inflation was to provide a robust explanation for the finely tuned conditions in the early universe—in particular, the remarkably uniform density of matter in widely separated regions. The acceleration driven by the temporary dark

energy smoothes out the universe almost perfectly. The prior distribution of matter and energy is irrelevant; once inflation starts, it removes any traces of the preexisting conditions, leaving us with a hot, dense, smooth early universe.

paradigm *n.* 模式，范例
invoke *v.* 唤起，引起，产生

21 The inflationary **paradigm** has been very successful in many ways. Its predictions of slight deviations from perfect uniformity agree with observations of density variations in the universe. As an explanation for time asymmetry, however, cosmologists increasingly consider it a bit of a cheat, for reasons that Roger Penrose[18] of the University of Oxford and others have emphasized. For the process to work as desired, the ultradense dark energy had to begin in a very specific configuration. In fact, its entropy had to be fantastically smaller than the entropy of the hot, dense gas into which it decayed. That implies inflation has not really solved anything: it "explains" a state of unusually low entropy (a hot, dense, uniform gas) by **invoking** a prior state of even lower entropy (a smooth patch of space dominated by ultradense dark energy).[19] It simply pushes the puzzle back a step: Why did inflation ever happen?

intuitively *adj.* 直觉地，直观地
stumble *v.* 摇摇晃晃地移动，步履蹒跚

22 One of the reasons many cosmologists invoke inflation as an explanation of time asymmetry is that the initial configuration of dark energy does not seem all that unlikely. At the time of inflation, our observable universe was less than a centimeter across. **Intuitively**, such a tiny region does not have many microstates, so it is not so improbable for the universe to **stumble** by accident into the microstate corresponding to inflation.

wind up 结束
generic *adj.* 一般的，普遍的

23 Unfortunately, this intuition is misleading. The early universe, even if it is only a centimeter across, has exactly the same number of microstates as the entire observable universe does today. According to the rules of quantum mechanics, the total number of microstates in a system never changes. (Entropy increases not because the number of microstates does but because the system naturally **winds up** in the most **generic** possible macrostate.) In fact, the early universe is the same physical system as the late universe. One evolves into the other, after all.

24 Among all the different ways the microstates of the universe can arrange themselves, only an incredibly tiny fraction corresponds

to a smooth configuration of ultradense dark energy packed into a tiny volume. The conditions necessary for inflation to begin are extremely specialized and therefore describe a very low entropy configuration. If you were to choose configurations of the universe randomly, you would be highly unlikely to hit on the right conditions to start inflation. Inflation does not, by itself, explain why the early universe has a low entropy; it simply assumes it from the start.

### A Time-Symmetric Universe

25 Thus, inflation is of no help in explaining why the past is different from the future. One bold but simple strategy is just to say: perhaps the very far past is not different from the future after all. Perhaps the distant past, like the future, is actually a high-entropy state. If so, the hot, dense state we have been calling "the early universe" is actually not the true beginning of the universe but rather just a transitional state between stages of its history.

26 Some cosmologists imagine that the universe went through a "bounce". Before this event, space was contracting, but instead of simply crashing to a point of infinite density, new physical principles—quantum gravity, extra dimensions, string theory[20] or other exotic phenomena—kicked in to save the day at the last minute, and the universe came out the other side into what we now perceive as the Big Bang. Though intriguing, bouncing cosmologies do not explain the arrow of time. Either entropy was increasing as the prior universe approached the **crunch**—in which case the arrow of time stretches infinitely far into the past—or the entropy was decreasing, in which case an unnatural low-entropy condition occurred in the middle of the universe's history (at the bounce). Either way, we have again **passed the buck** on the question of why the entropy near what we call the Big Bang was small.

crunch *n.* 关键时刻，转折点
pass the buck 推卸责任

27 Instead, let us suppose that the universe started in a high-entropy state, which is its most natural state. A good candidate for such a state is empty space. Like any good high-entropy state, the tendency of empty space is to just sit there, unchanging. So the problem is: How do we get our current universe out of a **desolate** and **quiescent** spacetime? The secret might lie in the existence of dark energy.

desolate *adj.* 荒凉的，无人烟的
quiescent *adj.* 不活动的，静态的

28 In the presence of dark energy, empty space is not completely

fluctuation *n.* 波动，起伏
disperse *v.* 散开，驱散
virtual *adj.* 虚的；虚拟的

empty. **Fluctuations** of quantum fields give rise to a very low temperature—enormously lower than the temperature of today's universe but nonetheless not quite absolute zero. All quantum fields experience occasional thermal fluctuations in such a universe. That means it is not perfectly quiescent; if we wait long enough, individual particles and even substantial collections of particles will fluctuate into existence, only to once again **disperse** into the vacuum. (These are real particles, as opposed to the short-lived "**virtual**" particles that empty space contains even in the absence of dark energy.)

pinch off 脱离；断裂
offspring *n.* 后代，子孙

29 Among the things that can fluctuate into existence are small patches of ultradense dark energy. If conditions are just right, that patch can undergo inflation and **pinch off** to form a separate universe all its own—a baby universe. Our universe may be the **offspring** of some other universe.

scenario *n.*（对可能出现的情况进行的）描述，推测
ignite *v.* 点燃，引发
bypass *v.* 绕过，越过

30 Superficially, this **scenario** bears some resemblance to the standard account of inflation. There, too, we posit that a patch of ultradense dark energy arises by chance, **igniting** inflation. The difference is the nature of the starting conditions. In the standard account, the patch arose in a wildly fluctuating universe, in which the vast bulk of fluctuations produced nothing resembling inflation. It would seem to be much more likely for the universe to fluctuate straight into a hot big bang, **bypassing** the inflationary stage altogether. Indeed, as far as entropy is concerned, it would be even more likely for the universe to fluctuate straight into the configuration we see today, bypassing the past 14 billion years of cosmic evolution.

31 In our new scenario, the preexisting universe was never randomly fluctuating; it was in a very specific state: empty space. What this theory claims—and what remains to be proved—is that the most likely way to create universes like ours from such a preexisting state is to go through a period of inflation, rather than fluctuating there directly. Our universe, in other words, is a fluctuation but not a random one.

Notes

1. **That may seem like a strange thing to say, given that cosmologists have very little standard for comparison:** 这种说法似乎很奇怪，因为宇宙学家没有什么可以对比的标准。given在句中是介词，意为taking (sth.) into account（鉴于，考虑到），在这里译为“因为”。
2. **time asymmetry:** 时间不对称。物理学在微观的层次几乎完全是时间对称的，这意味着物理学定律在时间流逝的方向倒转之后仍然保持不变。但是在宏观层次却并非如此：时间存在着明显的方向。时间箭头（the arrow of time）就是用于描述这种不对称现象的。
3. **the Big Bang:** 大爆炸。大爆炸理论是迄今关于宇宙形成的最有影响的学说。该理论认为，宇宙在遥远的过去曾处于一种极度高温和极大密度的状态，这种状态被形象地称为“原始火球”，即一个无限小的点，火球爆炸，宇宙开始膨胀，物质密度和温度逐渐降低，直到今天的状态。
4. **the arrow of time:** 时间箭头。时间箭头将过去和未来区别开来，使时间有了方向。已知至少存在三种不同的时间箭头：热力学时间箭头、心理学时间箭头、宇宙学时间箭头。
5. **spacetime:** 时空，即事件或物体存在的由一个时间坐标和三个空间坐标组成的四维空间。
6. **entropy:** 熵，热力学中表征物质状态的参量之一，通常用符号S表示。熵用来表示任何一种能量空间分布的均匀程度。能量分布得越均匀，熵就越大。该术语是德国物理学家克劳修斯（Rudolf Clausius, 1822–1888）在1850年提出的。
7. **second law of thermodynamics:** 热力学第二定律，热力学的基本定律之一。此定律认为，在有限的空间和时间内，一切和热运动有关的物理和化学过程都具有不可逆性，例如，在自然条件下，热量只能从高温物体向低温物体转移，另外，自然界中任何形式的能都很容易变成热，而反过来热却不能在不受其他影响的条件下变成其他形式的能。
8. **Ludwig Boltzmann:** 路德维希·波尔兹曼（1844–1906），奥地利物理学家。他提出这样的假设：无限小的世界中所发生的过程（这些过程与单个分子运动有关）全部可以逆转；相反，含有大量分子的宏观世界的过程都是不可逆转的。
9. **by the luck of the draw:** 完全取决于运气（purely by chance）。The luck of the draw是成语，原意为“碰运气的事”，其中draw的意思是“抽签、抽奖”。
10. **On average, the density differed from place to place by only about one part in 100,000:** 不同区域的密度差异平均只有十万分之一。part一词在此句中的意思是“……分之一”、“等分”。如A diet containing such a small amount of chlordane as 2.5 parts per million may eventually lead to storage of 75 parts per million in the fat of experimental animals.（即使受试动物的日常饮食中只含有百万分之二点五的氯丹，逐渐积累最终会导致这些动物脂肪内的氯丹含量达到百万分之七十五。）
11. **Stephen Hawking:** 史蒂芬·霍金（1942–2018），英国杰出的科学思想家和理论物理学家，剑桥大学应用数学及理论物理学系教授。他二十几岁时患卢伽雷氏症（肌萎缩性侧索硬化症），逐渐丧失行动和语言能力，但他依然成为当代最重要的广义相对论和宇宙论家。
12. **Jacob Bekenstein:** 雅各布·贝肯斯坦（1947–2015），以色列物理学家，1972年在普林斯顿大学物理系做研究生时，对黑洞产生兴趣，1973年发表论文《黑洞热力学》，他一反当时黑洞的熵值为零的说法，提出黑洞不但有很大的熵，而且它的熵不与体积成正比，却与表面积成正比。

13. **solar mass:** 太阳质量，天文学表达质量的标准单位，用来描述其他恒星与星系的质量。
14. **Hawking radiation:** 霍金辐射。史蒂芬·霍金发现所有宇宙黑洞至少释放少许能量，这种能量被命名为“霍金辐射”。
15. **It is not the only explanation for cosmic acceleration, but attempts to come up with a better idea have so far fallen short:** 以上这种说法并不是对宇宙膨胀速度加快的唯一解释，然而人们到目前为止还没有找到更具说服力的解释。短语come up with意为“提出”、“提供”，常与idea, proposal, answer, solution等词连用。短语fall short原意为“达不到”，在理解本句时应注意与上下文的联系。
16. **Huw Price:** 胡·普赖斯，生于英国剑桥，曾先后在爱丁堡大学和悉尼大学执教，发表专著 *Facts and the Function of Truth* (Blackwell, 1988), *Time's Arrow and Archimedes' Point* (Oxford, 1996) 等，以及多篇学术论文。
17. **…，or else we will be guilty of assuming the very thing we were trying to prove—that the past was special:** 否则我们就会因为假定我们过去试图证明的东西——过去的特殊性——而犯错误。句中的very是形容词，用于加强语气，意思是“正是那个”、“恰好的”。如：the very heart of the city（市区的正中心），at the very moment（恰恰在那个时刻）。
18. **Roger Penrose:** 罗杰·彭罗斯（1931– ），英国数学家、牛津大学数学教授，与著名理论物理学家斯蒂芬·霍金一起创立了现代宇宙论的数学结构理论。
19. **That implies inflation has not really solved anything: it "explains" a state of unusually low entropy (a hot, dense, uniform gas) by invoking a prior state of even lower entropy (a smooth patch of space dominated by ultradense dark energy):** 这表明宇宙膨胀的说法并不能提供真正的答案，它虽然“解释”了为什么熵处于极低的状态（即一种高温、密集、均衡的气体），却又引出了另一个问题，即在此之前还有一个熵值更低的状态（由密度超高的暗能量主导的平稳空间）。该句中的inflation指“宇宙膨胀这种说法”。
20. **string theory:** 弦理论、弦论，理论物理学理论。该理论的基本观点是，自然界的基本单元并非电子、光子、中微子和夸克之类的粒子，这些看似粒子的物质实际上都是很小的弦的闭合圈（称为闭合弦或闭弦），闭弦的不同振动和运动产生出各种不同的基本粒子。

## Exercises

### I. Answer the following questions.

1. How have cosmologists explained the evolution of the universe?
2. What do physicists say about entropy according to the second law of thermodynamics?
3. What did Jacob Bekenstein suggest about black holes?
4. What is the most straightforward explanation for the discovery that cosmic expansion is accelerating?
5. According to the author, why don't bouncing cosmologies explain the arrow of time?

## II. The following statements are incomplete. Search the missing information in the passage and fill in the blanks.

1. The universe we see does not qualify as we have developed __________ over the years for what counts as "natural".
2. Cosmologists are currently at an utter loss to explain the arguably most blatant feature of the universe: __________.
3. Ludwig Boltzmann explained entropy in terms of the distinction between two states of an object: __________ and __________.
4. It is clear that __________ has been very important to the evolution of the universe.
5. For many basic features of the universe, __________ is an attractive explanation.
6. Driven by the temporary __________, the acceleration smoothes out the universe almost perfectly.
7. The reason why many cosmologists invoke inflation as an explanation of time asymmetry is partly because __________ of dark energy does not seem all that unlikely.
8. If we are patient enough, individual particles and even substantial collections of particles will __________.

## III. Identify the implied meanings of the underlined parts of the following sentences according to the context of the passage, and translate the sentences into Chinese.

1. Every time you break an egg, you are doing observational cosmology.
2. High-entropy states greatly outnumber low-entropy ones; almost any change to the system will land it in a higher-entropy state, simply by the luck of the draw.
3. Regions with slightly more particles formed stars and galaxies, and regions with slightly fewer particles emptied out to form voids.
4. Whereas we can relate the entropy of a fluid to the behavior of the molecules that constitute it, we do not know what constitutes space, so we do not know what gravitational microstates correspond to any particular macrostate.
5. In the 1970s Stephen Hawking of the University of Cambridge confirmed a provocative suggestion of Jacob Bekenstein, now at the Hebrew University of Jerusalem, that black holes fit neatly into the second law.
6. Either we have to take the profound asymmetry of time as a blunt feature of the universe that escapes explanation, or we have to dig deeper into the workings of space and time.
7. It simply pushes the puzzle back a step: Why did inflation ever happen?
8. Intuitively, such a tiny region does not have many microstates, so it is not so improbable for the universe to stumble by accident into the microstate corresponding to inflation.
9. Either way, we have again passed the buck on the question of why the entropy near what we

call the Big Bang was small.

10. If conditions are just right, that patch can undergo inflation and pinch off to form a separate universe all its own—a baby universe.

## Translation Techniques (1)

### Conversion from English Adjectives into Chinese Verbs
### （英语形容词与汉语动词的转换）

Similar to the conversion between nouns and verbs in technical translation, sometimes it is also necessary to convert English adjectives into Chinese verbs to achieve smoothness in Chinese translation. For example, "It is no wonder that scientists around the world are racing to harness fusion. But the race is costly and timeconsuming." The Chinese translation of the adjectives "costly" and "timeconsuming" in the second sentence adopts this technique: 这就难怪全世界的科学家都在展开竞争，要驾驭聚变过程。但这一竞争既消耗资金又耗费时间。

More examples are provided as follows.

1. If *practical*, superconductors will become the basis for a gargantuan struggle, initially between the United States and Japan, over ways to exploit the new invention.

   如果具有实用价值，超导体将作为基础，导致美国和日本之间展开如何开发利用这项新发明的大规模竞争。

2. The ultimate cost of the disaster is only beginning to be *clear*: water and farmland in the Ukraine remain contaminated with the radioactive isotopes.

   这场灾难造成的最终代价只是初露端倪：乌克兰的水和农田都受到放射性同位素的污染。

3. Plastic bumpers and fuel tanks will soon be *commonplace* in new cars.

   塑料保险杠和燃料箱不久将在新式汽车上普遍采用。

4. Moreover, these signals are broadcast by satellites and are therefore not *available* to regular television viewers.

   此外，由于这些信号是通过卫星发射的，因此普通电视观众收看不到。

5. Scientists say proposals like Zenith's are but the first step on the digital highway, and that the transmission journey might be *complete* within a decade.

科学家认为，类似齐尼思公司的那些建议是在数字技术快速发展方面迈出的第一步，而数字传播的过程可能在十年内完成。

**IV. Translate the following sentences into Chinese. Pay attention to how the italicized adjectives in the sentences should be translated.**

1. Make no mistake: cosmologists have put together an incredibly *successful* picture of what the universe is made of and how it has evolved.
2. The asymmetry of time, the arrow that points from past to future, plays an *unmistakable* role in our everyday lives.
3. From this point of view, it is not *surprising* that entropy tends to increase with time.
4. In situations where gravity is *negligible*, such as a cup of coffee, a uniform distribution of particles has a high entropy.
5. Indeed, if we want to maximize the entropy of a volume when gravity is *active*, we know what we will get: a black hole.
6. But we do know that in an accelerating universe the entropy within the observable volume approaches a constant value *proportional* to the area of its boundary.
7. Inflation is an *attractive* explanation for many basic features of the universe.
8. This energy caused the expansion of the universe to accelerate at a fantastic rate, after which it decayed into matter and radiation, leaving behind a tiny wisp of dark energy that is becoming *relevant* again today.

## Translation Techniques (2)

### Translation of "and" (and的译法)

As a coordinate conjunction, the word "and" is used to connect words, phrases and clauses in a sentence, and is often translated as 和, 与 or 以及. However, if you translate the word "and" in the following sentence into those expressions, you may find the translation sounds somewhat awkward: "The process of oxidation in human body gives off heat slowly and regularly." This is because "and" in this sentence does not simply link the two adverbs "slowly" and "regularly", but also implies that two ways heat is given off exist side by side. Therefore, it would be better to translate the sentence as: 人体内的氧化过程缓慢而又有规律地放出热量。Obviously, the word 和 does not convey this subtle meaning exactly.

Here are a few more examples:

1. Several disadvantages tend to limit the use of hydraulic controls *and* they do offer many distinct advantages.
   液压控制虽有许多突出的优点，但也存在一些缺陷，使其应用范围受到限制。
2. Sound is carried by air, *and* without air there can be no sound.
   声音靠空气传播，因此没有空气也就没有声音。
3. Aluminium is used as the engineering material for planes and spaceships *and* it is both light and tough.
   铝用作制造飞机和宇宙飞船的工程材料，因为铝不但轻而且韧性好。

Sometimes, the word "and" is used to connect two clauses loosely with no substantial meaning, and therefore it may not be necessary to put it into Chinese. For example:

1. The small three-person submarine is less than eight meters long, *and* it can dive almost four kilometers under the ocean.
   三人小型潜水艇全长不足8米，可潜入水下约4公里。
2. The scientific and medical communities are beginning to rethink their ideas about how human body works, *and* gradually what was considered a minor science just a few years ago is being studied in major universities and medical centers around the world.
   科学界和医学界人士正在重新考虑关于人体如何运转的问题，这一学科仅仅在几年前还被认为是微不足道的，如今在全世界一些主要的大学和医学中心，都已逐步地展开研究。

**V. Discuss the translation technique and the ways of applying the technique to the translation of the following sentences. Complete each of the Chinese translations.**

1. We have come to the last *and* most important step of the experiment.（我们的实验现在已经到了最后__________最重要的阶段。）
2. This is extruded through minute holes in a nozzle, *and* the threads of filaments produced are solidified in various ways.（先将该物质从喷嘴的小孔中挤压出来，__________将产生的纤维丝利用各种方法固化。）
3. Thousands of the electric power generators, often installed on windfarms in North America and Europe, now total over 800 megawatts of rated capacity, *and* their numbers are continuing to grow.（数以千计的风力发电机大多安装在北美和欧洲的风力发电场，其总额

定功率现已达800兆瓦，__________数量仍在继续增加。）

4. Solid silicones serve, among other things, as a kind of artificial rubber, *and* liquid silicones have been used as hydraulic fluids.（固体硅酮的用途很广，其中之一是用作人工橡胶，__________液体硅酮则被用作各种液压流体。）
5. This fact can be demonstrated by rubbing a comb through one's hair to create static electricity *and* placing it by an electric fish's tank.（要证明这一点，可用梳子在头发上梳几下以产生静电，__________把梳子放在装有电鱼的水槽边。）
6. It took the genius of Newton to leap from there to the idea that it was the same force that caused the moon to drop from its natural motion in a straight line, to calculate at once the amount by which it dropped *and* from that to infer the inverse-square law.（牛顿的天才之处在于由此产生了一个思想上的飞跃，认识到就是相同的力使月球偏离它的正常直线运动，并且立即计算出其偏离的大小，__________推导出平方反比定律。）
7. The company says the homes are far more efficient than conventional houses *and* use less than a third as much power.（这家公司说，这种住宅的效益远远超过普通房屋，__________它耗电还不到普通房屋的三分之一。）
8. Another ministation could handle biomedical studies, *and* others could be used as assembly and take-off points for the Mars and subsequent missions.（另一个小型空间站可以从事生物医学研究，__________其他的小型空间站则可以用作火星和随后飞行的装配及发射地点。）
9. The area of a triangle is equal to half the product of the base *and* the perpendicular height.（三角形的面积等于二分之一底__________高。）
10. On the average, oceans are two *and* one third miles deep.（海洋的平均深度为二__________三分之一英里。）

**VI. Translate the following sentences into Chinese. Pay attention to how the conjunction "and" is translated properly.**

1. Some 14 billion years ago the cosmos was hotter and denser than the interior of a star, *and* since then it has been cooling off and thinning out as the fabric of space expands.
2. In other words, the real challenge is not to explain why the entropy of the universe will be higher tomorrow than it is today but to explain why the entropy was lower yesterday *and* even lower the day before that.
3. Gradually, as the universe expanded *and* cooled, the pull of gravity enhanced those differences.
4. The universe will be in equilibrium, *and* nothing much will ever happen.
5. Entropy requires microstates, *and* at first glance empty space does not have any.

## VII. Translate the following passage into Chinese.

Entropy is the only quantity in the physical sciences that "picks" a particular direction for time, sometimes called an arrow of time. As one goes "forward" in time, the second law of thermodynamics says that the entropy of an isolated system can only increase or remain the same; it cannot decrease. Hence, from one perspective, entropy measurement is thought of as a kind of clock.

The thermodynamic arrow is often linked to the cosmological arrow of time, because it is ultimately about the boundary conditions of the early universe. According to the Big Bang theory, the Universe was initially very hot with energy distributed uniformly. For a system in which gravity is important, such as the universe, this is a low-entropy state (compared to a high-entropy state of having all matter collapsed into black holes, a state to which the system may eventually evolve). As the Universe grows, its temperature drops, which leaves less energy available to perform useful work in the future than was available in the past. Additionally, perturbations in the energy density grow (eventually forming galaxies and stars). Thus the Universe itself has a well-defined thermodynamic time. But this doesn't address the question of why the initial state of the universe was that of low entropy. If cosmic expansion were to halt and reverse due to gravity, the temperature of the Universe would once again increase, but its entropy would continue to increase due to the continued growth of perturbations and eventually black hole formation.

## VIII. Translate the following passage into English.

### 宇宙大爆炸

大爆炸理论从诞生以来就不断受到人们的质疑，从而促使那些相信该理论的人去寻找更多的证据，来证明他们是对的。在本书完成时，许多人又做了更进一步的研究，而且的确有了多项发现，使人们更加全面地了解宇宙的起源。

最近美国国家航空航天局就做出了令人惊讶的发现，为证明大爆炸理论提供了证据。其中最重要的发现是，天文学家利用Astro-2天文台证实了宇宙通过大爆炸诞生的必要条件之一。1995年6月，科学家在宇宙深处探测到原生氦（如重氢）的存在。这一发现与大爆炸理论的一个重要方面相符，即在宇宙诞生之初就产生了氢与氦的混合物。

此外，哈勃望远镜（该望远镜以大爆炸理论之父的名字命名）也对解答宇宙形成后出现了哪些元素这一问题提供了某些线索，天文学家通过哈勃望远镜在极为古老的恒星内发现了硼元素。他们推测硼有可能是星系形成时能量爆发的残留物，也可能表明硼比宇宙更古老，可以追溯到大爆炸发生的时候。如果后一种推测成立的话，科学家不得不再次修改有关宇宙形成以及随后发生了什么的理论，因为根据目前的理论，重量如此之大、结构如此复杂的原子当时是不可能存在的。

## Section 2 Reading for Academic Purposes

*Reading is a process of thinking, which involves using strategies or approaches to understand and relate the information to other readings, ideas and themes from lectures, and to the goals of your research and learning. This is another important part of active reading. While reading, you should check your understanding, monitor for difficulties, and check for ways to correct difficulties. Therefore, active reading is brain intensive; that is, it involves thinking as you read and direct that thinking to achieve certain reading goals.*

*Some students wish to make their approach to reading such that they will always read without difficulty. However, no strategy can guarantee that reading will proceed without difficulty (some difficulty may be a sign that you're working at the understanding). So, as you develop your reading strategies, remind yourself that it is important to remain flexible in your approach to reading, for different kinds of information as well as for different purposes.*

*The following passage focuses on the reading strategies that are commonly adopted in reading for academic purposes. They will help you to eliminate the wasteful and often mindless repetition that is necessitated by forgetting what you have read. The active reading strategies also involve selecting information relevant to a purpose, which may mean that you are reading only a percentage of what others might be mindlessly reading and that you are reading with better results.*

### Reading a Scientific Paper
### 科技论文的阅读方法

Although it is **tempting** to read the paper straight through as you would do with most texts, it is more efficient to organize the way you read. Generally, you first read the Abstract in order to understand the major points of the work. The extent of background assumed by different authors, and allowed by the journal, also varies.

tempting *adj.* 诱人的，有吸引力的

One extremely useful habit in reading a paper is to read the Title and the Abstract and, before going on, review in your mind what you know about the topic. This serves several purposes. First, it clarifies whether you in fact know enough background to appreciate the paper. If not, you might choose to read the background in a review or textbook, as appropriate.

Second, it refreshes your memory about the topic. Third, and perhaps most importantly, it helps you as the reader **integrate** the new information into your previous knowledge about the topic. That is, it is used as a part of the self-education process that any professional must

integrate *v.* 使成一体，使结合，使合并

continue throughout his/her career.

If you are very familiar with the field, the Introduction can be skimmed or even skipped. As stated above, the logical flow of most papers goes straight from the Introduction to Results; accordingly, the paper should be read in that way as well, skipping Materials and Methods and referring back to this section as needed to clarify what was actually done. A reader familiar with the field who is interested in a particular point given in the Abstract often skips directly to the relevant section of the Results, and from there to the Discussion for interpretation of the findings. This is only easy to do if the paper is organized properly.

### Codewords

shorthand phrase（对某事的）简短且常为故意隐晦的表达方法
connotation *n.* 内涵，隐含意义
explicit *adj.* 明晰的，详尽的
document *v.* 记录，记载
preliminary *adj.* 初步的，预备的

Many papers contain **shorthand phrases** that we might term "codewords", since they have **connotations** that are generally not **explicit**. In many papers, not all the experimental data are shown, but referred to by "data not shown". This is often for reasons of space; the practice is accepted when the authors have **documented** their competence to do the experiments properly (usually in previous papers). Two other codewords are "unpublished data" and "**preliminary** data". The former can either mean that the data are not of publishable quality or that the work is part of a larger story that will one day be published. The latter means different things to different people, but one connotation is that the experiment was done only once.

### Difficulties in reading a paper

Several difficulties confront the reader, particularly one who is not familiar with the field. As discussed above, it may be necessary to bring yourself up to speed before beginning a paper, no matter how well written it is. Be aware, however, that although some problems may lie in the reader, many are the fault of the writer.

One major problem is that many papers are poorly written. Some scientists are poor writers. Many others do not enjoy writing, and do not take the time or effort to ensure that the prose is clear and logical. Also, the author is typically so familiar with the material that it is difficult to step back and see it from the point of view of a reader not familiar with the topic and for whom the paper is just another of a large **stack** of papers that need to be read.

stack *n.* (一) 堆，(一) 摊

Bad writing has several consequences for the reader. First, the logical connections are often left out. Instead of saying why an experiment was done, or what ideas were being tested, the experiment is simply described. Second, papers are often **cluttered** with a great deal of **jargon**. Third, the authors often do not provide a clear road-map through the paper; **side issues** and fine points are given equal **air time** with the main logical thread, and the reader loses this thread. In better writing, these side issues are **relegated** to Figure legends or Materials and Methods or clearly identified as side issues, so as not to distract the reader.

clutter *v.* 乱糟糟地堆满；把……弄得杂乱
jargon *n.* 行话，隐语
side issue 与正题无关的问题，枝节问题
air time 广播时间，播放时段
relegate *v.* 把……归类

Another major difficulty arises when the reader seeks to understand just what the experiment was. All too often, authors refer back to previous papers; these refer in turn to previous papers in a long chain. Often that chain ends in a paper that describes several methods, and it is unclear which was used. Or the chain ends in a journal with severe space limitations, and the description is so **compressed** as to be unclear. More often, the descriptions are simply not well-written, so that it is **ambiguous** what was done.

compress *v.* 压缩
ambiguous *adj.* 模棱两可的，模糊不清的

Other difficulties arise when the authors are **uncritical** about their experiments; if they firmly believe a particular model, they may not be **open-minded** about other possibilities. These may not be tested experimentally, and may even go unmentioned in the Discussion. Still another related problem is that many authors do not clearly distinguish between fact and **speculation**, especially in the Discussion. This makes it difficult for the reader to know how **well-established** are the "facts" under discussion.

uncritical *adj.* 不加批评的，不作批评的
open-minded *adj.* 虚心的，思想开明的
speculation *n.* 推测的结论，猜测
well-established *adj.* 已得到确认的，论证充分的

One final problem arises from the sociology of science. Many authors are ambitious and wish to publish in **trendy** journals. As a consequence, they **overstate** the importance of their findings, or put a speculation into the title in a way that makes it sound like a well-established finding. Another example of this approach is the "**Assertive** Sentence Title", which presents a major conclusion of the paper as a **declarative sentence** (such as "LexA is a **repressor** of the *recA* and *lexA* genes"). This trend is becoming **prevalent**; look at recent issues of *Cell* for examples. It's not so bad when the assertive sentence is

trendy *adj.* 时尚的，流行的
overstate *v.* 夸大
assertive *adj.* 断言的；肯定的
declarative sentence 陈述句
repressor *n.* 【生】阻遏因子
prevalent *adj.* 普遍的，盛行的

hasty *adj.* 草率的，轻率的

praiseworthy *adj.* 值得称颂的
novice *n.* 新手，初学者

well-documented (as it was in the example given), but all too often the assertive sentence is nothing more than a speculation, and the **hasty** reader may well conclude that the issue is settled when it isn't.

These last factors represent the public relations side of a competitive field. This behavior is understandable, if not **praiseworthy**. But when the authors mislead the reader as to what is firmly established and what is speculation, it is hard, especially for the **novice**, to know what is settled and what is not. Therefore, a careful evaluation is necessary.

## Exercises

**I. Read the passage and decide whether the following statements are true or false. Write T for True and F for False in the brackets.**

1. An efficient reader should read a paper straight through to acquire a whole idea. ( )
2. When reading a paper, it is a good habit to read the Title and the Abstract and review in your mind what you know about the topic. ( )
3. The paper should be read straight from the Introduction to Results. ( )
4. Codewords have connotations that are generally well-defined in the text. ( )
5. Typically the author familiar with the material will be more thoughtful of his reader. ( )
6. An author uncritical of his own experiment may refuse to explore other possibilities. ( )
7. It is easy for the reader to distinguish between facts and speculation in the paper. ( )
8. An Assertive Sentence Title is unacceptable even if it is well-documented. ( )
9. Trendy journals mislead authors to publish ill-documented papers. ( )
10. A careful evaluation is needed to distinguish speculation from established conclusion. ( )

**II. Read the passage again, and complete the following items.**

1. In order to understand the major points of the work, you should first read: ___________
2. Reading the Title and the Abstract serves three purposes: ___________
3. When reading in a familiar field, you can skim or even skip: ___________
4. The three typical codewords: ___________
5. The poorly written papers are often related to three types of writers: ___________
6. The three characteristics of "bad writing": ___________
7. In better writing, the side issues are dealt with in the following ways: ___________
8. Another problem faced by the readers is that when they seek to understand just what the experiment was, they may find: ___________

## Quotations from Great Scientists

*Science is a way of thinking much more than it is a body of knowledge.*

*—Carl Sagan*

科学是一种思想方法，因而远远超越了知识的范畴。

——卡尔·萨根

# Chapter 5 Computer Technology 计算机技术

## Section 1 Reading and Translation

### I What You Are Going to Read

The new inventions in computer technology bring us surprises constantly, and one of them—multi-touch screens—is introduced in the following article published in the June 2008 issue of *Scientific American*. From it, you can see rather than responding to the presence of a single finger, multi-touch computer screens can follow the instructions of many fingers simultaneously. For example, a wall-size screen developed by Perceptive Pixel can respond to as many as 10 fingers or multiple hands. Other companies such as Microsoft and Mitsubishi are offering smaller, specialized systems for hotels, stores, and engineering and design firms.

As the article indicates, multi-touch computing may one day free us from the mouse as our primary computer interface, the way the mouse freed us from keyboards.

### II About the Author

Stuart F. Brown is a science author of *Fortune* magazine. His writing involves chemistry, biotechnology and other scientific topics, and is highly appreciated for richness in analogies, metaphors and evocative images, which has helped make science understandable and exciting for millions of readers. In recognition of his achievements, the American Chemical Society, the world's largest scientific society, has selected him as winner of its 2007 James T. Grady-James H. Stack Award for Interpreting Chemistry for the Public.

### III Reading Passage

#### Hands-on Computing: How Multi-touch Screens Could Change the Way We Interact with Computers and Each Other

#### 触摸式计算机技术：多点触控式显示屏如何改变人机之间和人与人之间的交流方式

By Stuart F. Brown

*The iphone*[1] *and even wilder interfaces could improve collaboration without a mouse or keyboard.*

1 When Apple's iPhone hit the streets last year, it introduced so-called multi-touch screens to the general public.[2] Images on the screen can be moved around with a fingertip and made bigger or smaller by placing two fingertips on the image's edges and then either spreading those fingers apart or bringing them closer together. The **tactile** pleasure the interface provides beyond its utility quickly brought it **accolades**. The operations felt intuitive, even sensuous. But in laboratories around the world at the time of the iPhone's launch, multi-touch screens had vastly outgrown two-finger commands. Engineers have developed much larger screens that respond to 10 fingers at once, even to multiple hands from multiple people.

tactile *adj.*（有）触觉的；能触觉到的
accolade *n.* 赞赏；赞美

2 It is easy to imagine how photographers, graphic designers or architects—professionals who must manipulate lots of visual material and who often work in teams—would welcome this multi-touch computing. Yet the technology is already being applied in more **far-flung** situations in which anyone without any training can reach out during a brainstorming session and move or **mark up** objects and plans.[3]

far-flung *adj.* 分布很广的；范围广泛的
mark up 把……标出

**Perceptive Pixels**

3 Jeff Han, a consulting computer scientist at New York University and founder of Perceptive Pixel[4] in New York City, is at the forefront of multi-touch technology. Walking into his company's lobby, one is greeted by a three-by-eight-foot flat screen. Han steps up to the electronic wall and **unleashes** a world of images using nothing but the touch of his fingers. As many as 10 or more video feeds can run **simultaneously**, and there is no toolbar in sight. When Han wants the display to access different files he taps it twice, bringing up charts or menus that can also be tapped.

unleash *v.* 释放出
simultaneously *adv.* 同时发生地

4 Several early adopters have purchased complete systems, including intelligence agencies that need to quickly compare geographically coordinated **surveillance** images in their war rooms. News anchors on CNN[5] used a big Perceptive Pixel system during coverage of the presidential primaries that boldly displayed all 50 US states; to depict voting results, the anchors, standing in front of the screen, dramatically zoomed in and out of states, even counties, simply by moving their fingers across the map[6]. Looking ahead, Han expects the technology to find a home in graphically intense businesses such

surveillance *n.* 监视，监督

rudimentary *adj.* 初步的；早期的
fine-resolution *n.* 高清晰度；高分辨率

as energy trading and medical imaging.

5 **Rudimentary** work on multi-touch interfaces dates to the early 1980s, according to Bill Buxton[7], a principal researcher at Microsoft Research. But around 2000, at NYU[8], Han began a journey to overcome one of the technology's toughest hurdles: achieving **fine-resolution** fingertip sensing. The solution required both hardware and software innovations.

crisply *adv.* 轮廓分明地

6 Perhaps most fundamental was exploiting an optical effect known as frustrated total internal reflection (FTIR)[9], which is also used in fingerprint-recognition equipment. Han, who describes himself as "a very tactile person", became aware of the effect one day when he was looking through a full glass of water. He noticed how **crisply** his fingerprint on the outside of the glass appeared when viewed through the water at a steep angle. He imagined that an electronic system could optically track fingertips placed on the face of a clear computer monitor. Thus began his six-year absorption with multi-touch interfaces.

kiosk *n.* 公用电话亭
capacitance *n.* 【电】电容；电流容量
predefine *v.* 预先确定
functionality *n.* 功能（性）
rectangular *adj.* 矩形的，长方形的
acrylic *adj.* 丙烯酸的
waveguide *n.* 【电】波导
infrared *adj.* 【物】红外线的；产生红外辐射的
optical fiber 【物】光学纤维

7 He first considered building a very high resolution version of the single-touch screens used in automated teller machines and **kiosks**, which typically sense the electrical **capacitance** of a finger touching **predefined** points on the screen. But tracking a randomly moving finger would have required an insane amount of wiring behind the screen, which also would have limited the screen's **functionality**. Han ultimately devised a **rectangular** sheet of clear **acrylic** that acts like a **waveguide**, essentially a pipe for light waves. Light-emitting diodes (LEDs)[10] around the edges pump **infrared** light into the sheet. The light streams through, reflecting internally off the sheet walls, much as light flows through an **optical fiber**. No light leaks out. But when someone places a finger on one face of the sheet, some of the internally reflecting light beams hit it and scatter off, bouncing through the sheet and out the opposite face. Cameras behind the screen sense this leaking light, or FTIR[11], revealing the location being touched. The cameras can track this leakage from many points at once.

diffusion *n.* 【物】漫射；扩散

8 Han soon discovered that the acrylic panel could also serve as a **diffusion** screen; a projector behind the panel, linked to a computer, could beam images toward it, and they would diffuse through to the

other side. The screen could therefore serve as both an output of **imagery** and an input of touches made on that imagery.

imagery *n.* [总称] 像；画像

9 Sensing the exact location of fingers was one challenge. Devising software routines that could track the finger movements and convert them to instructions for what should be happening with images on the screen was tougher. The half a dozen software developers working with Han had to first write software that would function as a high-performance graphics engine, in part to give the display low **latency**, or **ghosting**, when fingers dragged objects quickly across the screen. Then they had to deal with the screen's **unorthodox** FTIR light output from fingertips sweeping around in random directions.

latency *n.* 【计】等待时间
ghosting *n.* 重像
unorthodox *adj.* 异端的；非正统的

10 Deep in the architecture of a computer's operating system is an assumption that a user's input will come either from a keyboard or a mouse. Keystrokes are unambiguous; a "q" means "q". The movement of a mouse is expressed as Cartesian coordinates[12]—*x* and *y* locations on a two-dimensional **grid**. Such methods for representing input belong to a general discipline known as the graphical user interface, or GUI[13]. Han's multi-touch screen generates 10 or more streams of *x* and *y* coordinates at the same time, and "the traditional GUIs are really not designed for that much **simultaneity**," he notes. The current operating systems—Windows, Macintosh, Linux—are so predicated on the single mouse **cursor** that "we had to tear up a lot of plumbing to make a new multi-touch graphical framework," Han says.

grid *n.* 【计】网格
simultaneity *n.* 同时发生
cursor *n.* （计）光标

11 During all this work, Han found that pressure sensing could be accomplished, too, by applying to the front of the acrylic screen a thin layer of **polymer** with microscopic ridges **engineered** into its surface. When a user presses harder or more softly on any spot on the polymer, it **flexes** slightly, and the fingerprint area becomes larger or smaller, causing the scattered light to become brighter or darker, which the camera can sense. By maintaining firm pressure on an object on the screen, a user can slide it behind an adjacent object.

polymer *n.* 【化】聚合物
engineer *v.* 设计；建造
flex *v.* （柔韧的或有弹性的东西）弯曲

12 Han's Perceptive Pixel team, formed in 2006, put all the elements together and demonstrated the system at the TED (for technology, entertainment and design) conference that year to an enthusiastic audience. Since then, orders for the system have steadily increased. Perceptive Pixel is not disclosing prices.

## Microsoft Scratches the Surface[14]

envision *v.* 想象，设想，展望
pinball *n.* 弹球游戏
browser *n.* 浏览器

13 While Han was perfecting his setup, engineers elsewhere were pursuing similar goals by different means. Software giant Microsoft is now rolling out a smaller multi-touch computer called Surface and is trying to brand this category of hardware as "surface computers". The initiative dates back to 2001, when Stevie Bathiche of Microsoft Hardware and Andy Wilson of Microsoft Research began developing an interactive tabletop that could recognize certain physical objects placed on it. The two innovators **envisioned** that the tabletop could function as an electronic **pinball** machine, a video puzzle or a photo **browser**.

14 More than 85 prototypes later, the pair ended up with a table that has a clear acrylic top and houses a projector on the floor below. The projector sends imagery up onto the horizontal, 30-inch screen. An infrared LED shines light up to the tabletop as well, which bounces off fingertips or objects on the other side, thus allowing the device to recognize commands from people's fingers. A Windows Vista[15] computer provides the processing.

atop *prep.* 在……顶上
cue *v.* 给……提示

15 Microsoft is shipping Surface table computers to four partners in the leisure, retail and entertainment industries, which it believes are most likely to apply the technology. Starwood Hotels' Sheraton chain[16], for example, will try installing surface computers in hotel lobbies that will let guests browse and listen to music, send home digital photographs, or order food and drinks. Customers in T-Mobile USA's[17] retail stores will be able to compare different cell phone models by simply placing them **atop** a surface screen; black-dotted "domino" tags[18] on the undersides of the phones will **cue** the system to display price, feature and phone plan details. Other Microsoft software will allow a wireless-enabled digital camera[19], when placed on a surface computer, to upload its photographic content to the computer without a cable.

16 First-generation surface systems are priced from $5,000 to $10,000. As with most electronic items, the company expects the price to decline as production volume increases. Microsoft says Surface computers should be available at consumer prices in three to five years.

## Mitsubishi Wired In, Too[20]

17 Technology developers might be interested in the DiamondTouch table[21] from a **start-up company** called Circle Twelve[22] in Framingham, Mass., that was recently **spun off** from Mitsubishi Electric Research Laboratories[23]. The table, developed at Mitsubishi, is **configured** so that outside parties can write software for applications they envision; several dozen tables are already in the hands of academic researchers and commercial customers.

start-up company 新公司，刚刚开始运营的公司
spin off （从现有公司中）派生出（一个部分独立的子公司）
configure *v.* 使成形

18 The purpose of DiamondTouch "is to support small-group collaboration," says Adam Bogue, Mitsubishi's vice president of marketing. "Multiple people can interact, and the system knows who's who." Several people sit in chairs that are positioned around the table and are linked to a computer below. When one of them touches the tabletop, an **array** of **antennas** embedded in the screen sends an extremely small amount of radio-frequency energy through the person's body and chair to a receiver in the computer, a scheme known as **capacitive coupling**. Alternatively, a special floor mat can be used to complete the circuit. The antennas that are **coupled** indicate the spot on the screen that the person is touching.

array *n.* 整齐的一批；大量；显眼的一系列
antenna *n.* 天线
capacitive *adj.*【电】电容的
coupling *n.*【电】耦合
couple *v.*【电】将……耦合

19 Though seemingly restrictive, this setup can keep track of who makes what input, and it can give control to whoever touches the screen first. In that case it will ignore other touches, sensed through the assigned seating, until the first user has completed his or her inputs. The system can also track who makes which **annotations** to images, such as blueprints.

annotation *n.* 注解，注释，评注

20 Parsons Brinckerhoff[24], a global engineering firm headquartered in New York City, has been experimenting with the tables and plans to acquire more. "We have thousands of meetings during the course of a big project," says Timothy Case, the company's **visualization** department regional manager. "We could have multiple tables in multiple locations, and everybody can be looking at the same thing."

visualization *n.* 想象，形象化

21 Both the DiamondTouch and Perceptive Pixel systems feature keyboard "**emulators**" that shine a virtual keyboard onto the screen so that people can type. But it seems unlikely that enthusiasts would prefer to use the dynamic systems for this **mundane** activity. The great strength of multi-touch is letting multiple people work together

emulator *n.*【计】仿真器，仿真程序
mundane *adj.* 现世的，世俗的

untether *v.* 解开束缚

on a complex activity. It is hard to remember how liberating the mouse seemed when it freed people from keyboard arrow keys some 25 years ago. Soon the multi-touch interface could help **untether** us from the ubiquitous mouse. "It's very rare that you come upon a really new user interface," Han says. "We're just at the beginning of this whole thing."

## Notes

1. **iphone:** (或iPhone）苹果公司于2007年1月推出的一款新型移动电话，它将移动电话、可触摸宽屏以及具有桌面级电子邮件、网页浏览、搜索和地图功能的突破性因特网通信设备融为一体。
2. **When Apple's iPhone hit the street last year, it introduced so-called multi-touch screens to the general public:** 去年，苹果公司的iPhone手机风靡大街小巷，它面向公众推出了一款所谓的多点触控显示屏。其中multi-touch screen指可以让触控板一次感应多个点的触摸式显示屏。短语hit the street的意思是"(某事）轰动一时或风行一时"。
3. **Yet the technology is already being applied in more far-flung situations in which anyone without any training can reach out during a brainstorming session and move or mark up objects and plans:** 但是该技术已在更加广泛的范围内得以应用，即使没有受过训练的人也可以在"头脑风暴"会上大显身手，调整或标注物品和计划。句中的短语reach out原意为"伸手触到或拿到"，在本句中用作比喻，意思是"施展自己的才能"。
4. **Perceptive Pixel:** 美国高新科技公司，专门生产和销售大尺寸多点触控显示屏。
5. **CNN:** (美国）有线电视新闻网（Cable News Network)。创办于1980年6月，通过卫星向有线电视网和卫星电视用户提供全天候的新闻节目，总部设在美国佐治亚州的亚特兰大。
6. **…；to depict voting results, the anchors, standing in front of the screen, dramatically zoomed in and out of states, even counties, simply by moving their fingers across the map:** 在报道投票结果时，节目主持人站在屏幕前，只要将手指在地图上滑动，就可以像魔术师一样穿梭于各州（甚至是县）之间。zoom原指"(飞行器、汽车等）急速移动（特别是伴有嗡嗡声或轰轰声)"，在本句中与副词dramatically连用，比喻手指的移动使屏幕上的画面快速变换的神奇情景。
7. **Bill Buxton:** 比尔·巴克斯顿，美国著名人机界面专家，微软高级研究员。早在20世纪80年代，巴克斯顿就提出了多手输入技术（multi-hand input）的原型。
8. **NYU:** 纽约大学（New York University)。
9. **frustrated total internal reflection (FTIR):** 受抑全内反射技术，用于指纹辨识系统。
10. **light-emitting diodes (LEDs):** 发光二极管。一种半导体二极管，可将实用电压转变成亮光，用于数字显示，如计算器等。
11. **FTIR:** 傅立叶转换红外线光谱（Fourier transform infrared spectrometry)。该项技术可以用来探测各种不同的化学分子，对同时出现的不同化学物质具有相当高的鉴别率。

12. **Cartesian coordinates:** 笛卡尔坐标系，直角坐标系和斜角坐标系的统称。
13. **graphical user interface (GUI):** 图形用户界面或图形用户接口，指的是采用图形方式显示的计算机操作环境用户接口。
14. **Microsoft Scratches the Surface:** Surface是微软公司推出的一款平面触摸式电脑。短语scratch the surface原意是"作肤浅的探讨"、"浅尝辄止"，本文作者奇妙地借用该短语作这部分的标题，其含义是"微软开发出了一款叫做Surface的新产品"，因此scratch the surface在这里用作双关语（pun）。
15. **Windows Vista:** Windows XP之后新一代的Windows操作系统。
16. **Starwood Hotel's Sheraton chain:** 喜达屋酒店集团下属的喜来登连锁酒店，世界顶级酒店之一。
17. **T-Mobile USA:** 美国第四大移动服务提供商，T-Mobile国际公司的子公司，而T-Mobile国际公司则是欧洲电信巨头德国电信公司的海外业务子公司。
18. **"domino" tags:** "骨牌"标签，一种电子装置，放在Surface平面触摸式电脑上，电脑即可根据标签上的圆点分布形状识别指令，使用者可自行设定不同的指令。
19. **a wireless-enabled digital camera:** 无线数码相机，可与无线网络连接，不必使用连线就可以将照片下载到计算机或手机上，使图像下载、打印、共享等程序得到简化。
20. **Mitsubishi Wired In, Too:** 三菱电子公司也在努力。wire in为口语表达方法，意思是"努力干"。如: Let's wire in and finish the job.（让我们加把劲把工作干完吧。）
21. **DiamondTouch table:** 三菱电子研究实验室研发的一种由触摸和姿势激活的显示屏，支持小范围群体协作，自桌面上方投射图像，使用电容耦合（类似于笔记本电脑中的触摸板）方式来跟踪指尖位移，该产品甚至能够识别触摸者。
22. **Circle Twelve:** 目前世界上唯一的DiamondTouch table生产商，该公司主要制作多人使用的计算机硬件接口和支持小范围群体使用的软件产品。
23. **Mitsubishi Electric Research Laboratories:** (MERL) 三菱电子研究实验室。
24. **Parsons Brinckerhoff:** (PB) 栢诚公司，1885年创立于美国纽约，是世界上历史最悠久的工程公司之一，在基础设施和楼宇设施的咨询、规划、工程设计、项目统筹管理、施工管理、运营和维护方面处于公认领先地位。

## Exercises

**I. Answer the following questions.**

1. Why professionals such as photographers, graphic designers, architects and those who often work in teams would welcome multi-touch computing?
2. What would Jeff Han's future customers of this multi-touch technology be like?
3. What was the toughest difficulty of multi-touch technology?
4. Which industries are more likely to apply Microsoft's "Surface computer" technology?
5. According to the author, what is the great advantage of multi-touch technology?

**II. The following statements are incomplete. Search the missing information in the passage and fill in the blanks.**

1. The tactile pleasure __________ provides quickly made it very popular.
2. When iPhone was introduced to the market, the engineers in laboratories around the world have developed much larger __________ that could respond to multiple hands from many people.
3. The rudimentary work of multi-touch interface can be dated back to __________.
4. After being inspired by the crisp image of his fingerprint on __________, Han began his absorption with multi-touch interfaces.
5. Designing __________ that could track the finger movements and convert them into __________ for what should be happening with images was more difficult than sensing the exact location of fingers.
6. Since __________ orders for Perceptive Pixel's system have steadily increased, but it did not disclose prices.
7. Microsoft expects the price of __________ computers to decline as production volume increases and they should be available at __________ price in the near future.
8. The DiamondTouch table that was designed by __________ is configured so that outside parties can write programs for applications they imagine.

**III. Identify the implied meanings of the underlined parts of the following sentences according to the context of the passage, and translate the sentences into Chinese.**

1. When Apple's iPhone hit the street last year, it introduced so-called multi-touch screens to the general public.
2. But in laboratories around the world at the time of the iPhone's launch, multi-touch screens had vastly outgrown two-finger commands.
3. Jeff Han, a consulting computer scientist at New York University and founder of Perceptive Pixel in New York City, is at the forefront of multi-touch technology.
4. Han steps up to the electronic wall and unleashes a world of images using nothing but the touch of his fingers.
5. But around 2000, at NYU, Han began a journey to overcome one of the technology's toughest hurdles: achieving fine-resolution fingertip sensing.
6. But tracking a randomly moving finger would have required an insane amount of wiring behind the screen, which also would have limited the screen's functionality.
7. Deep in the architecture of a computer's operating system is an assumption that a user's input will come either from a keyboard or a mouse.
8. The current operating systems—Windows, Macintosh, Linux—are so predicated on

the single mouse cursor that "we had to tear up a lot of plumbing to make a new multi-touch graphical framework," Han says.

9. Software giant Microsoft is now rolling out a smaller multi-touch computer called Surface and is trying to brand this category of hardware as "surface computers".
10. Soon the multi-touch interface could help untether us from the ubiquitous mouse.

## Translation Techniques (1)

### Conversion from English Prepositions into Chinese Verbs
### (英语介词与汉语动词的转换)

In English prepositions are also called "function words" (in contrast to the "content words", such as nouns, verbs and adjectives) which indicate the relationships with other words within a sentence. Prepositions can be used to express various meanings, and in many cases it is more appropriate to use verbs in Chinese translation where prepositions are used in the English versions.

Let's have a look at the following sentence from the reading passage: "Since then, orders for the system have steadily increased." The preposition "for" that follows the word "order" has the meaning of "buying something", and therefore the sentence should be translated as 自从那时以来，购买该系统的订单不断攀升.

Here are a few more examples:

1. *With* a mind toward practical applications, dozens of labs are working on wires and films thin enough to deposit on computer chips.
   由于对超导体的实际用途都抱有希望，几十家实验室都在研制可以固定在计算机芯片上的细线和薄膜。
2. If the resistance drops to zero *on* the digital meter, they proceed to slice the sample into little bits to explore its workings.
   如果数字电压表显示电阻降为零，他们就再将试样分割成小块来探索其工作原理。
3. Although doctors do not know all the causes of high blood pressure, they do know that overweight people and cigarette smokers have a tendency *to* high blood pressure.
   虽然医生们并不知道高血压的全部起因，但他们确实了解到超重者和吸烟者容易患高血压。

4. Today's thin, brittle photovoltaic cells are made from silicon, the second most abundant element on earth *after* oxygen.
   今天又薄又脆的光电池是由硅制成的，硅在地球上的蕴藏量仅次于氧。
5. About one in every 100 babies is born *with* a heart defect, the leading cause of heart disease deaths among infants.
   大约每100名婴儿中就有1名带有先天性心脏缺陷，这是婴儿死于心脏病的主要原因。

**IV. Translate the following sentences into Chinese. Pay attention to how italicized words in the sentences should be translated.**

1. Images on the screen can be moved around *with* a fingertip and made bigger or smaller by placing two fingertips on the image's edges and then either spreading those fingers apart or bringing them closer together.
2. Rudimentary work *on* multi-touch interfaces dates to the early 1980s, according to Bill Buxton, a principal researcher at Microsoft Research.
3. Han ultimately devised a rectangular sheet of clear acrylic that acts like a waveguide, essentially a pipe *for* light waves.
4. Devising software routines that could track the finger movements and convert them to instructions *for* what should be happening with images on the screen was tougher.
5. During all this work, Han found that pressure sensing could be accomplished, too, *by* applying to the front of the acrylic screen a thin layer of polymer with microscopic ridges engineered into its surface.
6. Other Microsoft software will allow a wireless-enabled digital camera, when placed on a surface computer, to upload its photographic content to the computer *without* a cable.
7. More than 85 prototypes later, the pair ended up *with* a table that has a clear acrylic top and houses a projector on the floor below.
8. Technology developers might be interested in the DiamondTouch table from a start-up company called Circle Twelve *in* Framingham, Mass.

## Translation Techniques (2)

### Translation of English Unanimated Nouns Used as Subjects
### （英语无灵名词作主语的翻译方法）

In English, nouns can be divided into animated and unanimated ones. The former are those connected with lives such as "doctor", "scientist" and "manager", while the latter with non-living objects such as "science", "scheme" and "universe". In technical English, the focus is usually placed on the objects that are studied instead of the people who study them, and therefore unanimated nouns are often put at the beginning of sentences as subjects. In addition, the use of unanimated nouns as subjects helps to make the language sound more formal.

However, when such kind of sentences in the English versions are translated into Chinese, some changes are often necessary so that the translated version will conform to Chinese ways of speaking. For example, in the sentence "Aminophylline does not share the usual objection to vasodilators", the subject is an unanimated noun "aminophylline", but in Chinese translation it would be more appropriate to change the structure of the sentence and start the sentence differently: 人们通常反对使用血管扩张剂，但并不反对使用氨茶碱。

Let's look at a few more examples.

1. *Mathematics and the military* are not new acquaintances, and the US government funding for math research has grown steadily since World War II.
   将数学用于军事领域早已不是新鲜事，自二战以来，美国政府对数学研究的拨款就一直在上升。
2. But *logic* alone does not make a first-class mind.
   然而仅凭逻辑还不能成为第一流的头脑。
3. *New manned and unmanned submarines* permit scientists to learn more about the ocean floor than ever before.
   科学家利用新型载人和不载人的潜水艇，能比以往任何时候更加深刻地了解海底的情况。
4. *Almost every use of electricity* suffers from resistance, an unavoidable phenomenon inside conventional wires and other conductors that turns part of any flow of electrical energy into useless heat.
   几乎所有用电的场合都存在着电阻，这是一种不可避免的现象，在常规导线和其他导体内将部分电能转变为无用的热量。
5. *Nature loves fusion*; it is the process that keeps our sun and other stars burning.

聚变是大自然的宠儿，这一过程使太阳和其他恒星不断燃烧。

**V. Discuss the translation technique and the ways of applying the technique to the translation of the following sentences. Complete each of the Chinese translations.**

1. *Their contributions* are important to the history of biology, and to modern science.（他们 __________ 都做出了重大贡献。）
2. *Nothing* better illustrates the broadening of the concern about pollution from a local affair to a global one than air pollution.（人们对于污染的关注已从局部地区扩展到全球范围，__________。）
3. *The view* is still common today that, initially, we should address local air pollution, then we should turn attention to regional issues like acid rain, and then, at some point in the future, we should address the global issue of greenhouse gases.（__________，认为开始时我们应关注当地的空气污染问题，然后 __________。）
4. Nevertheless, *efforts* in computational analysis and algorithm development must continue.（但是，我们 __________。）
5. *The potential of hardware advances* to make large-scale engineering and scientific computations practical is great.（通过硬件的进步使大规模工程计算与科学计算实际可行，__________。）
6. *Calculators* now slide into checkbooks, and stereo speakers no larger than bricks pack an audio wallop.（现在人们可以 __________，还可以 __________。）
7. *Plans* are in the works to launch a robot into space to locate and repair orbiting satellites.（人们 __________。）
8. *Reproduction* provides new generations and makes possible the continuation of race.（通过 __________。）

**VI. Translate the following sentences into Chinese. Pay attention to how the unanimated nouns as subjects of the sentences should be translated properly.**

1. *The solution* required both hardware and software innovations.
2. *The screen* could therefore serve as both an output of imagery and an input of touches made on that imagery.
3. *The initiative* dates back to 2001, when Stevie Bathiche of Microsoft Hardware and Andy Wilson of Microsoft Research began developing an interactive tabletop that could recognize certain physical objects placed on it.

4. The *table*, developed at Mitsubishi, is configured so that outside parties can write software for applications they envision; several dozen tables are already in the hands of academic researchers and commercial customers.
5. *The great strength of multi-touch* is letting multiple people work together on a complex activity.

## VII. Translate the following passage into Chinese.

Often the technologies that reshape daily life sneak up on us, until suddenly one day it's hard to imagine a world without them—instant messaging, for example, or microwave ovens. Other watershed technologies are visible a mile away, and when you contemplate their applications, the ultimate social impact looks enormous. A good example of the latter is radio frequency identification chips—RFID, for short.

An RFID chip is a tiny bit of silicon, smaller than a grain of rice, that carries information—anything from a retail price, to cooking instructions, to your complete medical records. A larger piece of equipment called an RFID "reader" can, without direct contact, pull that information off the chip and in turn deliver it to any electronic device—a cash register, a video screen, a home appliance, even directly onto the Internet. RFID is the technology used now to automate toll taking at bridges and tunnels; drivers are given a small plastic box with an RFID chip inside, allowing them to drive through the tollgates without stopping. An RFID reader in the tollbooth senses the information on the chip and the toll is automatically deducted from the driver's account.

The first wide-scale applications of RFID will be in retail. At a major industry conference next week, Wal-Mart is expected to urge its suppliers to adopt RFID—the same way that, 20 years ago, the giant retailer jump-started the use of bar codes. And some manufacturers are already on board. Gillette, for example, recently placed an order for half a billion RFID chips that they will begin to use to track individual packages of razors.

## VIII. Translate the following passage into English.

### 多点触控技术

多点触控指的是人机互动技术和应用该技术的硬件设施，利用这些设施，使用者不需要传统的输入设备（如鼠标、键盘）就可以进行运算。多点触控由触控式屏幕或触控垫，以及可以同时辨认多个触点的软件组成，它与标准的触摸屏不同，因为后者只能辨认一个触点。

多点触控技术始于1982年，当时多伦多大学首先研制出了指压多触点显示技术。同年，贝尔实验室发表了据信是将触摸屏技术用于接口装置的第一篇论文。21世纪初，该技术使各类公司得到发展，2007年苹果公司推出了iPhone，微软展示了桌面计算技术。尤其是iPhone引起了人们对多点触控计算技术的广泛兴趣，因为该技术规模虽小，却大大增加了用户的互动。随之出现了更强大的、可为用户量身定制的多点触控设备，例如Perceptive Pixel公司研制的壁式显示屏和台式设备可以允许20个手指同时操作。

预计多点触控技术会很快得到普遍应用。例如多点触控电话将从2006年出售的20万台增至2012年的2,100万台。事实上，开发商已提出应用多点触控技术的多种方法。

## Section 2 Reading for Academic Purposes

*Critical thinking in reading relates closely to active reading. Its purpose is to get you involved in a dialogue with the ideas obtained from reading so that you can summarize, analyze, hypothesize, and evaluate the ideas you encounter. As a graduate student, you are already doing some critical reading, and perhaps the most powerful thing you can do in furthering your abilities in this area is to become conscious in your application of a variety of questions to whatever you read. Even if you cannot always readily answer the questions you develop, you are beginning to think in a way which gets beyond there being just right and wrong answers, and which gets beyond you memorizing answers to the questions somebody else makes up. In fact, you are engaging in the practice which is often one of the most important aspects of critical thinking: evaluating a scientific paper.*

*The practice of evaluating a paper is probably not new to you, but you might be unsure of how to apply it to academic work in a strategic way. Taking the following essay written from the angle of biological research will provide you with some valuable ideas on this topic.*

### Evaluating a Scientific Paper
### 对科技论文的评估

*A thorough understanding and evaluation of a scientific paper involves answering several questions:*

1. What *questions* does the paper address?

2. What are the main *conclusions* of the paper?

3. What *evidence* supports those conclusions?

4. Do the data actually *support* the conclusions?

5. What is the *quality* of the evidence?

6. Why are the *conclusions* important?

## 1. What questions does the paper address?

Before addressing this question, we need to be aware that research can be of several different types: descriptive research, comparative research and analytical research.

*Descriptive research* often takes place in the early stages of our understanding of a system. We can't **formulate** hypotheses about how a system works, or what its interconnections are, until we know what is there. Typical descriptive approaches in **molecular biology**, for example, are DNA sequencing and DNA **microarray** approaches. In biochemistry, one could regard *x*-ray **crystallography** as a descriptive endeavor.

formulate *v.* 阐明，明确陈述
molecular biology 分子生物学
microarray *n.*【医】微点阵
crystallography *n.* 晶体学

*Comparative research* often takes place when we are asking how general a finding is. Is it specific to my particular **organism**, or is it broadly **applicable**? A typical comparative approach would be comparing the sequence of a gene from one organism with that from the other organisms in which that gene is found. One example of this is the observation that the **acting genes** from humans and **budding yeast** are 89% **identical** and 96% similar.

organism *n.* 生物体；有机体
applicable *adj.* 可应用的
acting gene【生】开放基因
budding yeast【生化】芽殖酵母
identical *adj.*（完全）相同的，一模一样的

*Analytical research* generally takes place when we know enough to begin formulating hypotheses about how a system works, about how the parts are interconnected, and what the **causal** connections are. A typical analytical approach would be to devise two (or more) alternative hypotheses about how a system operates. These hypotheses would all be consistent with current knowledge about the system. Ideally, the approach would devise a set of experiments to distinguish among these hypotheses. A classic example is the **Meselson-Stahl experiment**.

causal *adj.* 表示原因的；具有因果关系性质的
Meselson-Stahl experiment 梅瑟生–史达实验

Of course, many papers are a combination of these approaches. For instance, researchers might sequence a gene from their model organism; compare its sequence to **homologous** genes from other organisms; use this comparison to devise a hypothesis for the function of the gene product; and test this hypothesis by making a site-directed change in the gene and asking how that affects the **phenotype** of the organism and/or the biochemical function of the gene product.

homologous *adj.*【生】同源的
phenotype *n.* 显型；表现型（指有机体上可观察到的物理或生化特征）

orient *v.* 使……朝向

Being aware that not all papers have the same approach can **orient** you towards recognizing the major questions that a paper addresses.

*What are these questions?*

In a well-written paper, the Introduction generally goes from the general to the specific, eventually framing a question or set of questions. This is a good starting place. In addition, the results of experiments usually raise additional questions, which the authors may attempt to answer. These questions usually become evident only in the Results section.

**2. What are the main conclusions of the paper?**

This question can often be answered in a preliminary way by studying the abstract of the paper. Here the authors highlight what they think are the key points. This is not enough, because abstracts often have severe space constraints, but it can serve as a starting point. Still, you need to read the paper with this question in mind.

**3. What evidence supports those conclusions?**

Generally, you can get a pretty good idea about this from the Results section. The description of the findings points to the relevant tables and figures. This is easiest when there is one primary experiment to support a point. However, it is often the case that several different experiments or approaches combine to support a particular conclusion. For example, the first experiment might have several possible interpretations, and the later ones are designed to distinguish among these.

In the ideal case, the Discussion begins with a section of the form "Three lines of evidence provide support for the conclusion that... First, ... second, ..., etc." However, difficulties can arise when the paper is poorly written. The authors often do not present a concise summary of this type, leaving you to make it yourself. A skeptic might argue that in such cases the logical structure of the argument is weak and is omitted on purpose! In any case, you need to be sure that you understand the relationship between the data and the conclusions.

**4. Do the data actually support the conclusions?**

One major advantage of doing this is that it helps you to evaluate whether the conclusion is sound. If we assume for the moment that the data are believable, it still might be the case that the data do not actually support the conclusion the authors wish to reach. There are at least two different ways this can happen:

(1) The logical connection between the data and the interpretation is not sound.

(2) There might be other interpretations that might be consistent with the data.

One important aspect to look for is whether the authors take multiple approaches to answering a question. Do they have multiple lines of evidence, from different directions, supporting their conclusions? If there is only one line of evidence, it is more likely that it could be interpreted in a different way; multiple approaches make the argument more persuasive.

Another thing to look for is **implicit** or hidden assumptions used by the authors in interpreting their data. This can be hard to do, unless you understand the field thoroughly.

implicit *adj*. 内含的；含蓄的

## 5. What is the quality of that evidence?

This is the hardest question to answer, for novices and experts alike. At the same time, it is one of the most important skills to learn as a young scientist. It involves a major **reorientation** from being a relatively passive consumer of information and ideas to an active producer and critical evaluator of them. This is not easy and takes years to master. Beginning scientists often wonder, "Who am I to question these authorities? After all the paper was published in a top journal, so the authors must have a high standing, and the work must have received a critical review by experts."

reorientation *n*. 再定位，再调整

Unfortunately, that's not always the case. In any case, developing your ability to evaluate evidence is one of the hardest and most important aspects of learning to be a critical scientist and reader.

*How can you evaluate the evidence?*

First, you need to understand thoroughly the methods used in the experiments. Often these are described poorly or not at all. The details are often missing, but more importantly the authors usually assume

immunoblotting *n.* 免疫印迹；免疫印迹法
methodology *n.* 方法学；方法论

that the reader has a general knowledge of common methods in the field (such as **immunoblotting**, cloning, genetic methods, or DNase I footprinting). If you lack this knowledge, you have to make the extra effort to inform yourself about the basic **methodology** before you can evaluate the data.

Sometimes you have to go to the library, or to a lab that has a lot of back issues of common journals, to trace back the details of the methods if they are important. One new development that eventually will make this much easier is the increasing availability of journals on the Web. A comprehensive listing of journals relevant to the courses you are taking allows access to most of the listed volumes from any computer at the university.

linear *adj.* 线性的；直线的

Second, you need to know the limitations of the methodology. Every method has limitations, and if the experiments are not done correctly they can't be interpreted. For instance, an immunoblot is not a very quantitative method. Moreover, in a certain range of protein the signal increases (that is, the signal is at least roughly "**linear**"), but above a certain amount of protein the signal no longer increases. Therefore, to use this method correctly one needs a standard curve that shows that the experimental lanes are in a linear range. Often, the authors will not show this standard curve, but they should state that such curves were done. If you don't see such an assertion, it could of course result from bad writing, but it might also not have been done. If it wasn't done, a dark band might mean "there is this much protein or an indefinite amount more".

Third, you need to distinguish between what the data show and what the authors say they show. The latter is really an interpretation on the authors' part, though it is generally not stated to be an interpretation. Papers usually state something like "the data in Fig. x show that..." This is the authors' interpretation of the data. Do you interpret it the same way? You need to look carefully at the data to ensure that they really do show what the authors say they do. You can only do this effectively if you understand the methods and their limitations.

halftone *n.* 【印】网目版（画）
gel *n.* 凝胶，冻胶
autoradiogram *n.* 射线自显迹图

Fourth, it is often helpful to look at the original journal (or its electronic counterpart) instead of a photocopy. Particularly for **half-tone** figures in biology such as photos of **gels** or **autoradiograms**, the contrast is **distorted**, usually increased, by photocopying, so that the data are **misrepresented**.

distort *v.* 扭曲，歪曲
misrepresent *v.* 不如实地说明，歪曲

Fifth, you should ask if the proper **controls** are present. Controls tell us that nature is behaving the way we expect it to under the conditions of the experiment. If the controls are missing, it is harder to be confident that the results really show what is happening in the experiment. You should try to develop the habit of asking "where are the controls?" and looking for them.

control *n.* [常用复]控制的手段或措施

**6. Why are the conclusions important?**

Do the conclusions make a significant advance in our knowledge? Do they lead to new insights, or even new research directions?

Again, answering these questions requires that you understand the field relatively well.

Exercises

**I. Read the passage and decide whether the following statements are true or false. Write T for True and F for False in the brackets.**

1. Before we know what is there, we can propose hypotheses about how a system operates. ( )
2. When we are asking how general a finding is in a scientific research, it is a typical comparative study. ( )
3. A typical analytical approach will formulate two or more different hypotheses that are consistent with current knowledge about how a system works. ( )
4. Usually the results of experiments could raise new questions that usually become evident only in the results section. ( )
5. It is enough to search the main conclusions of a research paper in the abstract part in which the authors highlight them in detail. ( )
6. It is common that several different experiments or approaches are combined to support a certain conclusion in a research. ( )
7. Whether the authors take different approaches to answering a research question is crucial in evaluating the connection between the data and the conclusion. ( )
8. If there is only one line of evidence in the data, it is not necessary to interpret it in multiple ways which will only make it more complex. ( )
9. Since a paper was published in a top journal and the authors have a high standing, the work must have received critical review by experts and it is not necessary for novices

to question it. (   )

10. When we evaluate the evidence in a research paper, we should bear in mind that every method has limitations and if the experiments are not done correctly they can't be interpreted. (   )

**II. Read the passage again, and complete the following items.**

1. The three different types of research include: ___________
2. The characteristic of a well-written Introduction: ___________
3. The preliminary way of finding the main conclusions of a research paper: ___________
4. There are two different ways in which the data do not actually support the conclusion: ___________
5. The most difficult and important aspect of learning to be a critical scientist is: ___________
6. The first step in evaluating the evidence in a research paper is: ___________
7. The new development that will make searching back issues of journals easier is: ___________
8. The importance of the controls: ___________

## Quotations from Great Scientists

*The science of today is the technology of tomorrow.*

*—Edward Teller*

今天的科学就是明天的技术。

——埃德华·泰勒

# Chapter 6 Mathematics and Creativity 数学与独创性

## Section 1 Reading and Translation

### I What You Are Going to Read

Mathematics is a wonderful subject, full of imagination, fantasy and creativity that is not limited by the petty details of the physical world, but only by the strength of our inner light. Does this sound familiar? Probably not from the mathematics classes you may have attended. But consider the work of Professor William Byers. Traditionally, the essence of mathematical thinking and discourse is its precision, but Professor Byers argues that there are non-logical, ambiguous elements in math that are equally if not more important. He approaches math in a somewhat unexpected manner: through creativity.

In his publication *How Mathematicians Think: Using Ambiguity, Contradiction and Paradox to Create Mathematics* (Princeton University Press, 2007), Professor Byers argues that math is an ambiguous subject and cannot be ruled by logical and rational thought alone. According to Byers, mathematics is fraught with ambiguity that must be encouraged, not dismissed. In fact, many mathematicians describe their most important breakthroughs as creative and intuitive leaps in response to ambiguous situations that are not logical at all.

What you are going to read in this section is an excerpt from the Introduction of the book.

### II About the Author

William Byers is Professor of Mathematics and Statistics at Concordia University in Montreal. He has published widely in journals of mathematics.

### III Reading Passage

#### "The Light of Reason" or "the Light of Ambiguity"?
#### "理性之光"还是"模糊之光"?

By William Byers

[1] What is it that makes mathematics mathematics? What are the precise characteristics that make mathematics into a discipline that is so central to every advanced civilization, especially our own? Many

ultimate *n.* 终极；顶点
distinguishing *adj.* 区别性的
stringent *adj.* 严密的；有说服力的
deductive *adj.* 推论的，演绎的

explanations have been attempted. One of these sees mathematics as the **ultimate** in rational expression; in fact, the expression "the light of reason" could be used to refer to mathematics. From this point of view, the **distinguishing** aspect of mathematics would be the precision of its ideas and its systematic use of the most **stringent** logical criteria. In this view, mathematics offers a vision of a purely logical world. One way of expressing this view is by saying that the natural world obeys the rules of logic and, since mathematics is the most perfectly logical of disciplines, it is not surprising that mathematics provides such a faithful description of reality. This view, that the deepest truth of mathematics is encoded in its formal, **deductive** structure, is definitely not the point of view that this book assumes. On the contrary, the book takes the position that the logical structure, while important, is insufficient even to begin to account for what is really going on in mathematical practice, much less to account for the enormously successful applications of mathematics to almost all fields of human thought.

translogical *adj.* 超逻辑的
legitimate *adj.* 合理的，正统的
axiom *n.*【数】公理

2 This book offers another vision of mathematics, a vision in which the logical is merely one dimension of a larger picture. This larger picture has room for a number of factors that have traditionally been omitted from a description of mathematics and are **translogical**—that is, beyond logic—though not illogical. Thus, there is a discussion of things like ambiguity, contradiction, and paradox that, surprisingly, also have an essential role to play in mathematical practice. This flies in the face of conventional wisdom that would see the role of mathematics as eliminating such things as ambiguity from a **legitimate** description of the worlds of thought and nature.[1] As opposed to the formal structure, what is proposed is to focus on the central ideas of mathematics, to take ideas—instead of **axioms**, definitions, and proofs—as the basic building blocks of the subject and see what mathematics looks like when viewed from that perspective.

3 The phenomenon of ambiguity is central to the description of mathematics that is developed in this book. In his description of his own personal development, Alan Lightman[2] says, "Mathematics contrasted strongly with the ambiguities and contradictions in people. The world of people had no certainty or logic." For him, mathematics is the domain of certainty and logic. On the other hand, he is also a novelist who "realized that the ambiguities and complexities of the

human mind are what give fiction and perhaps all art its power." This is the usual way that people divide up the arts from the sciences: ambiguity in one, certainty in the other. I suggest that mathematics is also a human, creative activity. As such, ambiguity plays a role in mathematics that is **analogous** to the role it plays in art—it **imbues** mathematics with depth and power.

analogous *adj.* 类似的，可比拟的
imbue *v.* 使充满

4 Ambiguity is **intrinsically** connected to creativity. In order to make this point, I propose a definition of ambiguity that is derived from a study of creativity. The description of mathematics that is to be **sketched** in this book will be a description that is grounded in mathematical practice—what mathematicians actually do—and, therefore, must include an account of the great creativity of mathematics. We shall see that many creative insights of mathematics arise out of ambiguity, that in a sense the deepest and most revolutionary ideas come out of the most profound ambiguities. Mathematical ideas may even arise out of contradiction and paradox. Thus, eliminating the ambiguous from mathematics by focusing exclusively on its logical structure has the unwanted effect of making it impossible to describe the creative side of mathematics. When the creative, **open-ended** dimension is lost sight of, and, therefore, mathematics becomes identified with its logical structure, there develops a view of mathematics as rigid, inflexible, and unchanging.[3] The truth of mathematics is mistakenly seen to come exclusively from a rigid, deductive structure. This rigidity is then transferred to the domains to which mathematics is applied and to the way mathematics is taught, with unfortunate consequences for all concerned.

intrinsically *adv.* 本质上地，固有地
sketch *v.* 勾画，简略描绘
open-ended *adj.* 无限制的，没有限度的

5 Thus, there are two visions of mathematics that seem to be **diametrically** opposed to one another. These could be characterized by emphasizing "the light of reason", the **primacy** of the logical structure, on the one hand, and the light that Wiles[4] spoke of, a creative light that I maintain often emerges out of ambiguity, on the other (this is itself an ambiguity!). My job is to demonstrate how mathematics **transcends** these two opposing views: to develop a picture of mathematics that includes the logical and the ambiguous, that situates itself equally in the development of vast deductive systems of the most **intricate** order and in the birth of the extraordinary leaps of creativity that have changed the world and our understanding of the world.

diametrically *adv.* 直接地
primacy *n.* 首位，首要状态
transcend *v.* 超越，胜过
intricate *adj.* 错综复杂的

6 This is a book about mathematics, yet it is not your average mathematics book. Even though the book contains a great deal of mathematics, it does not systematically develop any particular mathematical subject. The subject is mathematics as a whole—its methodology and conclusions, but also its culture. The book puts forward a new vision of what mathematics is all about. It concerns itself not only with the culture of mathematics in its own right, but also with the place of mathematics in the larger scientific and general culture.

rigor *n.* 严密，精确
descendant *n.* 继承者，后代
beset *v.* 遭到困扰
predictability *n.* 可预言性，可预知
subscribe *v.* 赞成，同意
impediment *n.* 妨碍；阻碍物
rationality *n.* 理性，合理性
equate *v.* 等同
simplistic *adj.* 过分简单化的
aesthetic *adj.* 美学的；审美的
originality *n.* 创意，独创性
innovate *v.* 创新，革新
potent *adj.* 有力的，有效的
amend *v.* 修改，修正

7 The perspective that is being developed here depends on finding the right way to think about mathematical **rigor**, that is, logical, deductive thought. Why is this way of thinking so attractive? In our response to reason, we are the true **descendants** of the Greek mathematicians and philosophers[5]. For us, as for them, rational thought stands in contrast to a world that is all too often **beset** with chaos, confusion, and superstition. The "dream of reason" is the dream of order and **predictability** and, therefore, of the power to control the natural world. The means through which we attempt to implement that dream are mathematics, science, and technology. The desired end is the emergence of clarity and reason as organizational principles of the entire cosmos, a cosmos that of course includes the human mind. People who **subscribe** to this view of the world might think that it is the role of mathematics to eliminate ambiguity, contradiction, and paradox as **impediments** to the success of **rationality**. Such a view might well **equate** mathematics with its formal, deductive structure. This viewpoint is incomplete and **simplistic**. When applied to the world in general, it is mistaken and even dangerous. It is dangerous because it ignores one of the most basic aspects of human nature—in mathematics or elsewhere—our **aesthetic** dimension, our **originality** and ability to **innovate**. In this regard let us take note of what the famous musician, Leonard Bernstein[6], had to say: "Ambiguity...is one of art's most **potent** aesthetic functions. The more ambiguous, the more expressive." His words apply not only to music and art, but surprisingly also to science and mathematics. In mathematics, we could **amend** his remarks by saying, "The more ambiguous, the more potentially original and creative."

plumb *v.* 探求

8 If one wishes to understand mathematics and **plumb** its depths,

one must reevaluate one's position toward the ambiguous (as I shall define it in Chapter 1) and even the paradoxical. Understanding ambiguity and its role in mathematics will hint at a new kind of organizational principle for mathematics and science, a principle that includes classical logic[7] but goes beyond it. This new principle will be **generative**—it will allow for the dynamic development of mathematics. As opposed to the **static** nature of logic with its absolute dichotomies, a generative principle will allow for the existence of mathematical creativity, be it in research or in individual acts of understanding.[8] Thus "ambiguity" will force a reevaluation of the essence of mathematics.

generative *adj.* 有生产力的，能生成的
static *adj.* 静态的

9 Why is it important to reconsider mathematics? The reasons vary from those that are internal to the discipline itself to those that are external and affect the applications of mathematics to other fields. The internal reasons include developing a description of mathematics, a philosophy of mathematics if you will, that is consistent with mathematical practice and is not merely a set of a **priori** beliefs.[9] Mathematics is a human activity; this is a triumph, not a **constraint**. As such, it is potentially accessible to just about everyone. Just as most people have the capacity to enjoy music, everyone has some capacity for mathematics appreciation. Yet most people are fearful and **intimidated** by mathematics. Why is that? Is it the mathematics itself that is so frightening? Or is it rather the way in which mathematics is viewed that is the problem?

priori *adj.* 先验的
constraint *n.* 约束，束缚
intimidate *v.* 使胆怯，使害怕

10 Beyond the valid "internal" reasons to reconsider the nature of mathematics, even more **compelling** are the external reasons—the impact that mathematics has, one way or another, on just about every aspect of the modern world. Since mathematics is such a central discipline for our entire culture, reevaluating what mathematics is all about will have many implications for science and beyond, for example, for our conception of the nature of the human mind itself. Mathematics provided humanity with the ideal of reason and, therefore, a certain model of what thinking is or should be, even what a human being should be. Thus, we shall see that a close investigation of the history and practice of mathematics can tell us a great deal about issues that arise in philosophy, in education, in cognitive science, and in the sciences in general. Though I shall **endeavor** to remain within

compelling *adj.* 令人信服的
endeavor *v.* 努力

the boundaries of mathematics, the larger implications of what is being said will not be ignored.

reverence *n.* 崇敬，敬重
occasion *v.* 引起，为……提供场合
intricacy *n.* 错综复杂，难以理解

11 Mathematics is one of the most profound creations of the human mind. For thousands of years, the content of mathematical theories seemed to tell us something profound about the nature of the natural world—something that could not be expressed in any way other than the mathematical. How many of the greatest minds in history, from Pythagoras[10] to Galileo[11] to Gauss[12] to Einstein[13], have held that "God is a mathematician." This attitude reveals a **reverence** for mathematics that is **occasioned** by the sense that nature has a secret code that reveals her hidden order. The immediate evidence from the natural world may seem to be chaotic and without any inner regularity, but mathematics reveals that under the surface the world of nature has an unexpected simplicity—an extraordinary beauty and order. There is a mystery here that many of the great scientists have appreciated. How does mathematics, a product of the human intellect, manage to correspond so precisely to the **intricacies** of the natural world? What accounts for the "extraordinary effectiveness of mathematics"?

irreducible *adj.* 不能简化的，无法缩减的
propensity *n.* 倾向，习性
hard-wired *adj.* 永久连接的
implicit *adj.* 固有的，内含的

12 Beyond the content of mathematics, there is the fact of mathematics. What is mathematics? More than anything else, mathematics is a way of approaching the world that is absolutely unique. It cannot be reduced to some other subject that is more elementary in the way that it is claimed that chemistry can be reduced to physics.[14] Mathematics is **irreducible**. Other subjects may use mathematics, may even be expressed in a totally mathematical form, but mathematics has no other subject that stands in relation to it in the way that it stands in relation to other subjects. Mathematics is a way of knowing—a unique way of knowing. When I wrote these words I intended to say "a unique human way of knowing". However, it now appears that human beings share a certain **propensity** for number with various animals. One could make an argument that a tendency to see the world in a mathematical way is built into our developmental structure, **hard-wired** into our brains, perhaps **implicit** in elements of the DNA structure of our genes. Thus mathematics is one of the most basic elements of the natural world.

span *v.* 跨越

13 From its roots in our biology, human beings have developed mathematics as a vast cultural project that **spans** the ages and all

civilizations. The nature of mathematics gives us a great deal of information, both direct and indirect, on what it means to be human. Considering mathematics in this way means looking not merely at the content of individual mathematical theories, but at mathematics as a whole. What does the nature of mathematics, viewed globally, tell us about human beings, the way they think, and the nature of the cultures they create? Of course, the latter, global point of view can only be seen clearly **by virtue of** the former. You can only speak about mathematics with reference to actual mathematical topics. Thus, this book contains a fair amount of actual mathematical content, some very elementary and some less so. The reader who finds some topic obscure is advised to skip it and continue reading. Every effort has been made to make this account self-contained, yet this is not a mathematics textbook—there is no systematic development of any large area of mathematics. The mathematics that is discussed is there for two reasons: first, because it is intrinsically interesting, and second, because it contributes to the discussion of the nature of mathematics in general. Thus, a subject may be introduced in one chapter and returned to in subsequent chapters.

by virtue of 凭借，由于

14 It is not always appreciated that the story of mathematics is also a story about what it means to be human—the story of being blessed (some might say cursed) with self-consciousness and, therefore, with the need to understand the natural world and themselves. Many people feel that such a human perspective on mathematics would **demean** it in some way, diminish its claim to be revealing absolute, objective truth. To anticipate the discussion in Chapter 8, I shall claim that mathematical truth exists, but is not to be found in the content of any particular **theorem** or set of theorems. The intuition that mathematics accesses the truth is correct, but not in the manner that it is usually understood. The truth is to be found more in the fact than in the content of mathematics. Thus it is consistent, in my view, to talk simultaneously about the truth of mathematics and about its **contingency**.

demean *v.* 贬低
theorem *n.* 定理，法则
contingency *n.* 偶然性，突发性

15 The truth of mathematics is to be found in its human dimension, not by avoiding this dimension. This human story involves people who find a way to transcend their limitations, about people who dare to do what appears to be impossible and is impossible by any reasonable

aftermath *n.* 后果；结果

standard. The impossible is rendered possible through acts of genius—this is the very definition of an act of genius, and mathematics boasts genius in abundance. In the **aftermath** of these acts of genius, what was once considered impossible is now so simple and obvious that we teach it to children in school. In this manner, and in many others, mathematics is a window on the human condition. As such, it is not reserved for the initiated, but is accessible to all those who have a fascination with exploring the common human potential.

construct *n.* 概念
formulation *n.* 公式化的表述
allude *v.* 暗指；间接提到

16 We do not have to look very far to see the importance of mathematics in practically every aspect of contemporary life. To begin with, mathematics is the language of much of science. This statement has a double meaning. The normal meaning is that the natural world contains patterns or regularities that we call scientific laws and mathematics is a convenient language in which to express these laws. This would give mathematics a descriptive and predictive role. And yet, to many, there seems to be something deeper going on with respect to what has been called "the unreasonable effectiveness of mathematics in the natural sciences". Certain of the basic **constructs** of science cannot, in principle, be separated from their mathematical **formulation**. An electron is its mathematical description via the Schrödinger equation[15]. In this sense, we cannot see any deeper than the mathematics. This latter view is close to the one that holds that there exists a mathematical, Platonic substratum[16] to the real world. We cannot get closer to reality than mathematics because the mathematical level is the deepest level of the real. It is this deeper level that has been **alluded** to by the brilliant thinkers that were mentioned above. This deeper level was also what I meant by calling mathematics irreducible.

mathematization *n.* 数学化，数字处理

17 Our contemporary civilization has been built upon a mathematical foundation. Computers, the Internet, CDs, and DVDs are all aspects of a digital revolution that is reshaping the world. All these technologies involve representing the things we see and hear, our knowledge, and the contents of our communications in digital form, that is, reducing these aspects of our lives to a common numerical basis. Medicine, politics, and social policy are all increasingly expressed in the language of the mathematical and statistical sciences. No area of modern life can escape from this **mathematization** of the world.

18 If the modern world stands on a mathematical foundation, it **behooves** every thoughtful, educated person to attempt to gain some familiarity with the world of mathematics. Not only with some particular subject, but with the culture of mathematics, with the manner in which mathematicians think and the manner in which they see this world of their own creation.

behoove *v.* 【主语用it】对（某人）来说应该（做），有必要

## Notes

1. **This flies in the face of conventional wisdom that would see the role of mathematics as eliminating such things as ambiguity from a legitimate description of the worlds of thought and nature:** 这正好与传统认识相反，传统上人们认为数学的作用就是把模糊性从思想界和自然界的合理描述中消除。句中to fly in the face of sth. 是习语，意思是to be the opposite of what is usual or accepted（与通常的、已被接受的观点相反）。conventional wisdom意为“传统上被认为是真知灼见的看法”。
2. **Alan Lightman:** 阿兰·莱特曼（1948– ），美国物理学家、小说家、随笔作家，现任麻省理工学院物理学与人文学教授，他创作的小说《爱因斯坦的梦》（*Einstein's Dreams*, 1992）畅销全球。
3. **When the creative, open-ended dimension is lost sight of, and, therefore, mathematics becomes identified with its logical structure, there develops a view of mathematics as rigid, inflexible, and unchanging:** 人们一旦忽略了数学中创造性的、非确定性的一面，并将数学等同于其逻辑结构，就会把数学看成呆板的、僵化的、一成不变的。本句中的状语从句When...，含有“如果、一旦”的意思。为了与从句连接得更加紧密，主句采用了倒装的方式，主语是a view of mathematics，develops是谓语动词。
4. **Wiles:** 安德鲁·维尔斯（Andrew Wiles，1953– ），英国数学家，普林斯顿大学数学教授，因证明费马大定理而闻名于世。
5. **Greek mathematicians and philosophers:** 希腊数学家和哲学家，指的是古希腊为西方文化奠定理论实践基础的数学家和哲学家。数学家有毕达哥拉斯、泰勒斯、欧几里得、阿基米德、德谟克利特、埃拉托斯特尼等；哲学家有苏格拉底、柏拉图、亚里士多德、安提斯泰尼等。
6. **Leonard Bernstein:** 伦纳德·伯恩斯坦（1918–1990），美国指挥家、作曲家和作家。曾任纽约爱乐乐团指挥，创作音乐剧《西区故事》等。
7. **classical logic:** 古典逻辑，又称传统逻辑，指从古希腊亚里士多德开创至19世纪现代发展阶段的形式逻辑体系和理论。
8. **As opposed to the static nature of logic with its absolute dichotomies, a generative principle will allow for the existence of mathematical creativity, be it in research or in individual acts of understanding:** 与绝对二分法静止的逻辑本质相反，生成原则使数学拥有创造性，这无论对于研究还是个人对数学的理解来说都是如此。句中be it...起状语的作用，相

当于whether it is...。此外，absolute dichotomies（绝对二分法）认为，事物划分为因矛盾而彼此排斥的两部分，没有中间物。

9. **The internal reasons include developing a description of mathematics, a philosophy of mathematics if you will, that is consistent with mathematical practice and is not merely a set of a priori beliefs:** 内部原因包括对数学的描述，也可以称作“数学的哲学”，它与数学的实际应用一致，因此不仅仅是一套预先树立的信念。句中if you will的意思是“如果你愿意那样说的话”，阅读中应注意此类表达方法的含义。
10. **Pythagoras:** 毕达哥拉斯（580?–500?BC），古希腊哲学家和数学家，在意大利南部创立学派，强调对音乐和谐及几何的研究，他证明了毕达哥拉斯定理（直角三角形勾股定理）的广泛有效性，被认为是世界上第一位真正的数学家。
11. **Galileo:** 伽利略（Galileo Galilei, 1564–1642），文艺复兴后期伟大的意大利天文学家、力学家、哲学家、物理学家、数学家，也是近代实验物理学的开拓者，被誉为“近代科学之父”。
12. **Gauss:** 高斯（Karl Friedrich Gauss, 1777–1855），德国数学家和天文学家，因对代数、微积分几何、或然率理论和数字理论的贡献而著称。
13. **Einstein:** 爱因斯坦（Albert Einstein, 1879–1955），德裔美国理论物理学家，他创立的狭义和广义相对论使现代关于时间和时间性质的想法有了突破性进展，并为原子能的利用提供了理论基础。因其对光电效应的解释获1921年诺贝尔奖。
14. **It cannot be reduced to some other subject that is more elementary in the way that it is claimed that chemistry can be reduced to physics:** 我们不能把它（数学）简化为另一个更具基础性的学科，就像人们所说的可以把化学纳入物理学。reduce sth. to sth. 在本句中的意思是change sth. to a more general or basic form（将某事物概括或简化为某种形式），如：We can reduce this problem to two main issues.（我们可以把这个问题归纳成两个要点。）
15. **the Schrödinger equation:** 薛定谔方程，量子力学基本方程。1926年由奥地利物理学家薛定谔（Erwin Schrödinger, 1887–1961）得出。
16. **Platonic substratum:** 柏拉图实体。柏拉图（Plato, 427–347BC），古希腊哲学家。这里的实体是一个哲学概念，指的是支持现实特点的无特征体。根据柏拉图的理论，千变万化的事物具有稳固的本质，这种本质就是殊相中的共相，特殊中的一般。柏拉图把这种共相从事物中间提取出来，成为独立的实体，即理念。

## Exercises

### I. Answer the following questions.

1. What view does the book present towards the logical structure of mathematics?
2. What is the new vision this book puts forward of what mathematics is all about?
3. Why does the author think the “dream of reason” is dangerous?
4. What does it mean that mathematics is irreducible?
5. Why does the book contain a fair amount of actual mathematical content?

II. **The following statements are incomplete. Search the missing information in the passage and fill in the blanks.**

1. In a point of view, the distinguishing aspect of mathematics would be __________.
2. __________ is essential to the description of mathematics that is developed in this book.
3. It is wrong to see the truth of mathematics coming exclusively from __________.
4. The primacy of the logical structure is __________.
5. According to Leonard Bernstein, one of art's most potent aesthetic functions is __________.
6. For thousands of years, the content of mathematical theories seemed to tell us something profound about __________.
7. The author thinks it is consistent to talk simultaneously about __________ and about __________.
8. According to the author, there is no area of modern life that can escape from __________.

III. **Identify the implied meanings of the underlined parts of the following sentences according to the context of the passage, and translate the sentences into Chinese.**

1. One of these sees mathematics as the ultimate in rational expression; in fact, the expression "the light of reason" could be used to refer to mathematics.
2. This book offers another vision of mathematics, a vision in which the logical is merely one dimension of a larger picture.
3. My job is to demonstrate how mathematics transcends these two opposing views: to develop a picture of mathematics that includes the logical and the ambiguous, that situates itself equally in the development of vast deductive systems of the most intricate order and in the birth of the extraordinary leaps of creativity that have changed the world and our understanding of the world.
4. In our response to reason, we are the true descendents of the Greek mathematicians and philosophers.
5. If one wishes to understand mathematics and plumb its depths, one must reevaluate one's position toward the ambiguous (as I shall define it in Chapter 1) and even the paradoxical.
6. Though I shall endeavor to remain within the boundaries of mathematics, the larger implications of what is being said will not be ignored.
7. One could make an argument that a tendency to see the world in a mathematical way is built into our developmental structure, hard-wired into our brains, perhaps implicit in elements of the DNA structure of our genes.
8. What does the nature of mathematics, viewed globally, tell us about human beings, the way they think, and the nature of the cultures they create?

9. The impossible is rendered possible through acts of genius—this is the very definition of an act of genius, and mathematics boasts genius in abundance.
10. To begin with, mathematics is the language of much of science.

## Translation Techniques (1)

### Conversion Between Passive Voice in English and Active Voice in Chinese
### （被动语态和主动语态的转换）

Technical English is noted for its wide use of the passive voice, which reflects the impersonal and objective attitude of the scientists and technologists toward the scientific studies. The impersonal structure in technical writing emphasizes the matters and processes in discussion instead of the agent of action. However, owing to the different nature of the two languages, it is often preferable to convert the passive voice in English into the active voice in Chinese.

From the following example, we can see that the passive voice in the English sentence is not directly translated as被……; otherwise, the sentence would sound strange: "In the past, scientists simply reshuffled genes within a particular species; corn could not be crossed with soybeans, nor cows with pigs."（过去，科学家只是在某一特定物种内对基因进行重组，因此玉米不能和大豆杂交，牛也不能和猪杂交。）

Here are more examples:

1. When the ionization anemometer *was exposed* to an airstream, the ion flow was disturbed and current changes occurred.
   把电离速度计置于气流中，里边的离子流就会受到干扰，从而使电流发生变化。
2. Should the blood become either too diluted or too viscous, the kidneys retain or excrete surplus water until the proper balance *is reached*.
   如果血液太稀或太稠，肾脏就要保存或排除多余的水分，直到恢复平衡。
3. This suggests that cosmic rays do not come from all over the Universe, but *are produced* within galaxies.
   这种情况说明，宇宙射线并不是来自整个宇宙，而是产生于星系内部。

Similarly, it is often necessary to convert the active voice into the passive voice when Chinese sentences are translated into English. For example:

1. 虽然计算机主要是为了满足科学的需要而设计的，但商业及政府对数据处理的要求却是这种机器获得进一步发展的主要推动因素。
   Although the computer *was devised* mainly in response to scientific needs, the requirements of business and government data handling have been a major stimulant to the further development of this machinery.

2. 当前的问题是软件价格已远高于硬件，而要使计算机成为普遍使用的工具，还需要更为复杂的软件。

The problem is that software cost more than hardware already, and to turn the computer into an everyman tool much more complex software *is needed* than is available now.

**IV. Translate the following sentences. Pay attention to how the technique of conversion between the passive voice in English and the active voice in Chinese should be used.**

1. The phenomenon of ambiguity is central to the description of mathematics that is developed in this book.
2. For thousands of years, the content of mathematical theories seemed to tell us something profound about the nature of the natural world—something that could not be expressed in any way other than the mathematical.
3. Other subjects may use mathematics, may even be expressed in a totally mathematical form, but mathematics has no other subject that stands in relation to it in the way that it stands in relation to other subjects.
4. The reader who finds some topic obscure is advised to skip it and continue reading.
5. The mathematics that is discussed is there for two reasons: first, because it is intrinsically interesting, and second, because it contributes to the discussion of the nature of mathematics in general.
6. The intuition that mathematics accesses the truth is correct, but not in the manner that it is usually understood.
7. Our contemporary civilization has been built upon a mathematical foundation.
8. Medicine, politics, and social policy are all increasingly expressed in the language of the mathematical and statistical sciences.

## Translation Techniques (2)

### Translation of English Adverbs（英语副词的译法）

English adverbs are often located before or after verbs and used widely to indicate how actions are conducted, but in Chinese translation English adverbs are handled more flexibly. Sometimes an adverb is translated as a phrase and occasionally even as an adjective.

Take the following sentence for example, "In many systems, the power output above the rated wind speed is *mechanically* maintained at a constant level, allowing better system control."（在许多系统中，当风速超过额定风速时，则利用机械装置使输出功率维持在恒定的水平，以更好地控制整个系统。）From this example we can see that the adverb "mechanically" is changed into the phrase 利用机械装置 in Chinese translation.

Here are more examples:

1. The worldwide use of electricity is expected to grow dramatically over the next decades, and wind technology, in particular, is *ideally* suited to meet this need.
   预计今后几十年中全世界的用电量将急剧增加，而风力技术对满足这种需求是特别理想的。
2. First, the raw materials are treated *chemically*, in some cases melted by heating, to form a viscous liquid.
   先用化学方法处理原料，有时通过加热使其融化，以制成黏性液体。
3. Picturephone service, which will become available *commercially* in the next few years, will at first probably be used by large business corporations.
   电视电话服务将在今后几年进入寻常百姓家，在此之前可能首先在大型商业公司使用。
4. Experimental stress analysis is the *strictly* practical branch of stress analysis and for most problems it relies heavily on strain measurement techniques.
   实验应力分析是应力分析一门实用性极强的分支，而且就大多数问题来说，它在很大程度上依赖于应变测量技术。

## V. Discuss the translation technique and the ways of applying the technique to the translation of the following sentences. Complete each of the Chinese translations.

1. Using Einstein's relativity equations, Hawking developed new techniques to prove *mathematically* that at the heart of black holes were singularities.（霍金运用爱因斯坦的相对论等式__________，证明在黑洞的中心存在着奇点。）
2. This relationship is usually expressed *graphically* in a power curve.（这个关系__________来表示。）
3. The ranges of plants and animal species could change *regionally*, endangering protected areas and many species whose habitats are now few and confined.（__________，从而危及动植物保护区，以及许多生存地区所剩无几且面积有限的动植物。）
4. In machining contours, numerical control can *mathematically* translate the defined curve into a finished product, saving time and eliminating templates.（在加工外形轮廓时，__________，既节省时间又无需样板。）
5. Patients with Karnofsky scores greater than 50 did not perform *differently* than controls.（卡诺夫斯基评分高于50的患者的答题正确率__________。）
6. The Hummer, or the M998 series, as it is *officially* known, is being exhibited by LTV Aerospace and Defense Company's AM General Division.（蜂鸟__________正由LTV宇航与防务公司的AM通用分公司展出。）
7. It seems desirable to choose the science which is most responsible for the attitude and viewpoint of the scientific age and which is today influencing scientific thought most *profoundly*, namely, physics.（看来，我们需要选一门对于科学时代态度和观点的形成最具决定性，__________，这就是物理学。）
8. This buildup is *largely* a consequence of the use of fossil fuels and CFCs, deforestation, and various agricultural activities, and it now threatens societies with far-reaching climate change.（__________，它以影响深远的气候变化威胁着人类社会。）

## VI. Translate the following sentences into Chinese. Pay attention to how the adverbs are translated properly.

1. On the contrary, the book takes the position that the logical structure, while important, is insufficient even to begin to account for what is *really* going on in mathematical practice.
2. This larger picture has room for a number of factors that have *traditionally* been omitted from a description of mathematics and are translogical—that is, beyond logic—though not illogical.
3. Thus, there is a discussion of things like ambiguity, contradiction, and paradox that, *surprisingly*, also have an essential role to play in mathematical practice.
4. Ambiguity is *intrinsically* connected to creativity.

5. The truth of mathematics is *mistakenly* seen to come exclusively from a rigid, deductive structure.

**VII. Translate the following passage into Chinese.**

Carl Friedrich Gauss referred to mathematics as "the Queen of the Sciences". In the original Latin *Regina Scientiarum*, as well as in German *Königin der Wissenschaften*, the word corresponding to *science* means (field of) knowledge. Indeed, this is also the original meaning in English, and there is no doubt that mathematics is in this sense a science. The specialization restricting the meaning to *natural* science is of later date. If one considers science to be strictly about the physical world, then mathematics, or at least pure mathematics, is not a science. Albert Einstein has stated that "as far as the laws of mathematics refer to reality, they are not certain; and as far as they are certain, they do not refer to reality."

Many philosophers believe that mathematics is not experimentally falsifiable, and thus not a science according to the definition of Karl Popper. However, in the 1930s important work in mathematical logic showed that mathematics cannot be reduced to logic, and Karl Popper concluded that "most mathematical theories are, like those of physics and biology, hypothetico-deductive: pure mathematics therefore turns out to be much closer to the natural sciences whose hypotheses are conjectures, than it seemed even recently". Other thinkers, notably Imre Lakatos, have applied a version of falsificationism to mathematics itself.

**VIII. Translate the following passage into English.**

**世界是数学的一面镜子**

伽利略在1623年发表的《试金者》（*Assayer*）中对数学的批评性评论从根本上给予了答复，他写道："宇宙是由数学的语言写成的。"伽利略所表达的也是当时学者广泛持有的观点，即数学远远不像批评者所说的那样不具有任何实质性内容，相反，数学的研究目的就是整个自然界。然而，虽然大多数人认为数学与物质世界密切相关，但这种关系的性质仍然是激烈争论的焦点。

其中一种主要的看法与传统观点一致，即数学是表示数字和数量的一门严密推理性学科。根据这种观点，数学的普遍定律是制约物质现实的基本定律。因此，人们研究数学和几何学的关系时，实际上研究的是物质的基本结构。

这种观点的主要倡导者是勒奈·笛卡尔（René Descartes），他将数学看作是上帝为了创造宇宙而制定的基本的、推理性的定律。当神圣的建筑师——上帝——使完全理性化的宇宙开始运转，它就会根据数学原理永远运转下去。因此，数学研究是按照神的旨意对自然界的研究，而世界则是抽象的数学原理的直接表达。

*The rapid development of computer science and the Internet has brought about many changes to the academic world, and one of them is the use of electronic journals (e-journals) for gaining information.*

*You may find many advantages in consulting electronic journals, and one of the most obvious ones is that the data made available can be kept constantly up-to-date and relevant. Moreover, owing to speedy publication, the information you need is immediately accessible through the Internet, and therefore time-consuming inter-library loans are no longer necessary. Compared with the printed publications, e-journals have virtually no limitation to the quantity of material that could be included. This is particularly important when it comes to the photographs illustrating the inscriptions, their monuments, and their research context, and an electronic publication can contain a larger selection of digital photographs, in color as well as the conventional black and white to which a paper publication may be limited. Finally, Internal hyperlinks make it easier for you to move from one part of the publication to another, follow cross-references, and forge multiple paths through the data provided.*

*More detailed introduction to e-journals and their archiving is provided in the following essay.*

## Electronic Journal
## 电子期刊

Electronic journals are **scholarly** journals or magazines that can be accessed via electronic transmission. They are a specialized form of electronic document: they have the purpose of providing material for academic research and study, they are **formatted** approximately like printed journal articles, the **metadata** is entered into specialized databases, such as **DOAJ** or **OACI** as well as the databases for the discipline, and they are **predominantly** available through academic libraries and special libraries.

scholarly *adj.* 学术的，具有学术特点的
format *v.* 安排格局，按照……格式
metadata *n.* 元数据
DOAJ (Directory of Open Access Journals) 开放存取期刊目录
OACI (Open Archive Citation Index) 开放性过刊资料库引文索引
predominantly *adv.* 最主要地，最普遍地

Some electronic journals are online-only journals; some are online versions of printed journals, and some consist of the online equivalent of a printed journal, but with additional online-only material.

Most commercial sites are subscription-based, or allow **pay-per-view** access. Many universities subscribe to electronic journals to provide access to their students and faculty, and it is generally also possible for individuals to subscribe. An increasing number of journals are now available as open access journals, requiring no subscription.

pay-per-view *adj.* 按次计费的
archive *n.* [常作~s]档案

Most working paper **archives** and articles on personal homepages are free, as are collections in institutional repositories and subject repositories.

HTML (Hypertext Markup Language) 超文本标示语言
PDF (Portable Document Format) 移植文件格式
ASCII (American Standard Code for International Interchange) 美国国家标准资讯交换码

Most electronic journals are published both in **HTML** and **PDF** formats, but some are available in only one of the two. Some early electronic journals were first published in **ASCII** text, and some informally published ones continue in that format.

## E-Journal archiving

segment *n.* 组成部分

In the past two or three years, e-journals have become the largest and fastest growing **segment** of the digital collections for most libraries. Collections that a few years ago numbered in the few hundreds of titles now number in the thousands, and the rate of growth continues to increase.

redundancy *n.* 过剩，多余
binding *n.* 装订
shelving *n.*（图书资料等）上架
usability *n.* 可用性
differentiate *v.* 区别，区分
microfilming *n.* 缩微胶片
coordination *n.* 调整，协调

In many ways, archiving and preserving e-journals will be dramatically different from what has been done for paper-based journals. In the paper era, there was large-scale **redundancy** in the storage of journals. Many different institutions collected the same titles. The copies of journals being saved for future generations were the same copies being read by the current generation of users. Many of the things that helped maintain journals for the long term (**binding**, repair, sound handling and **shelving** practices, environmental control, reformatting when **usability** was threatened) were not **differentiated** from what a library did to provide current services. Other than in the case of preservation **microfilming** and the odd instance of shared book storage facilities, there was little conscious **coordination** of preservation activities, and in fact a level of redundancy was expected and thought useful.

replication *n.* 复制

The common service model for e-journals is quite different than that for paper journals. Most e-journal access is through a single delivery system maintained either by the publisher or its agent. There is little **replication**, and only a few institutions actually hold copies of journals locally. Libraries can fulfill their current service requirements without facing the issues involved in the preservation of the resources. Further, in the digital realm the issues involved in day-to-day service are quite different from those involved in long-term preservation.

The issue of long-term archiving and preservation of e-journal content has become one of increasing importance. Specifically because of archiving concerns, many research libraries continue to collect paper copies at the same time they pay for access to the electronic versions. This dual expense is not likely to be sustainable over time. Publishers are finding that authors, editors, scholarly societies, and libraries frequently resist moving to electronic-only publication because of concern that long-term preservation and access to the electronic version is uncertain. Of perhaps even greater long-term concern, while libraries continue to rely on the paper copy as the archival version, from the viewpoint of publishers it is increasingly the electronic versions of titles that are the version of record, containing content not available in the print version.

These tensions and concerns led to a series of meetings over the past few years among publishers, librarians, and technologists sponsored by a variety of organizations, including the Society of Scholarly Publishers, the National Science Foundation, the Council on Library and Information Resources, and the **Coalition** for Networked Information. While these meetings helped to identify many of the issues, they did not result in any specific **follow-up** action. Finally, in the summer of 2000, the **Andrew W. Mellon Foundation**, working with the Council on Library and Information Resources (CLIR), took the initiative to move beyond exchanges of viewpoint to **experimentation** and **implementation.**

coalition *n.* 联盟，联合会
follow-up *adj.* 后续的
Andrew W. Mellon Foundation（美国）梅隆基金会
experimentation *n.* 试验
implementation *n.* 执行

**What content is archived?**

At first hearing, most people assume that e-journal archiving is basically concerned with the content of journal articles. Indeed, while articles are the intellectual core of journals, in fact e-journals contain many other kinds of materials. Some examples of commonly found content are:

- **Editorial boards**
- Rights and usage terms
- Copyright statements
- Journal descriptions
- Advertisements

editorial board 编委会

- Reprint information
- Editorials
- Events lists
- **Errata**
- Conference announcements
- Various sorts of digital files related to individual articles (**datasets**, images, tables, videos, models, etc.).

errata *n.* 勘误表

dataset *n.* 不依赖于数据库的独立数据集合

asset *n.* 资产，财产
ephemeral *adj.* 短命的，短暂的
masthead *n.* 刊头

Which of these content types need to be archived and preserved for the future? Some of these types of materials will pose issues for publishers. Not all of these items are controlled in publishers' **asset** management systems. Some are treated as **ephemeral**, "**masthead**" information and are simply handled as website content. When such information changes, the site is updated and earlier information is lost. For example, few if any e-journals provide a list of who was on the editorial board for an issue published a year or two ago. Another difficult content type is advertisements. Advertisements are, of course, frequently not tied to any given issue, and they change over time with the business arrangements of the publisher. In some cases, advertisements are specific to certain populations, and what advertisements you see depend on who or where you are. (For instance, drug ads are frequently regulated at the national level.) Deciding what of all that is seen on e-journal sites today should be archived and maintained will require careful consideration by archives, publishers, and scholars.

## Exercises

**I. Read the passage and decide whether the following statements are true or false. Write T for True and F for False in the brackets.**

1. Electronic journals articles are formatted in a way different from printed ones. (  )
2. Not all the electronic journals can be found in printed form. (  )
3. Large-scale redundancy in the storage of journals was considered a sheer waste of resources in the paper era. (  )
4. The common service model for e-journals is quite similar to that for paper journals. (  )
5. Libraries don't have to face the issues involved in the preservation of the resources when

fulfilling their current service requirements on e-journals. ( )

6. Many of those who resist moving to electronic-only publication are concerned with the uncertainty of long-term preservation and access to the electronic version. ( )
7. E-journal archiving is exclusively concerned with the content of journal articles. ( )
8. Conference announcements may be included in the content of e-journals. ( )
9. Advertisements should never appear in electronic journals. ( )
10. What should be archived and maintained on e-journal sites depends on readers. ( )

**II. Read the passage again, and complete the following items:**

1. The way electronic journals are accessed: ___________
2. How most commercial sites are paid: ___________
3. The formats in which most electronic journals are published: ___________
4. Some early electronic journals were first published in: ___________
5. The delivery system through which most e-journals are accessed is maintained by: ___________
6. The intellectual core of electronic journals: ___________
7. Simply handled as website content, some e-journals are treated as: ___________
8. Deciding what should be archived and maintained must be carefully considered by: ___________

**Quotations from Great Scientists**

*The important thing in science is not so much to obtain new facts as to discover new ways of thinking about them.*

*—William Bragg*

对于科学来说，不仅需要获取新的事实，更重要的是找到解释这些事实的新的思维方法。

——威廉·布拉格

# Chapter 7 Climate Change 气候变化

## Section 1 Reading and Translation

### I What You Are Going to Read

Climate change is already happening and represents one of the greatest environmental, social and economic threats facing the planet. The warming of the climate system is unequivocal, as is now evident from observations of increases in global average air and ocean temperatures, widespread melting of snow and ice, and rising global average sea level. The Earth's average surface temperature has risen by 0.76°C since 1850. Most of the warming that has occurred over the last 50 years is very likely to have been caused by human activities.

The impact on climate from 200 years of industrial development is an everyday fact of life, but did humankind's active involvement in climate change really begin with the industrial revolution, as commonly believed? Prof. William Ruddiman's provocative book *Plows, Plagues, and Petroleum: How Humans Took Control of Climate* (published by Princeton University Press, 2005) argues that humans have actually been changing the climate for some 8,000 years—as a result of the earlier discovery of agriculture.

What you are going to read is the excerpt of the first chapter of the book, in which Prof. Ruddiman will show you the general development of climate science, different disciplines involved in the study of climate, and the relationship between climate and human history.

### II About the Author

Before retirement, William F. Ruddiman worked as Professor of Environmental Sciences at the University of Virginia, following many years as a Doherty Senior Research Scientist at Lamont-Doherty Earth Observatory of Columbia University. He has published many articles in *Scientific American*, *Nature*, and *Science* as well as various scientific journals.

### III Reading Passage

#### Climate and Human History
#### 气候与人类史

By William F. Ruddiman

1 When I started my graduate student career in the field of

climate science almost 40 years ago, it really was not a "field" as such. Scattered around the universities and laboratories of the world were people studying **pollen grains**, shells of marine **plankton**, records of ocean temperature and **salinity**, the flow of ice sheets, and many other parts of the climate system, both in their modern form and in their past **manifestations** as suggested by evidence from the geologic record. A half-century before, only a few dozen people were doing this kind of work, mostly university-based or self-taught "gentleman" geologists and geographers in Western Europe and the eastern United States. Now and then, someone would organize a conference to bring together 100 or so colleagues and compare new findings across different disciplines.

pollen grain *n.* 花粉粒
plankton *n.* 浮游生物
salinity *n.* 盐度，盐分
manifestation *n.* 显示，表现

2 Today, this field has changed beyond recognition. Thousands of researchers across the world explore many aspects of the climate system, using aircraft, ships, satellites, **innovative** chemical and biological techniques, and high-powered computers. Geologists measure a huge range of processes on land and in the ocean. **Geochemists** trace the movement of materials and measure rates of change in the climate system. **Meteorologists** use numerical models to **simulate** the circulation of the atmosphere and its interaction with the ocean. **Glaciologists** analyze how ice sheets flow. Ecologists and biological **oceanographers** investigate the roles of vegetation on land and plankton in the ocean. **Climatologists** track trends in climate over recent decades. Hundreds of groups with shorthand **acronyms** for their longer names hold meetings every year on one or another aspect of climate. I am certain there are now more groups with acronyms in the field of climate science than there were people when I began.[1]

innovative *adj.* 创新的，革新的
geochemist *n.* 地球化学家
meteorologist *n.* 气象学家
simulate *v.* 模仿，模拟
glaciologist *n.* 冰河学家
oceanographer *n.* 海洋学家
climatologist *n.* 气象学家
acronym *n.* 首字母缩略词

3 Studies of Earth's climate history utilize any material that contains a record of past climate: deep-ocean cores collected from sea-going research vessels, ice cores drilled by fossil-fuel machine power in the Antarctic or Greenland ice sheets or by hand or solar power in mountain **glaciers**; soft-sediment cores hand-driven into lake muds; **hand-augered drills** that extract thin wood cores from trees; **coral** samples drilled from tropical **reefs**. The intervals investigated vary from the geological past many tens of millions of years ago to the recent historical past and changes occurring today.

glacier *n.* 冰河，冰川
hand-augered drill 手摇钻
coral *n.* 珊瑚
reef *n.* 礁；暗礁

4 These wide-ranging investigations have, over the last half-

precipitation *n.* 降雨量，雨量
plate-tectonic *adj.*【地】板块构造的
plateau *n.* 高地，高原
isthmus *n.* 地峡，地颈
tilt *n.* 倾斜
axis *n.* 轴；轴线

century or so, produced enormous progress in understanding climate change on every scale. For intervals lying in the much more distant past, tens or hundreds of millions of years ago, changes in global temperature, regional **precipitation**, and the size of Earth's ice sheets have been linked to **plate-tectonic** reorganizations of Earth's surface such as movements of continents, uplift and erosion of mountains and **plateaus**, and opening and closing of **isthmus** connections between continents. Over somewhat shorter intervals, cyclic changes in temperature, precipitation, and ice sheets over tens of thousands of years have been linked to subtle changes in Earth's orbit around the Sun, such as the **tilt** of its **axis** and the shape of the orbit. At still finer resolution, changes in climate over centuries or decades have been tied to large volcanic explosions and small changes in the strength of the Sun.

archbishop *n.* 大主教
patriarch *n.*（基督教的）早期的主教

5 Some scientists regard the results of this ongoing study of climate history as the most recent of four great revolutions in earth science, although advances in understanding climate have come about gradually, as in most of the earlier revolutions. In the 1700s James Hutton[2] concluded that Earth is an ancient planet with a long history of gradually accumulated changes produced mainly by processes working at very slow rates. Only after a century or more did Hutton's concept of an ancient planet displace the careful calculations of an **archbishop** in England who had added up the life spans of the **patriarchs** in the Bible and calculated that Earth was formed on October 26 in 4004 BC. Today chemistry, physics, biology, and astronomy have all provided critical evidence in support of the geology-based conclusion that our Earth is very old indeed, in fact several billions of years old.

meteorite *n.* 陨星

6 In 1859 Charles Darwin[3] published his theory of natural selection, based in part on earlier work showing that organisms have appeared and disappeared in an ever-changing but well-identified sequence throughout the immense interval of time for which we have the best fossil record (about 600 million years). Darwin proposed that new species evolve as a result of slow natural selection for attributes that promote reproduction and survival. Although widely accepted in its basic outline, Darwin's theory is still being challenged and enlarged by new insights. For example, only recently has it become clear that very rare collisions of giant **meteorites** with Earth's surface

also play a role in evolution by causing massive extinctions of most living organisms every few hundred million years or so. Each of these **catastrophes** opens up a wide range of environmental **niches** into which the surviving species can evolve with little or no competition from other organisms (for a while).

catastrophe *n.* 大灾难
niche *n.* 【生】小生态环境

7 The third great revolution, the one that eventually led to the theory of plate tectonics[4], began in 1912 when Alfred Wegener[5] proposed the concept of continental drift. Although this idea attracted attention, it was widely rejected in North America and parts of Europe for over 50 years. Finally, in the late 1960s, several groups of scientists realized that marine **geophysical** data that had been collected for decades showed that a dozen or so chunks of Earth's crust and outer **mantle**, called "plates", must have been slowly moving across Earth's surface for at least the last 100 million years. Within three or four years, the power of the plate tectonic theory to explain this wide range of data had convinced all but the usual handful of reflex contrarians that the theory was basically correct.[6] This revolution in understanding is not finished; the mechanisms that drive the motions of the plates remain unclear.

geophysical *adj.* 地球物理学的
mantle *n.* 【地】地幔

8 As with the three earlier revolutions, the one in climate science has come on slowly and in fact is still under way. Its oldest roots lie in field studies dating from the late 1700s and explanatory hypotheses dating from the late 1800s and early 1900s. Major advances in this field began in the late 1900s, continue today, and seem **destined** to go on for decades.

destine *v.* [常用被动语态] 注定，命定

9 Research into the history of humans is not nearly as large a field as climate science, but it attracts a nearly comparable amount of public interest.[7] This field, too, has expanded far beyond its intellectual boundaries of a half-century ago. At that time, the fossil record of our distant **precursors** was still extremely **meager**. Humans and our precursors have always lived near sources of water, and watery soils contain acids that dissolve most of the bones overlooked by **scavenging** animals. The chance of preservation of useful remains of our few ancestors living millions of years ago is tiny. When those opposed to the initial Darwinian hypothesis of an evolutionary descent from apes to humans cited "missing links" as a counterargument, their

precursor *n.* 先驱；先锋；前辈
meager *adj.* 贫乏的，不足的
scavenging *adj.* 以（腐肉、腐物）为食的

anthropologist *n.* 人类学者
doggedly *adv.* 顽强地；固执地
outcrop *n.* 【地】露出地面的岩层；(矿脉等的）露头
skeletal *adj.* 骨骼的；骸骨的

criticisms were at times difficult to refute.[8] The gaps in the known record were indeed immense. Now the missing links in the record of human evolution are at most missing minilinks. Gaps that were as much as a million years in length are generally now less than 1/10 that long, filled in by a relatively small number of **anthropologists** and their assistants **doggedly** exploring **outcrops** in Africa and occasionally stumbling upon fossil **skeletal** remains.

sandwich *v.* 夹入，挤进
basalt *n.* 【地】玄武岩

10 Suppose that skeletal remains are found in ancient lake sediments **sandwiched** between two layers of lava that have long since turned into solid rock (**basalt**). The basalt layers can be dated by the radioactive decay of key types of minerals enclosed within. If the dating shows that the two layers were deposited at 2.5 and 2.3 million years ago, respectively, then the creatures whose remains were found in the lake sediments sandwiched in between must have lived within that time range. With dozens of such dated skeletal remains found over the last half-century, the story of how our remote precursors changed through time has slowly come into focus.

intermediate *adj.* 中间的
australopithecine *n.* 南（方古）猿
marginally *adv.* 在边缘地
hominid *n.* 原始人类，类人动物
hominine *adj.* 人的，类人的，有人类特征的

11 Even though the details of the pathway from apes to modern humans still need to be worked out, the basic trend is clear, and no credible scientist that I know of has any major doubts about the general sequence. Creatures **intermediate** between humans and apes (**australopithecines**, or "southern apes") lived from 4.5 to 2.5 million years ago, around which time they gave way to beings (the genus Homo[9], for "man") that we would consider **marginally** human, but not fully so. Today anthropologists refer to everything that has followed since 2.5 million years ago as the **hominid** (or **hominine**) line. By 100,000 years ago, or slightly earlier, fully modern humans existed. This long passage was marked by major growth in brain size; progressively greater use of stone tools for cutting, crushing, and digging; and later by control of fire.

archeology *n.* (=archaeology) 考古学
encompass *v.* 包含，包括

12 Knowledge of the more recent history of humans has increased even more remarkably. Decades ago the field of **archeology** was focused mainly on large cities and buildings and on the cultural artifacts found in the tombs of the very wealthy; today this field **encompasses** or interacts with disciplines such as historical ecology[10] and environmental geology[11] that explore past human activities across

the much larger fraction of Earth's surface situated well away from urban areas. **Radiocarbon** dating (also based on radioactive decay) has made it possible to place even tiny organic fragments with a time framework. The development of cultivated cereals in the Near East[12] nearly 12,000 years ago and their spread into previously forested regions of Europe from 8,000 to 5,500 years ago can be dated from **trace amounts** of crops found in lake sediments. On other research fronts, archeologists unearthing mud-brick and stone foundations of houses have been able to estimate population densities thousands of years ago. Others examining photos taken from the air in early morning at low sun angles find distinct patterns of field cultivation created by farmers centuries before the present. Geochemists can tell from the kind of carbon preserved in the teeth and bones of humans and other animals the mixture of plants and animals they ate. From these and other explorations, the developing pattern of human history over the last 12,000 years has come into much sharper focus.

radiocarbon *n.* 放射性碳，碳的放射性同位素
trace amount 微量

13 Because both of these research fields—climate and human history—concentrate on the past, they have much in common with the field of crime solving. Imagine that a breaking and entering and a murder have been committed. The detectives arrive and examine the crime scene, searching for evidence that will point to the guilty person. How and when did the criminal enter the house? Was anything stolen? Were muddy footprints or fibers or other evidence left behind? Based on all the evidence, and the **modus operandi** of the possible **perpetrators**, the detectives gradually **zero in** on the identity of the criminal. Was the crime the work of a family member, an outsider who knew the family, or a complete stranger? A list of possible suspects emerges, the detectives check out where they were at the time of the murder, and a primary suspect is identified.

modus operandi <拉>（罪犯的）一贯手法，惯技
perpetrator *n.* 犯罪者，作恶者
zero in 把（枪炮、抨击的火力等）对准……，对……集中火力

14 **By analogy**, students of climate and human history also arrive on the scene after the event has occurred, but in this case hundreds, thousands, or even many millions of years later.[13] And, as in the crime scene, the first thing these scientists encounter is evidence that something of importance has happened. Twenty thousand years ago, an ice sheet more than a mile high covered the area of the present-day city of **Toronto**. Ten thousand years ago, grasslands with streams and abundant wildlife existed in regions now covered by blowing sand in

by analogy 用类推的方法
Toronto *n.* 多伦多（加拿大城市）

the southern Sahara Desert[14].

invoke *v.* 援引（法规、条文等）
paradigm *n.* 范例，示例
immune *adj.* 免除……的；豁免的

15 Natural curiosity drives scientists to wonder how such striking changes could have happened, and for some scientists this process of wondering leads to hypotheses that are first attempts at explanations. Soon after a major discovery is made, other scientists challenge the initial hypothesis or propose competing explanations. Over many years and even decades, these ideas are evaluated and tested by a large community of scientists. Some of the hypotheses are found to be inadequate or simply wrong, most often because additional evidence turns out to be inconsistent with specific predictions made in the initial hypotheses. If any hypothesis survives years of challenges and can explain a large amount of old and new evidence, it may become recognized as a theory. Some theories become so familiar that they are **invoked** almost without conscious thought and called **paradigms**. But even the great paradigms are not **immune** from continual testing. Science takes nothing for granted and draws no protective shield around even its time-honored "successes".

diagnostic *adj.* 特征的；有助于诊断的
incriminate *v.* 显示……有罪
prosecutor *n.* 原告，起诉人；检举人
plausible *adj.* 似乎有理的，似乎可能的
contention *n.* 论点

16 Only rarely do scientists studying climate history manage to isolate one causal explanation for any specific piece of evidence. By analogy to a crime scene, the detectives might be lucky enough to find totally **diagnostic** and **incriminating** evidence near the murder victim or at the point of the break-in, such as high-quality fingerprints or blood samples with DNA that match evidence from a suspect. If so, the perpetrator of the crime is convicted (unless the **prosecutors** are totally incompetent). In climate science, the explanation for an observation (the presence of ice sheets where none exist today, or of ancient streambeds in modern-day deserts) more commonly ends up with several contributing factors in **plausible contention**.

monsoon *n.* 季风
amplitude *n.* 广大，广阔
inherently *adv.* 内在地，固有地

17 But sometimes nature can be more cooperative in revealing cause-and-effect connections. The changes in Earth's orbit mentioned earlier occur at regular cycles of tens of thousands of years. These same cycles have occurred during many of Earth's major climate responses, including changes in the size of its ice sheets and in the intensity of its tropical **monsoons**. Because "cycles" are by definition regular in both length (duration in time) and size (**amplitude**), they are **inherently** predictable. This gives climate scientists like me a major

opportunity. We can look at past records of climate and see where and when the natural cycles were behaving "normally", but if we then find a trend developing that doesn't fit into the long-term "rules" set by the natural system, we are justified in concluding that the explanation for this departure from the norm cannot be natural.

18 Several years ago, just before I retired from university life, I noticed something that didn't seem to fit into what I knew about the climate system. What bothered me was this: the amount of **methane** in the atmosphere began going up around 5,000 years ago, even though everything I had learned about the natural cycles told me it should have kept going down. It has occurred to me since then that this was like an early scene in every episode of Peter Falk's *Columbo*[15] television series when he has just begun to investigate a recently committed crime. After he finishes an initial talk with the person whom he will eventually accuse of the crime, he starts to leave. Halfway out of the room, he stops, turns back, scratches his head, and says: "There's just this one thing that's bothering me..." That's how it all started, with just this one thing that bothered me—a trend that went up instead of down.

methane *n.* 甲烷

19 During the rest of the every *Columbo* show, Falk gradually pieces the story together and figures out what really happened. And that's how this new hypothesis came about. Having noticed the mystery of the wrong-way methane trend, I wondered what might explain it and eventually found an answer in the literature of early human history that convinced me. Just about the time the methane trend began its **anomalous** rise, humans began to irrigate for rice in Southeast Asia. I concluded that the irrigation created unnatural wetlands that emitted methane and explained the **anomaly**.

anomalous *adj.* 反常的，异常的
anomaly *n.* 异常（现象），反常（事物）

20 That first "Columbo moment" and the subsequent investigation has been followed by other, similar mysteries: the cause of a similarly anomalous rise in atmospheric $CO_2$ in the last 8,000 years, the reason why new ice sheets have failed to appear in northeast Canada when the natural cycles of Earth's orbit predict that they should have, and the origin of brief drops in $CO_2$ that again cannot be easily explained by natural processes but that appear to **correlate** with the greatest **pandemics** in human history. But before these "Columbo moments"

correlate *v.* 相互关联
pandemic *n.*（较大范围的）传染病

can be explored, we need to go back in time to see where humans came from, and to find out how and why climate has changed during our time on Earth.

## Notes

1. **Hundreds of groups with shorthand acronyms for their longer names hold meetings every year on one or another aspect of climate:** 数百个组织用类似速记的首字母缩略词代表全称，每年举办会议研讨各个气候专题。介词for在句中意为“代替，代表”。
2. **James Hutton:** 詹姆斯·霍顿（1726–1797），近代地质学奠基人之一。在200多年前提出了著名的地质学基本原理（即均变原理）。
3. **Charles Darwin:** 查尔斯·达尔文（1809–1882），英国博物学家，《物种起源》（*The Origin of Species by Means of Natural Selection*，1859）的作者。此后他又完成了一系列补充性的论述，如《人类的起源及性的选择》（*The Descent of Man, and Selection in Relation to Sex*, 1871）。
4. **the theory of plate tectonics:** 板块构造说（the plate tectonic theory），20世纪60年代在大陆漂移说和海底扩张说的基础上发展起来的全球大地构造学说。该学说认为，地球表层是由为数不多、大小不等的岩石圈板块构成，彼此独立运动，并相互挤压、摩擦。大陆是板块的一部分，而且同板块一起运动。
5. **Alfred Wegener:** 阿尔弗雷德·韦格纳（1880–1930），德国地球物理学家、气象学家，提出大陆漂移理论，著有《海陆的起源》等。
6. **Within three or four years, the power of the plate tectonic theory to explain this wide range of data had convinced all but the usual handful of reflex contrarians that the theory was basically correct:** 在三四年内，板块构造说对上述范围广泛的数据所做的解释，使大多数人（除了少数习惯与大家思维相反的人们）相信该理论基本上是正确的。句中的contrarian指“想法或行为总是与大家相反的人”，reflex原意为“反射（作用）”，在此意为“习惯性思维（或行为）方式”。
7. **Research into the history of humans is not nearly as large a field as climate science, but it attracts a nearly comparable amount of public interest:** 人类历史的研究规模远远比不上气候学的研究规模，但是人们对它的兴趣却与气候学相差无几。nearly一词在与否定词not连用时，其含义为“远非，相差很远”。如：It's not nearly so easy as you think.（这远不是你所想的那么容易。）
8. **When those opposed to the initial Darwinian hypothesis of an evolutionary descent from apes to humans cited "missing links" as a counterargument, their criticisms were at times difficult to refute:** 当那些反对达尔文进化论假设的人们提出在进化过程中有缺失环节而对达尔文的理论——人是由猿进化而来的——进行反驳时，他们的评判有时确实很难被驳倒。在时间状语中，opposed to the initial Darwinian hypothesis of an evolutionary descent from apes to humans为形容词短语，相当于一个定语从句who were opposed to...

9. **the genus Homo:** 人属，灵长目人科中的一个属。人属最大的特点是大脑发达，在200万年的进化过程中其脑的含量扩大了3倍。
10. **historical ecology:** 历史生态学。该领域探讨文化和环境在不同的历史阶段如何相互作用，这些研究具有历时性质。历史生态学从全面论的角度断言，人类的生存一定伴随着文化，并试图论证土地是人类活动的产物这一命题。
11. **environmental geology:** 环境地质学，是研究人类活动和地质环境相互作用的学科，是地质学的一个分支，也是环境地学的组成部分，其研究内容包括自然和人为引起的环境地质问题。环境地质学以地质学作为学科基础，在基础理论和研究方法上带有地学、生态学、物理学和化学等学科相互渗透、融合的特色。
12. **Near East:** 近东，通常指地中海东部沿岸地区，包括非洲东北部和亚洲西南部，有时还包括巴尔干半岛。早期近代西方地理学者还以"近东"指邻近欧洲的东方。
13. **By analogy, students of climate and human history also arrive on the scene after the event has occurred, but in this case hundreds, thousands, or even many millions of years later:** 做个比喻，事件发生后，到达现场的还有气候和人类史的研究者，但是时间却是在几百年、几千年，甚至几百万年之后。本句中的students of...意为"某领域的研究者，研究……的学者"，常用于正式文体。
14. **Sahara Desert:** 撒哈拉沙漠，世界上除南极洲之外最大的荒漠，位于非洲北部，东西长达5,600公里，南北宽约1,600公里，总面积约9,065,000平方公里，约占非洲总面积的32%。撒哈拉沙漠气候条件极其恶劣，是地球上最不适合生物生长的地方之一。
15. ***Columbo*:**《神探可伦坡》，由彼得·佛克（Peter Falk）主演的美国电视剧。在剧中，可伦坡为洛杉矶凶杀调查组副组长，其办案特点是对证人或疑犯提出一个接一个的问题，问到人家头皮发麻，有时甚至会不自觉地自曝线索或掉入陷阱，他的口头禅是Sir, one more question...

## Exercises

### I. Answer the following questions.

1. What do studies of the Earth's climate history involve?
2. What are the four great revolutions contributing to the ongoing study of climate history?
3. According to the author, what is basalt and how can the basalt layers be dated?
4. Why does the author say that researches on climate and human history share similarity with crime solving?
5. How does a hypothesis become recognized as a theory eventually?

### II. The following statements are incomplete. Search the missing information in the passage and fill in the blanks.

1. The exploration of various aspects of the climate system is carried out by thousands of

researchers across the world, including __________, __________, __________, __________, __________, __________, and __________.

2. Studies of Earth's climatic history make use of any material that contains a record of past climate, such as __________, __________, __________, __________, and __________.
3. Today chemistry, physics, biology, and astronomy have all provided critical evidence in support of the geology-based conclusion that our Earth is __________ years old.
4. Darwin proposed that new species evolve as a result of __________ for attributes that promote reproduction and survival.
5. In the late 1960s, several groups of scientists realized that __________ data that had been collected for decades showed that a dozen or so chunks of Earth's crust and outer mantle must have been slowly moving across Earth's surface for at least the last __________ years.
6. The oldest roots of the revolution in climate science lie in field studies dating from __________ and explanatory hypotheses dating from __________ and __________.
7. The development of modern humans was marked by major growth in __________; progressively greater use of __________; and later by __________.
8. Geochemists can tell from the kind of carbon preserved in __________ of humans and other animals the mixture of plants and animals they ate.

## III. Identify the implied meanings of the underlined parts of the following sentences according to the context of the passage, and translate the sentences into Chinese.

1. A half-century before, only a few dozen people were doing this kind of work, mostly university-based or self-taught "gentleman" geologists and geographers in Western Europe and the eastern United States.
2. Although widely accepted in its basic outline, Darwin's theory is still being challenged and enlarged by new insights.
3. This field, too, has expanded far beyond its intellectual boundaries of a half-century ago.
4. Gaps that were as much as a million years in length are generally now less than one-tenth that long, filled in by a relatively small number of anthropologists and their assistants doggedly exploring outcrops in Africa and occasionally stumbling upon fossil skeletal remains.
5. Suppose that skeletal remains are found in ancient lake sediments sandwiched between two layers of lava that have long since turned into solid rock (basalt).
6. Creatures intermediate between humans and apes (australopithecines, or "southern apes") lived from 4.5 to 2.5 million years ago, around which time they gave way to beings (the genus Homo, for "man") that we would consider marginally human, but not

fully so.

7. From these and other explorations, the developing pattern of human history over the last 12,000 years has come into much sharper focus.
8. Based on all the evidence, and the modus operandi of the possible perpetrators, the detectives gradually zero in on the identity of the criminal.
9. But even the great paradigms are not immune from continual testing.
10. During the rest of the every *Columbo* show, Falk gradually pieces the story together and figures out what really happened.

## Translation Techniques (1)

### The Translation of Comparatives（比较级的译法）

Owing to the differences between English and Chinese, conversion is commonly used in translation. Besides the conversion techniques introduced in the previous chapters of this book, there is another case in which conversion is used—the translation of the comparatives in English. In order to achieve accuracy and smoothness in the translated versions, there is sometimes the need to convert the comparatives in English into other forms of expressions in Chinese.

Let's have a look at an example: "The expert system generated both higher performance and simpler operation." The conversion of the comparatives in this sentence *higher* and *simpler* into the verbs 提高 and 简化 in Chinese translation conforms better to the way of the Chinese speakers in expressing the meaning: 该专家系统既提高了性能，又简化了操作程序.

Here are more examples:

1. Owing to biological rhythms, some people may begin to eat *more* and often feel sleepy in autumn.

   由于生物节律，有些人一到秋天食欲就开始增强，而且往往昏昏欲睡。

2. The space shuttle is by far the most complex flying machine ever built—vastly *more sophisticated* than the Apollo rockets that sent men to the moon in capsules.

   航天飞机是迄今为止所建造的最为复杂的飞行器，其复杂程度远远超过通过密闭舱把人送上月球的阿波罗火箭。

3. Explosions of the sunspots result in *stronger* ultraviolet radiation.
   太阳黑子的爆炸增强了紫外线辐射。
4. The experiment shows that in case of electronic scannings the beamwidth gets *broader* by a factor of two.
   实验表明，在电子扫描时，波束宽度增加一倍。
5. According to the report, *tighter* tolerances are possible, but with *lower* yield and *greater* cost.
   根据这份报告，缩小公差是可能的，但这会降低成品率，使成本增加。

## IV. Translate the following sentences into Chinese. Pay attention to the appropriate translation of the comparatives in English.

1. I am certain there are now *more* groups with acronyms in the field of climate science than there were people when I began.
2. At still *finer* resolution, changes in climate over centuries or decades have been tied to large volcanic explosions and small changes in the strength of the Sun.
3. Some scientists regard the results of this ongoing study of climate history as the most recent of four great revolutions in earth science, although advances in understanding climate have come about gradually, as in most of the *earlier* revolutions.
4. By 100,000 years ago, or slightly *earlier*, fully modern humans existed.
5. Today this field encompasses or interacts with disciplines such as historical ecology and environmental geology that explore past human activities across the much *larger* fraction of Earth's surface situated well away from urban areas.
6. The changes in Earth's orbit mentioned *earlier* occur at regular cycles of tens of thousands of years.
7. In climate science, the explanation for an observation (the presence of ice sheets where none exist today, or of ancient streambeds in modern-day deserts) *more commonly* ends up with several contributing factors in plausible contention.
8. In 1859 Charles Darwin published his theory of natural selection, based in part on *earlier* work showing that organisms have appeared and disappeared in an everchanging but well-identified sequence throughout the immense interval of time for which we have the best fossil record (about 600 million years).

## Translation Techniques (2)

### Conversion of Clauses（从句的转换）

Complex sentences in the English language are made up of related units called clauses. According to their functions in the sentences, the clauses usually play the roles of subjects and objects (noun clauses), or can be used as adjectives and adverbs (adjective and adverbial clauses). The Chinese translation of clauses normally depends on their roles in the sentences. Meanwhile, it should be noted that the difference between English and Chinese in the use of specific expressions sometimes makes it necessary to change one type of clause into another one.

Take the following sentence as an example, "*When parts must be inspected in large numbers*, 100% inspection of each part is not only slow and costly, but in addition does not eliminate all of the defective pieces."（"如果待检验的零件数量很大，对每个零件都做百分之百的全面检查不仅费时费钱，而且也不一定能把所有的次品全都检查出来。"）The clause in this sentence starts with "when" to indicate time, but when it is translated into Chinese, the words 如果 are used in stead of 当……的时候.

More examples are provided as follows so that this point can be demonstrated more clearly.

1. *When all parameters are in limits* the on conditions light is switched to green and a momentary audible tone is sounded.
   如果全部参数都处于要求的范围之内，试验条件灯即切换成绿色，并发出短暂的响声。
2. By observing such a flow pattern, *which might be stationary or variable with time*, one can get an idea of the whole development of the flow.
   通过观察这样的流谱（无论是稳态流谱还是随时间变化的流谱），人们可以了解整个流动过程。
3. *If a computerized cutting machine could be used to design a replacement hip*, they reasoned, why not use a computerized machine to do the replacement?
   他们不禁要问：既然能够用计算机控制切割机来设计髋关节置换品，为什么不能用计算机控制机器来做置换手术？
4. Stress during labor comes from the periodic reduction in the oxygen supply *when the pressure of the contraction stops blood flow through the placenta.*
   由于子宫收缩阻止血液通过胎盘，引起间歇性供氧的减少而导致分娩紧张。

**V. Discuss the translation technique and the ways of applying the technique to the translation of the following sentences. Complete each of the Chinese translations.**

1. *When the thermometer becomes warmer* the mercury expands, and the amount of expansion can be noted on a scale engraved on, or attached to, the stem. (__________，水银就会膨胀，其膨胀程度可以从刻在管柱上或附在管柱上的刻度看出。)
2. As a substitute for chlorine, *which is a poisonous gas*, tablets of the organic compound hexachlorethane can be used. (__________，因此可以用有机化合物六氯乙烷片代替氯气。)
3. What are the consequences of multiple stresses—a variety of pollutants, heat waves and climate changes, increased ultraviolet radiation—*when realized together*? (__________，将会造成什么后果呢?)
4. *Where a vessel has vertical sides*, the pressure on the bottom is equal to the height of the liquid times its density. (__________，则容器底部的压强等于液体高度乘以液体的密度。)
5. Living standards have been found to improve substantially *when electricity is introduced for such purposes as lighting, refrigeration, irrigation, and communication.* (__________，人们的生活水准已经大大提高。)
6. Adrenalin in the fetus shunts blood toward vital internal organs *that may be damaged by a reduction in oxygen supply.* (胎儿体内的肾上腺素会使血液流向体内的重要器官，__________。)
7. Moreover, iron possesses magnetic properties, *which have made the development of electrical power possible.* (而且，铁具有磁性，__________。)
8. The other scale in general use nowadays is the binary, *in which numbers are expressed by combinations of only two digits, 0 and 1.* (当今普遍采用的另一种进位制是二进制，__________。)
9. For example, *if a bowl were filled with many different kinds of fruit*, you could tell the robot, "Get me an orange", and using its shape-recognizing "eyes" it could travel to the fruit bowl and pick out an orange from the rest of the fruit. (例如，__________，你可以告诉机器人："给我一个橘子"，机器人随即利用能够辨认形状的"眼睛"，走到水果盘边，从众多水果里挑出一个橘子来。)
10. Researchers found that physicians who had computers ordered nine percent fewer tests *when the software told them beforehand the patient's logistical probability of having an abnormal result.* (研究人员发现，对于使用计算机的医生来说，__________，他们开出的检验项目可减少9%。)

## VI. Translate the following sentences into Chinese. Pay attention to how the conversion of the clauses should be made.

1. *When these organisms died*, they decayed and over millions of years layers of slime formed on the ocean bed; with the pressure of movements in the earth's crust and the heat this pressure caused, these layers were converted into petroleum.
2. Electronics is a field *in which scientific research and technical development is still intensive*, and so many more manifestations of the versatility of the electron can be expected.
3. Researchers have long suspected that heredity plays a role in some *if not all cases*, and the Amishes present an ideal setting in which to test that hypothesis.
4. Pressure transducers are generally precision instruments and often yield signals that are consistent to 0.1 percent of design pressure or better *when used properly*.
5. As water gets colder it contracts *until it reaches the temperature of 4℃ above freezing point*; after that, as it gets colder, it expands.

## VII. Translate the following passage into Chinese.

The European Union has long been at the forefront of international efforts to combat climate change and has played a key role in the development of the two major treaties addressing the issue, the 1992 United Nations Framework Convention on Climate Change (UNFCCC) and its Kyoto Protocol, agreed in 1997.

The EU has been taking serious steps to address its own greenhouse gas emissions since the early 1990s. In 2000 the European Commission launched the European Climate Change Programme (ECCP). The ECCP has led to the adoption of a wide range of new policies and measures. These include the pioneering EU Emissions Trading System, which has become the cornerstone of EU efforts to reduce emissions cost-effectively, and legislation to tackle emissions of fluorinated greenhouse gases.

Monitoring data and projections indicate that the 15 countries that were EU members at the time of the EU's ratification of the Kyoto Protocol in 2002 will reach their Kyoto Protocol target for cutting greenhouse gas emissions. This requires emissions in 2008-2012 to be 8% below 1990 levels.

However, Kyoto is only a first step and its targets expire in 2012. International negotiations are now taking place under the UNFCCC with the goal of reaching a global agreement governing action to address climate change after 2012.

In January 2007, as part of an integrated climate change and energy policy, the European Commission set out proposals and options for an ambitious global agreement in its Communication "Limiting Global Climate Change to 2 degrees Celsius: The way ahead for 2020 and beyond". EU leaders endorsed this vision in March 2007. They committed the EU to cutting its greenhouse gas emissions by 30% of 1990 levels by 2020 provided other developed countries commit to making comparable reductions under a global agreement.

**VIII. Translate the following passage into English.**

## 人类对气候变化的影响

人为因子指改变环境、影响气候的人类活动。在某些情况下因果关系链直接而明确（例如灌溉对温度和湿度的影响），而在其他情况下则不那么明显。多年来人们一直在争论人类造成气候变化的各种假设，然而重要的是我们应当注意到，科学争论已经从怀疑论中走了出来，因为对于气候变化人们已经在科学上达成一致，即人类活动毫无疑问是造成当今全球气候迅速变化的主要原因。

### 矿物燃料

19世纪80年代，随着工业革命的开始以及此后的迅速发展，人类矿物燃料的消耗使二氧化碳浓度由280ppm增加到今天的387ppm，而且估计到21世纪末其浓度将会达到535ppm–983 ppm。我们知道，今天二氧化碳的浓度比过去75万年中的任何一个时期都高。预计这些变化与甲烷气体的增加会使1990年到2100年间的温度上升1.4–5.6摄氏度。

### 土地利用

在矿物燃料得到广泛应用之前，人类对当地气候最大的影响可能来自于土地的使用。灌溉、砍伐森林和农业活动从根本上改变了环境。例如，上述因素使流入和流出某个地区的水量发生变化，并可能影响到地表植被，造成阳光吸收量的改变，从而改变地表反照率。例如，有证据表明，公元前700年到公元1年之间大面积的森林砍伐（砍伐的木材用于造船、建筑，并用作燃料），使希腊和其他地中海国家的气候产生了永久性的变化，结果，该地区当前的气候要比过去热得多，也干燥得多，而古代用于造船的树种也从该地区消失了。

2007年的一项研究发现，进入现代社会以来，加利福尼亚州的平均温度在过去的50年中上升了大约2度，而城市地区的温度增加得更多。这种变化主要是由于人们对自然景观的开发所致。

*For many people, the image associated with a "library" is a building with a large quantity of books and other publications stacked on rows upon rows of bookshelves in the reading rooms, where the librarians move about quietly keeping the books in order. However, with the appearance of the Internet, a kind of "invisible" library—digital library—began to develop. With a strong focus on network information sources, the vast array of topics found on the World Wide Web is accessible through digital libraries. Now sitting in front of your computer and clicking the mouse, you may travel to many different digital libraries throughout the world to retrieve the information you need for your research.*

*A variety of tools are available at digital libraries to help you retrieve information efficiently and conveniently. They can be used to explore tens of millions of Web pages, and the information data can be effectively collected, stored, searched, and mined. Moreover, with these tools, you can also build unique indexes into the Web, and have the system deliver Web pages to feature analysis programs at very high data rates. This digital information resource facilitates and provides access to materials such as manuscripts, photographs, and works of art held in libraries, museums, archives, and other institutions across the world.*

## Digital Library
## 数字图书馆

A digital library is a library in which collections are stored in digital formats (as opposed to print, microform, or other media) and accessible by computers. The digital content may be stored locally, or accessed remotely via computer networks. A digital library is a type of information **retrieval** system.

retrieval *n.* 检索

The first use of the term *digital library* in print may have been in a 1988 report to the **Corporation for National Research Initiatives**. The term *digital libraries* was first popularized by the **NSF/DARPA/NASA Digital Libraries Initiative** in 1994. The older names *electronic library* or *virtual library* are also occasionally used, though electronic library nowadays more often refers to **portals**, often provided by government agencies, as in the case of the Florida Electronic Library.

Corporation for National Research Initiatives（美国）全国研究创新联合会
NSF (the National Science Foundation)（美国）国家科学基金会
DARPA (Defense Advanced Research Projects Agency)（美国）国防部高级研究计划署
NASA (National Aeronautics and Space Administration)（美国）国家航空航天局
Digital Libraries Initiative 数字图书馆创新计划
portal *n.*（网络）入口网站，门户

### Types of digital libraries

The term digital library is **diffuse** enough to be applied to a wide range of collections and organizations, but, to be considered a digital library, an online collection of information must be managed

diffuse *adj.* 散开的；弥漫的

Project Perseus 伯尔修斯计划（数字图书馆）
Project Gutenberg 古藤堡计划（最早的数字图书馆）
ibiblio（北卡罗来纳大学主办的）数字图书馆网站
audiobook *n.* 有声读物

by and made accessible to a community of users. Thus, some web sites can be considered digital libraries, but far from all. Many of the best known digital libraries are older than the web including **Project Perseus, Project Gutenberg**, and **ibiblio**. Nevertheless, as a result of the development of the internet and its search potential, digital libraries such as the European Library and the Library of Congress are now developing in a Web-based environment. Public, school and college libraries are also able to develop digital download websites, featuring eBooks, **audiobooks**, music and video, through companies like OverDrive, Inc.

born-digital *adj.* 原生数字形态的
digitize *n.* 将（资料）数字化
hybrid library 混合程序库，复合图书馆
ePrint arXiv 电子预印本文献库

A distinction is often made between content that was created in a digital format, known as **born-digital**, and information that has been converted from a physical medium, e.g., paper, by **digitizing**. The term **hybrid library** is sometimes used for libraries that have both physical collections and digital collections. They consist of a combination of traditional preservation efforts such as microfilming and new technologies involving digital projects. For example, American Memory is a digital library within the Library of Congress. Some important digital libraries also serve as long term archives, for example, the **ePrint arXiv**, and the Internet Archive.

### Academic repositories

corporate *adj.* 共同的；全体的

Many academic libraries are actively involved in building institutional repositories of the institution's books, papers, theses, and other works which can be digitized or were "born digital". Many of these repositories are made available to the general public with few restrictions, in accordance with the goals of open access. Institutional, truly free, and **corporate** repositories are often referred to as digital libraries.

### Digital archives

Archives differ from libraries in several ways. Traditionally, archives were defined as:

1. Containing primary sources of information (typically letters and papers directly produced by an individual or organization) rather than the secondary sources found in a library (books, etc.);

2. Having their contents organized in groups rather than individual items. Whereas books in a library are cataloged individually, items in an archive are typically grouped by **provenance** (the individual or organization who created them) and original order (the order in which the materials were kept by the creator);

3. Having unique contents. Whereas a book may be found at many different libraries, depending on its **rarity**, the records in an archive are usually one-of-a-kind, and cannot be found or consulted at any other location except at the archive that holds them.

provenance *n.* 出处

rarity *n.* 稀有，罕见

The technology used to create digital libraries has been even more revolutionary for archives since it breaks down the second and third of these general rules. The use of search engines, **Optical Character Recognition** and metadata allow digital copies of individual items (i.e. letters) to be cataloged, and the ability to remotely access digital copies has removed the necessity of physically going to a particular archive to find a particular set of records. The Oxford Text Archive is generally considered to be the oldest digital archive of academic primary source materials.

Optical Character Recognition 光学字符识别（技术）

## Searching

Most digital libraries provide a search interface which allows resources to be found. These resources are typically deep Web (or invisible Web) resources since they frequently cannot be located by search engine **crawlers**. Some digital libraries create special pages or sitemaps to allow search engines to find all their resources. Digital libraries frequently use the **Open Archives Initiative Protocol for Metadata Harvesting** (OAI-PMH) to expose their metadata to other digital libraries, and search engines like Google Scholar, Google, Yahoo! and Scirus can also use OAI-PMH to find these deep web resources.

crawler *n.* 网络蜘蛛，网络爬虫（搜索引擎软件）

Open Archives Initiative Protocol for Metadata Harvesting 开放档案元数据获取协定

There are two general strategies for searching a **federation** of digital libraries:

1. distributed searching, and

2. searching previously harvested metadata.

federation *n.* 联邦式（数据搜索方法）

query *v.* 查询
algorithm *n.* 运算法则

Distributed searching typically involves a client sending multiple search requests in parallel to a number of servers in the federation. The results are gathered, duplicates are eliminated or clustered, and the remaining items are sorted and presented back to the client. Protocols like Z39.50 are frequently used in distributed searching. A benefit to this approach is that the resource-intensive tasks of indexing and storage are left to the respective servers in the federation. A drawback to this approach is that the search mechanism is limited by the different indexing and ranking capabilities of each database, making it difficult to assemble a combined result consisting of the most relevant found items.

Searching over previously harvested metadata involves searching a locally stored index of information that has previously been collected from the libraries in the federation. When a search is performed, the search mechanism does not need to make connections with the digital libraries it is searching—it already has a local representation of the information. This approach requires the creation of an indexing and harvesting mechanism which operates regularly, connecting to all the digital libraries and **querying** the whole collection in order to discover new and updated resources. OAI-PMH is frequently used by digital libraries for allowing metadata to be harvested. A benefit to this approach is that the search mechanism has full control over indexing and ranking **algorithms**, possibly allowing more consistent results. A drawback is that harvesting and indexing systems are more resource-intensive and therefore expensive.

Exercises

**I. Read the passage and decide whether the following statements are true or false. Write T for True and F for False in the brackets.**

1. A digital library is a library in which the information is collected and stored through digital technology and can be obtained by means of computers. (    )
2. The term digital library can be applied to any kind of collections and organizations. (    )
3. The term hybrid library is only used for the libraries that have digital collections. (    )
4. American Memory is a hybrid library, which belongs to the Library of Congress. (    )
5. Archives differ from libraries in at least three ways. (    )
6. All digital libraries provide a search interface which allows resources to be found. (    )

7. Distributed searching typically involves a client sending multiple search requests in parallel to a number of servers in the federation. ( )
8. When a search is performed, the search mechanism needs to make connections with the digital libraries it is searching. ( )
9. Searching over previously harvested metadata involves searching a locally stored index of information that has previously been collected from the libraries in the city. ( )
10. Harvesting and indexing systems are more resource-extensive and costly. ( )

## II. Read the passage again, and complete the following items.

1. The occasion on which the term *digital library* was first used in print: ___________
2. The year when the term *digital libraries* was first popularized by the NSF/DARPA/NASA Digital Libraries Initiative: ___________
3. The prerequisite for a digital library: ___________
4. The distinction made between two types of collections of the hybrid library: ___________
5. The reason why many of the institutional repositories are made available to the general public with few restrictions: ___________
6. The repositories which are often referred to as digital libraries: ___________
7. The three ways in which archives differ from libraries: ___________
8. The two strategies for searching a federation of digital libraries: ___________

### Quotations from Great Scientists

*You cannot teach a man anything; you can only help him discover it in himself.*

—*Galileo Galilei*

你无法把所有知识教给他人，你只能帮助他自己去发现。

——伽利略·伽利莱

# Chapter 8 Energy Conservation 节能

## Section 1 Reading and Translation

### I What You Are Going to Read

Energy conservation is the practice of decreasing the quantity of energy used. It may be achieved through efficient energy use, in which case energy use is decreased while achieving a similar outcome, or by reduced consumption of energy services. Electrical energy conservation is an important element of energy policy. In this field, energy conservation reduces the energy consumption and energy demand per capita, and thus offsets the growth in energy supply needed to keep up with population growth. This reduces the rise in energy costs, and can reduce the need for new power plants and energy imports.

*A Solar Grand Plan*, an essay published in *Scientific American* (December, 2007), puts forward a proposal concerning the use of solar power as the energy source in the United States by 2050. According to the essay, a massive switch from coal, oil, natural gas and nuclear power plants to solar power plants could supply 69 percent of the US's electricity and 35 percent of its total energy by 2050. The following excerpt of the essay will tell you more about this grand plan.

### II About the Authors

Ken Zweibel, James Mason and Vasilis Fthenakis met a decade ago while working on life-cycle studies of photovoltaics. Zweibel is president of PrimeStar Solar in Golden, Colo., and for 15 years was manager of the National Renewable Energy Laboratory's Thin-Film PV Partnership. Mason is director of the Solar Energy Campaign and the Hydrogen Research Institute in Farmingdale, NY. Fthenakis is head of the Photovoltaic Environmental Research Center at Brookhaven National Laboratory and is a professor in and director of Columbia University's Center for Life Cycle Analysis.

### III Reading Passage

**A Solar Grand Plan**

**太阳能利用的长远规划**

By Ken Zweibel, James Mason and Vasilis Fthenakis

*By 2050 solar power could end US dependence on foreign oil and slash greenhouse gas emissions.*

1 High prices for gasoline and home heating oil are here to stay. The US is at war in the Middle East at least in part to protect its foreign oil interests. And as China, India and other nations rapidly increase their demand for fossil fuels[1], future fighting over energy **looms** large. In the meantime, power plants that burn coal, oil and natural gas, as well as vehicles everywhere, continue to pour millions of tons of pollutants and greenhouse gases into the atmosphere annually, threatening the planet.

loom *v.* 隐约呈现，朦胧出现；（危险、忧虑等）逼近

2 **Well-meaning** scientists, engineers, economists and politicians have proposed various steps that could slightly reduce fossil-fuel use and emissions. These steps are not enough. The US needs a bold plan to free itself from fossil fuels. Our analysis convinces us that a massive switch to solar power is the logical answer.

well-meaning *adj.* 善意的，好心的

3 Solar energy's potential is off the chart. The energy in sunlight striking the earth for 40 minutes is equivalent to global energy consumption for a year. The US is lucky to be endowed with a vast resource; at least 250,000 square miles of land in the Southwest alone are suitable for constructing solar power plants, and that land receives more than 4,500 **quadrillion British thermal units** (Btu)[2] of solar radiation a year. Converting only 2.5 percent of that radiation into electricity would match the nation's total energy consumption in 2006.

quadrillion *n.* < 美、法 > $10^{15}$（1000 的 5 次幂）
British thermal unit 英制热量单位

4 To convert the country to solar power, huge **tracts** of land would have to be covered with **photovoltaic** panels and solar heating troughs. A direct-current (DC) transmission backbone would also have to be erected to send that energy efficiently across the nation.[3]

tract *n.* 一片土地，地带
photovoltaic *adj.*【物】光电（池）的，光致电压的

5 The technology is ready. On the following pages we present a grand plan that could provide 69 percent of the US's electricity and 35 percent of its total energy (which includes transportation) with solar power by 2050. We project that this energy could be sold to consumers at rates equivalent to today's rates for conventional power sources, about five cents per **kilowatt-hour** (kwh). If wind, **biomass** and **geothermal** sources were also developed, renewable energy could provide 100 percent of the nation's electricity and 90 percent of its energy by 2100.

kilowatt-hour *n.*【电】千瓦（特）时；一度（电）（能量单位）
biomass *n.* 生物量；用作燃料或能源的植物材料、蔬菜或农业废弃物
geothermal *adj.* 地热的；地热产生的

infrastructure *n.* 基础设施；基础
deficit *n.* 赤字；亏空（额）
grid *n.*（高压）输电网

6 The federal government would have to invest more than $400 billion over the next 40 years to complete the 2050 plan. That investment is substantial, but the payoff is greater. Solar plants consume little or no fuel, saving billions of dollars year after year. The **infrastructure** would displace 300 large coal-fired power plants and 300 more large natural gas plants and all the fuels they consume. The plan would effectively eliminate all imported oil, fundamentally cutting US trade **deficits** and easing political tension in the Middle East and elsewhere. Because solar technologies are almost pollution-free, the plan would also reduce greenhouse gas emissions from power plants by 1.7 billion tons a year, and another 1.9 billion tons from gasoline vehicles would be displaced by plug-in hybrids refueled by the solar power **grid**.[4] In 2050 US carbon dioxide emissions would be 62 percent below 2005 levels, putting a major brake on global warming.

## Photovoltaic Farms

module *n.* 组件；模块
deployment *n.* 展开，调度，部署
cadmium telluride 【化】碲化镉

7 In the past few years the cost to produce photovoltaic cells and **modules** has dropped significantly, opening the way for large-scale **deployment**. Various cell types exist, but the least expensive modules today are thin films made of **cadmium telluride**. To provide electricity at six cents per kwh by 2020, cadmium telluride modules would have to convert electricity with 14 percent efficiency, and systems would have to be installed at $1.20 per watt of capacity. Current modules have 10 percent efficiency and an installed system cost of about $4 per watt. Progress is clearly needed, but the technology is advancing quickly; commercial efficiencies have risen from 9 to 10 percent in the past 12 months. It is worth noting, too, that as modules improve, rooftop photovoltaics will become more cost-competitive for homeowners, reducing daytime electricity demand.

gigawatt *n.* 【电】十亿瓦特

8 In our plan, by 2050 photovoltaic technology would provide almost 3,000 **gigawatts** (GW), or billions of watts, of power. Some 30,000 square miles of photovoltaic arrays would have to be erected. Although this area may sound enormous, installations already in place indicate that the land required for each gigawatt-hour of solar energy produced in the Southwest is less than that needed for a coal-powered plant when factoring in land for coal mining.[5] Studies by the National

Renewable Energy Laboratory[6] in Golden, **Colo**. show that more than enough land in the Southwest is available without requiring use of environmentally sensitive areas, population centers or difficult terrain. Jack Lavelle, a spokesperson for Arizona's Department of Water Conservation, has noted that more than 80 percent of his state's land is not privately owned and that Arizona is very interested in developing its solar potential. The **benign** nature of photovoltaic plants (including no water consumption) should keep environmental concerns to a minimum.

Colo. (=Colorado)（美国）科罗拉多州
benign *adj.* 有利的，易于产生有益影响的

9 The main progress required, then, is to raise module efficiency to 14 percent. Although the efficiencies of commercial modules will never reach those of solar cells in the laboratory, cadmium telluride cells at the National Renewable Energy Laboratory are now up to 16.5 percent and rising. At least one manufacturer, First Solar[7] in Perrysburg, Ohio, increased module efficiency from 6 to 10 percent from 2005 to 2007 and is reaching for 11.5 percent by 2010.

## Pressurized Caverns

10 The great limiting factor of solar power, of course, is that it generates little electricity when skies are cloudy and none at night. Excess power must therefore be produced during sunny hours and stored for use during dark hours. Most energy storage systems such as batteries are expensive or inefficient.

11 Compressed-air energy storage has emerged as a successful alternative. Electricity from photovoltaic plants compresses air and pumps it into vacant underground caverns, abandoned mines, **aquifers** and depleted natural gas wells. The pressurized air is released on demand to turn a **turbine** that generates electricity, aided by burning small amounts of natural gas.[8] Compressed-air energy storage plants have been operating reliably in Huntorf, Germany, since 1978 and in McIntosh, **Ala**., since 1991. The turbines burn only 40 percent of the natural gas they would if they were fueled by natural gas alone, and better heat recovery technology would lower that figure to 30 percent.

aquifer *n.* 含水土层，地下蓄水层
turbine *n.* 涡轮机
Ala. (=Alabama)（美国）亚拉巴马州

12 Studies by the Electric Power Research Institute[9] in Palo Alto, **Calif**., indicate that the cost of compressed-air energy storage today is about half that of lead-acid batteries. The research indicates that these facilities would add three or four cents per kwh to photovoltaic

Calif. (=California)（美国）加利福尼亚州

generation, bringing the total 2020 cost to eight or nine cents per kWh.

reservoir *n.* 贮气槽

13 Electricity from photovoltaic farms in the Southwest would be sent over high-voltage DC transmission lines to compressed-air storage facilities throughout the country, where turbines would generate electricity year-round. The key is to find adequate sites. Mapping by the natural gas industry and the Electric Power Research Institute shows that suitable geologic formations exist in 75 percent of the country, often close to metropolitan areas.[10] Indeed, a compressed-air energy storage system would look similar to the US natural gas storage system. The industry stores eight trillion cubic feet of gas in 400 underground **reservoirs**. By 2050 our plan would require 535 billion cubic feet of storage, with air pressurized at 1,100 pounds per square inch. Although development will be a challenge, plenty of reservoirs are available, and it would be reasonable for the natural gas industry to invest in such a network.

## Hot Salt

14 Another technology that would supply perhaps 1/5 of the solar energy in our vision is known as concentrated solar power. In this design, long, metallic mirrors focus sunlight onto a pipe filled with fluid, heating the fluid like a huge magnifying glass might. The hot fluid runs through a heat exchanger, producing steam that turns a turbine.

insulated *adj.* 绝缘的，隔热的

15 For energy storage, the pipes run into a large, **insulated** tank filled with molten salt, which retains heat efficiently. Heat is extracted at night, creating steam. The molten salt does slowly cool, however, so the energy stored must be tapped within a day.

megawatt *n.* 【电】兆瓦特

16 Nine concentrated solar power plants with a total capacity of 354 **megawatts** (MW) have been generating electricity reliably for years in the US. A new 64-MW plant in Nevada came online in March 2007. These plants, however, do not have heat storage. The first commercial installation to incorporate it—a 50-MW plant with seven hours of molten salt storage—is being constructed in Spain, and others are being designed around the world. For our plan, 16 hours of storage would be needed so that electricity could be generated 24 hours a day.

17 Existing plants prove that concentrated solar power is practical,

but costs must decrease. Economies of scale and continued research would help. In 2006 a report by the Solar Task Force of the Western Governors' Association[11] concluded that concentrated solar power could provide electricity at 10 cents per kWh or less by 2015 if 4 GW of plants were constructed. Finding ways to boost the temperature of heat exchanger fluids would raise operating efficiency, too. Engineers are also investigating how to use molten salt itself as the heat-transfer fluid, reducing heat losses as well as capital costs. Salt is **corrosive**, however, so more **resilient** piping systems are needed.

corrosive *adj.* 腐蚀的；腐蚀性的
resilient *adj.* 有弹性的，弹回的

18 Concentrated solar power and photovoltaics represent two different technology paths. Neither is fully developed, so our plan brings them both to large-scale deployment by 2020, giving them time to mature. Various combinations of solar technologies might also evolve to meet demand economically. As installations expand, engineers and accountants can evaluate the **pros and cons**, and investors may decide to support one technology more than another.

pros and cons 优缺点；赞成与反对

## Direct Current, Too

19 The geography of solar power is obviously different from the nation's current supply scheme. Today coal, oil, natural gas and nuclear power plants dot the landscape, built relatively close to where power is needed. Most of the country's solar generation would stand in the Southwest. The existing system of **alternating-current** (AC) power lines is not **robust** enough to carry power from these centers to consumers everywhere and would lose too much energy over long **hauls**. A new high-voltage, direct-current (HVDC) power transmission backbone would have to be built.

alternating current 交流电
robust *adj.* 强健的，强大的
haul *n.* 拖运（距离）

20 Studies by Oak Ridge National Laboratory[12] indicate that long-distance HVDC lines lose far less energy than AC lines do over equivalent spans. The backbone would radiate from the Southwest toward the nation's borders. The lines would **terminate** at **converter stations** where the power would be switched to AC and sent along existing regional transmission lines that supply customers.

terminate *v.* 停止，结束，终止
converter station 变电站

21 The AC system is also simply out of capacity, leading to noted shortages in California and other regions; DC lines are cheaper to build and require less land area than equivalent AC lines. About 500 miles of HVDC lines operate in the US today and have proved

reliable and efficient. No major technical advances seem to be needed, but more experience would help refine operations. The Southwest Power Pool of Texas[13] is designing an integrated system of DC and AC transmission to enable development of 10 GW of wind power in western Texas. And TransCanada, Inc.[14], is proposing 2,200 miles of HVDC lines to carry wind energy from Montana and Wyoming south to Las Vegas and beyond.

## Stage One: Present to 2020

subsidy *n.* 补助金，津贴
cumulative *adj.* 累积的，渐增的

22 We have given considerable thought to how the solar grand plan can be deployed. We foresee two distinct stages. The first, from now until 2020, must make solar competitive at the mass-production level. This stage will require the government to guarantee 30-year loans, agree to purchase power and provide price-support **subsidies**. The annual aid package would rise steadily from 2011 to 2020. At that time, the solar technologies would compete on their own merits.[15] The **cumulative** subsidy would total $420 billion.

in parallel 除此之外
right-of-way *n.* （道路或铁路）用地
hurdle *n.* 障碍，栏
Fla.（=Florida）（美国）佛罗里达州

23 About 84 GW of photovoltaics and concentrated solar power plants would be built by 2020. **In parallel**, the DC transmission system would be laid. It would expand via existing **rights-of-way** along interstate highway corridors, minimizing land-acquisition and regulatory **hurdles**. This backbone would reach major markets in Phoenix, Las Vegas, Los Angeles and San Diego to the west and San Antonio, Dallas, Houston, New Orleans, Birmingham, Ala., Tampa, **Fla.**, and Atlanta to the east.

scale up 按比例增加或提高
apiece *adv.* 每，每个

24 Building 1.5 GW of photovoltaics and 1.5 GW of concentrated solar power annually in the first five years would stimulate many manufacturers to **scale up**. In the next five years, annual construction would rise to 5 GW **apiece**, helping firms optimize production lines. As a result, solar electricity would fall toward six cents per kwh. This implementation schedule is realistic; more than 5 GW of nuclear power plants were built in the US each year from 1972 to 1987. What is more, solar systems can be manufactured and installed at much faster rates than conventional power plants because of their straightforward design and relative lack of environmental and safety complications.

## Stage Two: 2020 to 2050

25 It is **paramount** that major market **incentives** remain in effect through 2020, to set the stage for self-sustained growth thereafter. In extending our model to 2050, we have been conservative. We do not include any technological or cost improvements beyond 2020. We also assume that energy demand will grow nationally by 1 percent a year. In this scenario, by 2050 solar power plants will supply 69 percent of US electricity and 35 percent of total US energy. This quantity includes enough to supply all the electricity consumed by 344 million plug-in hybrid vehicles, which would displace their gasoline counterparts, key to reducing dependence on foreign oil and to **mitigating** greenhouse gas emissions. Some three million new domestic jobs—notably in manufacturing solar components—would be created, which is several times the number of US jobs that would be lost in the then **dwindling** fossil-fuel industries.

paramount *adj.* 极为重要的
incentive *n.* 动机，刺激，鼓励
mitigate *v.* 使缓和，减轻
dwindle *v.* 缩小

26 The huge reduction in imported oil would lower trade balance payments by $300 billion a year, assuming a crude oil price of $60 a barrel (average prices were higher in 2007). Once solar power plants are installed, they must be maintained and repaired, but the price of sunlight is forever free, **duplicating** those fuel savings year after year. Moreover, the solar investment would enhance national energy security, reduce financial burdens on the military, and greatly decrease the societal costs of pollution and global warming, from human health problems to the ruining of coastlines and farmlands.

duplicate *v.* 复制，复写

27 **Ironically**, the solar grand plan would lower energy consumption. Even with one percent annual growth in demand, the 100 quadrillion Btu consumed in 2006 would fall to 93 quadrillion Btu by 2050. This unusual offset arises because a good deal of energy is consumed to extract and process fossil fuels, and more is wasted in burning them and controlling their emissions.

ironically *adv.* 讽刺地

28 To meet the 2050 projection, 46,000 square miles of land would be needed for photovoltaic and concentrated solar power installations. That area is large, and yet it covers just 19 percent of the suitable Southwest land. Most of that land is barren; there is no competing use value. And the land will not be polluted. We have assumed that only 10 percent of the solar capacity in 2050 will come from distributed

photovoltaic installations—those on rooftops or commercial lots throughout the country. But as prices drop, these applications could play a bigger role.

Notes

1. **fossil fuels:** 矿物燃料，碳氢化合物的沉淀，如石油、煤、天然气等，由古代生物衍生而成，经过开发后用作燃料。
2. **British thermal units (Btu):** （亦可写作British Thermal Unit/BTU）英制热量单位，相当于将1磅水升高1华氏摄氏度所需的总热量，常用作美国的能量单位。
3. **A direct-current (DC) transmission backbone would also have to be erected to send that energy efficiently across the nation:** 还需要建立一条直流电传输主干线，将能量高效地传送到全国各地。backbone原意为"支柱，脊椎"，在这里转义为"(电力）主干线"。
4. **Because solar technologies are almost pollution-free, the plan would also reduce greenhouse gas emissions from power plants by 1.7 billion tons a year, and another 1.9 billion tons from gasoline vehicles would be displaced by plug-in hybrids refueled by the solar power grid:** 由于太阳能发电技术几乎不产生任何污染，因此该计划每年还可使发电厂排放的温室气体减少17亿吨。此外，用太阳能电力网充电的外接充电式混合动力车，还可避免汽车排放的19亿吨温室气体。句中pollution-free指"没有污染的"。-free意为"没有，免除"，如duty-free (免税的)，ice-free (没有冰的)。plug-in hybrids指外接充电式混合动力车，hybrid常用于生物学，意为"杂交品种"，在本句中指既可以使用汽油也可以使用直流电作为驱动力的汽车。
5. **Although this area may sound enormous, installations already in place indicate that the land required for each gigawatt-hour of solar energy produced in the Southwest is less than that needed for a coal-powered plant when factoring in land for coal mining:** 虽然上述面积看起来也许非常庞大，但已建立起来的设施表明，在美国西南部每生产10亿瓦特/小时的太阳能所需要的土地比火力发电厂占用的土地要少，因为后者还包括了开采煤矿所占的土地。在本句中factor用作动词，与in或into连用，意为"把……考虑在内"。
6. **National Renewable Energy Laboratory:** （美国）国家可再生能源实验室（简称NREL)，美国能源部直属的国家级实验室，主要进行可再生能源和节能技术的研究和实践，并对国家可再生能源和节能规划等工作予以指导。
7. **First Solar:** 第一太阳能公司，美国薄膜光伏电池领域的标志性厂商，也是在纳斯达克上市的高科技企业。
8. **The pressurized air is released on demand to turn a turbine that generates electricity, aided by burning small amounts of natural gas:** 压缩空气依据需求被释放出来，推动涡轮产生电力，同时燃烧少量天然气加以辅助。on demand意为"按要求，应需求"。
9. **Electric Power Research Institute:** （美国）电力科学研究院（简称EPRI)，成立于1973年，其任务是进行发、输、配、用电方面的科学研究，组织协调公用电业的新技术开发，组织电力技术的信息交流。

10. **Mapping by the natural gas industry and the Electric Power Research Institute shows that suitable geologic formations exist in 75 percent of the country, often close to metropolitan areas:** 依据天然气产业和电力科学研究院的勘查，全国75%的地区拥有合适的地质结构，而且通常靠近大城市。map在句中用作动词，意为“绘制地图，(为制地图而）勘测”。
11. **Solar Task Force of the Western Governors' Association:** （美国）西部州长协会太阳能工作小组。
12. **Oak Ridge National Laboratory:** （美国）橡树岭国家实验室（简称ORNL)，美国能源部所属最大的科学和能源研究实验室，成立于1943年，在许多科学领域中都处于国际领先地位。它主要从事六个科学领域的研究，包括中子科学、能源、高性能计算、复杂生物系统、先进材料和国家安全。
13. **The Southwest Power Pool of Texas:** （美国）德克萨斯州西南部电力联营体。
14. **TransCanada, Inc.:** （加拿大）泛加股份有限公司。
15. **At that time, the solar technologies would compete on their own merits:** 到那时，太阳能技术将依靠自身的优越性进行竞争。短语on one's own merits常常与consider或judge连用，意思是“根据某人或某物本身的品质、价值来考虑或判断”，在本句中指2010年政府资助停止后，太阳能本身的优越性将成为它与其他技术进行竞争的本钱。

## Exercises

### I. Answer the following questions.

1. According to the passage, what can US basically profit in the future if it develops solar power step by step?
2. What is the major disadvantage of solar power?
3. Why should a new direct current power transmission backbone be built for the application of solar power?
4. What would be achieved by 2020 according to the solar grand plan?
5. Why can the solar grand plan cut down the energy consumption under the background of annual growth in energy demand?

### II. The following statements are incomplete. Search the missing information in the passage and fill in the blanks.

1. ___________ and ___________ should be installed on large tracts of land in order to convert the whole nation to solar power US.
2. ___________ refueled by the solar power grid can reduce green houses gas emissions from gasoline vehicles by 1.9 billion tons.

3. To drop electricity price to six cents per kWh by 2020, the cadmium telluride module efficiency should be raised to __________.
4. In the solar grand plan __________ is regarded as an important technology which would supply perhaps 20% of the solar energy.
5. The first commercial power plant with heat storage equipment is being constructed in __________.
6. According to the Solar Task Force of the Western Governors' Association, __________ could improve operating efficiency.
7. In the first stage of the solar grand plan, the overall subsidy the US government should provide will be up to __________.
8. To meet the 2050 projection, 46,000 square miles of land which will take up 19% of __________ would be prepared for photovoltaic and concentrated solar power installations.

**III. Identify the implied meanings of the underlined parts of the following sentences according to the context of the passage, and translate the sentences into Chinese.**

1. And as China, India and other nations rapidly increase their demand for fossil fuels, future fighting over energy <u>looms large</u>.
2. Solar energy's potential is <u>off the chart</u>.
3. In 2050 US carbon dioxide emissions would be 62 percent below 2005 levels, <u>putting a major brake</u> on global warming.
4. In the past few years the cost to produce photovoltaic cells and modules has dropped significantly, <u>opening the way for</u> large-scale deployment.
5. Studies by the National Renewable Energy Laboratory in Golden, Colo. show that more than enough land in the Southwest is available without requiring use of <u>environmentally sensitive areas</u>, population centers or difficult terrain.
6. A new 64-MW plant in Nevada <u>came online</u> in March 2007.
7. Today coal, oil, natural gas and nuclear power plants <u>dot the landscape</u>, built relatively close to where power is needed.
8. The existing system of alternating-current (AC) power lines <u>is not robust enough</u> to carry power from these centers to consumers everywhere and would lose too much energy over long hauls.
9. It would expand via existing <u>rights-of-way</u> along interstate highway corridors, minimizing land-acquisition and regulatory hurdles.
10. It is paramount that major market incentives remain in effect through 2020, to set the stage for <u>self-sustained growth</u> thereafter.

## Translation Techniques (1)

### Negation in Technical Translation（科技英语中否定的译法）

In English, "negation" means saying "no" or expressing denial. According to linguists, "Every language has its peculiarities in negation." As English and Chinese belong to different language families, each has its own ways of expressing negation, and what is affirmative in form in one language may often mean something negative in another language. This is especially important in technical translation because the errors in this respect may lead to unclearness and distortion of the meaning of the original version.

There are generally two cases in which negation in translation should be handled cautiously: when it is necessary to add negative words in translation and when it is necessary to omit them.

Some English sentences do not have the words of "no" or "not", but there are words which imply negative meanings. When this kind of sentences are translated into Chinese, negative words may be needed to add to the translation. For example, "In the muddy waters of South American rivers these fishes' eyes are of little use to them; instead of eyes they use extremely accurate electric sense organs." In this sentence there are two expressions which have negative meanings—"little" and "instead of", and in translation the meanings are expressed explicitly: 在南美洲浑浊的河水里，这些鱼的眼睛几乎没有多少用处；它们不靠眼睛，而是使用极为准确的电感器官。

Another example:

**原文** A radical departure from A-bomb and H-bomb are a new generation of nuclear weapons that focus the power of nuclear explosions, *rather than* letting the force escape in all directions.

**译文** 新一代核武器与原子弹和氢弹有很大差别，新的核武器集中核爆炸的威力，而不是让它四处扩散。

On the other hand, although some English sentences do possess negative words, it would be better to use other expressions in translation. For example, "Under conditions on earth ammonia does *not* become a liquid until it is cooled to thirty degrees below zero Fahrenheit and does *not* freeze until a temperature of 100 degrees below zero is reached." The appropriate translation should be "在地球上，氨要冷却到华氏零下30摄氏度才会变成液体，而只有当温度达到零下100度时氨才会冻结。"

**IV. Translate the following sentences. Pay attention to how the translation technique of negation should be used.**

1. Adding negative words in translation

(1) The US needs a bold plan to free itself from fossil fuels.

(2) The benign nature of photovoltaic plants (including no water consumption) should keep environmental concerns to a minimum.

(3) The great limiting factor of solar power, of course, is that it generates little electricity when skies are cloudy and none at night.

(4) The AC system is also simply out of capacity, leading to noted shortages in California and other regions.

(5) Instead of using pumps for the cooling process, the system relies on natural circulation.

(6) Whether a computer will be able to exercise free will or originate anything is still an open argument.

2. Omitting negative words in translation

(1) Ice is *not* as dense as water and it therefore floats.

(2) There is *no* material but will deform more or less under the action of force.

(3) Owing to rigidity of the spindle and bearings, the fluid bearings *never* lose their accuracy.

(4) Microprogramming was *not* used much until the advent of relatively low-cost ROS.

## Translation Techniques (2)

### Conversion Between Clauses and Phrases（从句和短语的转换）

Within a clause, besides the linking word that connects the clause with other parts of the sentence, there are also such elements as a subject, verbs, and an object, etc. If the Chinese translation rigidly follows the structure of the sentences and includes all the elements in the clauses, in some cases the Chinese sentences may sound wordy because some parts in original versions are not needed in the Chinese sentences. Therefore, the technique of conversion between the clauses in English and the phrases in Chinese is applied.

For example, the *if*-clause in the following sentence "*If you melt two or more metals together*, you can get a new metal" can be changed into a phrase in Chinese translation, which makes the sentence compact and natural: 将两种或两种以上的金属熔化在一起就可产生一种新的金属。

There are more examples to illustrate this point:

1. Dr. John Nuckolls, head of physics at the Livermore laboratory, says humans suffer confusion and disorientation *when subjected to long wavelength radiation of great strength.*
   利弗莫尔实验室物理学负责人约翰·纳科尔斯博士说，在高强度的长波辐射下人们会产生心绪混乱和定向障碍。
2. Some atoms are so constructed *that they lose electrons easily.*
   有些原子的结构使其很容易失去电子。
3. Electromagnetic waves carry the energy *which is received from the current flowing in the conductor.*
   电磁波携带的能量是从导体传送的电流中获得的。
4. This circuit has the same number of natural modes *as does the circuit in Fig. 2.*
   该电路与图2电路的自然模数相同。
5. A good understanding has been obtained of evaporation processes *which are carried out at ordinary pressure.*
   人们对于常压下的蒸发过程已有了清楚的了解。

Meanwhile, the phrases in Chinese sentences may also be transferred into English clauses when necessary. For example, 半导体材料既不是良导体，也不是好的绝缘体。(A semiconductor is a material *that is neither a good conductor nor a good insulator*.)

**V. Discuss the translation technique and the ways of applying the technique to the translation of the following sentences. Complete each of the Chinese translations.**

1. Without all four forces working *as they do*, matter could not exist, stars and planets could not exist, human beings could not exist. (__________，就不存在物质，不存在恒星和行星，不存在人类。)
2. The ultimate type of resistance is that the insect changes its normal physiology *so that it is no longer sensitive to the insecticide*. (抗药性的最终形式是__________。)
3. The distance *light travels in one second* is 300,000 kilometers. (__________是30万公里。)
4. As many instruments *as are in the laboratory* have been made most use of. (__________得到充分的利用。)
5. The purpose *for which electricity is used* is numerous. (__________多种多样。)
6. Under this policy, the government agency stated that software would be subject to regulation *if it met the following two criteria*. (该政府部门说明，根据政策规定，__________。)
7. Smoke is a mixture of gases, vaporized chemicals, minute particles of ash, and other solids;

there is also nicotine, *which is a powerful poison,* and black tar.（吸烟产生的烟雾是一种混合物，包括各种气体、汽化的化学物质、极小的灰末微粒和其他固体，还有__________和黑焦油。）

8. High blood pressure cannot be cured; however, it can be brought under control, *so that it may not cause damage to the heart and other body systems.*（高血压病无法根治，但可予以控制，__________。）

## VI. Translate the following sentences into Chinese. Pay attention to how the technique of conversion between clauses and phrases should be used.

1. *It is worth noting*, too, that as modules improve, rooftop photovoltaics will become more cost-competitive for homeowners, reducing daytime electricity demand.
2. For our plan, 16 hours of storage would be needed *so that electricity could be generated 24 hours a day.*
3. Well-meaning scientists, engineers, economists and politicians have proposed various steps *that could slightly reduce fossil-fuel use and emissions.*
4. The turbines burn only 40 percent of the natural gas they would *if they were fueled by natural gas alone*, and better heat recovery technology would lower that figure to 30 percent.
5. For energy storage, the pipes run into a large, insulated tank filled with molten salt, *which retains heat efficiently.*

## VII. Translate the following passage into Chinese.

Tides, like the sun, always will be with us, and people have utilized the energy in their ebbs and flows since at least the 11th century. For centuries the moving waters turned mills that ground grains. Utilizing tidal power on a grand scale to generate electricity did not become a reality until 1966. That year France began full operation of the world's first tidal-power plant.

Such plants require tides with a large range flowing in and out of a narrow bay or river that can be closed off by a dam. High tides raise water in the bay or river, which is closed by the dam before the water begins to ebb. During low tide, the water level outside the dam drops below the level in the bay. Gates are opened, and as the stored water falls to a lower level, it drives turbines that generate electricity.

Tides come and go, but people need a continuous flow of electricity. To get it, the falling water is made to operate pumps that put water into storage ponds for release between tidal cycles.

North America's first tidal-power plant was built at the Annapolis Basin on the shore of Nova Scotia. Here tides with a maximum range of 8.7 meters bring water into the Annapolis River. As it flows out again into the Bay of Fundy, the water spins turbine generators to produce 50 million kilowatt-hours of electricity per year. If successful, this plant could pave the way for larger tidal-power plants at the head of the Bay of Fundy where tides range 15 meters. Such power plants, however, are not likely to become commonplace because they can be constructed only in a few places.

**VIII. Translate the following passage into English.**

### 光电池

光电池（亦称太阳能电池）是一种利用光电效应将光转变为直流电的装置。第一个光电池于19世纪80年代问世。虽然当时的硒电池模型只能将1%的入射光转变为电能，但研究者已认识到这个发现的重要性。

光电池最早的重要用途是作为“先锋1号”卫星的后备电源，在化学蓄电池耗尽以后，它使卫星得以继续传送信息达一年多之久。在光电池成功地应用于这次任务后，其他许多卫星也使用了这种电池，到20世纪60年代后期已成为卫星的常用电源。此后，光电池继续在早期商业卫星所取得的成功中起着必不可少的作用，直到今天对于电信的基础设施仍至关重要。

1973年的石油危机刺激了70年代和80年代初期光电池生产的迅速增加。20世纪90年代中期以来，光电池业的领导地位已从美国转移到日本和德国。1992至1994年，日本增加了这方面的研发经费，并出台了补贴计划，鼓励在居民区建立光电池系统，结果日本的光电池发电量由1994年的31.2兆瓦上升至1999年的318兆瓦。德国成为世界领先的光电池市场后，其光电池发电量从2000年的100兆瓦增加到约4,150兆瓦。2004年，西班牙成为光电池第三大市场，最近由于各种激励机制和本地市场条件，法国、意大利、韩国和美国也取得了快速发展。

## Section 2 Reading for Academic Purposes

*The introduction to digital libraries in the previous chapter has provided you with the basic idea of this new source of information, which opens another channel through which you may gain access to huge collection of network information. However, in spite of the advantages of digital libraries, there is the concern about the possibility of the violation of intellectual property rights.*

*What are intellectual property rights? They are the rights over creations of the mind, both artistic and commercial. The former is covered by copyright laws, which protect creative works, such as books, movies, music, paintings, photographs, and software, and gives the copyright holder exclusive right to control reproduction or adaptation of such works for a certain period of time. However, with the development of network information sources such as digital libraries, the protection of intellectual property rights faces new challenges.*

## Network Information Sources and Intellectual Property Rights

## 网上信息源与知识产权

The advantages of digital libraries as a means of easily and rapidly accessing books, archives and images of various types are now widely recognized by commercial interests and public bodies alike. You can also access any user account simply through a link like this.

do away with 废除，去掉

Traditional libraries are limited by storage space; digital libraries have the potential to store much more information, simply because digital information requires very little physical space to contain it. As such, the cost of maintaining a digital library is much lower than that of a traditional library. A traditional library must spend large sums of money paying for staff, book maintenance, rent, and additional books. Digital libraries **do away with** these fees.

wiki *n.* 维基网；维客网（一种互联网多人协作的写作工具）
conversion *n.* 变换，转化
patron *n.* 顾客（尤指老主顾）
affiliation *n.* 联系，从属关系

Digital libraries can immediately adopt innovations in technology providing users with improvements in electronic and audio book technology as well as presenting new forms of communication such as **wikis** and blogs. An important advantage to digital **conversion** is increased accessibility to users. There is also availability to individuals who may not be traditional **patrons** of a library, due to geographic location or organizational **affiliation**.

The other advantages include:

(1) No physical boundary. The users of a digital library need not go to the library physically; people from all over the world can gain access to the same information, as long as an Internet connection is available.

(2) Round the clock availability. A major advantage of digital libraries is that people can gain access to the information at any time, night or day.

(3) Multiple access. The same resources can be used simultaneously by a number of institutions and patrons

retrieval *n.* 【计】检索（从内存或其他存储设备中获取信息的过程）
user-friendly *adj.* 方便用户的
clickable *adj.* 可点击的

(4) Information **retrieval**. The user is able to use any search term (word, phrase, title, name, subject) to search the entire collection. Digital libraries can provide very **user-friendly** interfaces, giving **clickable** access to its resources.

(5) Preservation and conservation. **Digitization** is not a long-term preservation solution for physical collections, but does succeed in providing access copies for materials that would otherwise fall to **degradation** from repeated use. Digitized collections and born-digital objects pose many preservation and conservation concerns that **analog** materials do not.

digitization *n.* 【计】数字化
degradation *n.* 降级，退化
analog *n.* 模拟（格式）

(6) Space. Whereas traditional libraries are limited by storage space, digital libraries have the potential to store much more information, simply because digital information requires very little physical space to contain them and media storage technologies are more affordable than ever before.

(7) Added value. Certain characteristics of objects, primarily the quality of images, may be improved. Digitization can enhance **legibility** and remove visible flaws such as stains and **discoloration**.

legibility *n.* 易读性，易辨认
discoloration *n.* 变色，褪色

However, some people have criticized that digital libraries are **hampered** by copyright law, because works cannot be shared over different periods of time in the manner of a traditional library. There is a **dilution** of responsibility that occurs as a result of the spread-out nature of digital resources. Complex intellectual property matters may become involved since digital material isn't always owned by a library. The content is, in many cases, public domain or self-generated content only. Some digital libraries, such as Project Gutenberg, work to digitize out-of-copyright works and make them freely available to the public. An estimate of the number of distinct books still existent in library catalogs from 2000 BC to 1960 has been made.

hamper *v.* 妨碍，牵制
dilution *n.* 冲淡，稀释

Other digital libraries accommodate copyright concerns by **licensing** content and distributing it on a commercial basis, which allows for better management of the content's reproduction and the payment (if required) of **royalties**. The Fair Use Provisions under copyright law provide specific guidelines under which circumstances libraries are allowed to copy digital resources. Four factors that constitute fair use are purpose of use, nature of the work, market impact, and amount or substantiality used.

license *v.* 许可，特许
royalty *n.* 版税

The ease of digital reproduction has pushed art institutions into a more conservative **stance** on the issue of intellectual property rights (much in the same vein as it did the music industry). But to whom

stance *n.* 立场；态度

public domain works 进入公共领域的作品
milieu *n.* <法> 周围环境

do what the J. Paul Getty Museum's Kenneth Hamma labels "**public domain works** of art" belong? Public domain works that are too old to be included in ongoing copyright protection, yet still remain out of the public's view and also "of educators, and of the general **milieu** of creativity".

assertion *n.* 主张，断言，声明
undermine *v.* 破坏
proponent *n.* 支持者；辩护者
revenue *n.* 收入，税收
accrue *v.* 增加， 增多
effectuate *v.* 实行，完成

A museum's decision to guard this information hinders research and halts digitization efforts. If the Internet fosters the free-exchange of ideas for the sake of educational access, then these **assertions** **undermine** the mission of the museum and more general of digitization. Still, some **proponents** of intellectual property claim that controlling ownership of public domain artwork generates **revenue** through licensing fees, and this **accrued** funding directly benefits institutional operation, but no publicly available statistical data to **effectuate** this belief exists.

fiscal *adj.* 财政的，会计的
Renaissance *n.* （欧洲14-16世纪的）文艺复兴
decree *v.* 颁布
curatorial *adj.* （博物馆、图书馆等）馆长的

Excluding **fiscal** concerns, institutions often assume the role of guide and seek control over educational usage, thereby influencing the creative processes that result from these visual resources. For instance, an art historian developing critical theory on a particular issue in the Dutch **Renaissance** may be granted usage of certain works and not others as **decreed** by the **curatorial** interpretations of museum staff. This practice narrows the possibility for a unique, fresh perspective somewhat by limiting the intellectual freedom of the scholar.

overlap *v.* （与……）重叠
initiate *v.* 开始，发动
Impressionist *n.* 印象主义者，印象派作家
tote bag 大手提袋（或购物袋）

The issue **overlaps** other digitization obstacles that **initiate** debate in the arts community. For example, the usage of digital images for physical reproduction of artwork presents a common problem and also provides a more relatable model for those just beginning to grasp digitization. Few people take offense to **Impressionist** works featured on stationery sets, umbrellas, canvas **tote bags** and the like.

ethical *adj.* 道德的；合乎道德的（尤指合乎职业道德或规矩的）
validate *v.* 【律】使生效；使合法化
relinquish *v.* 放弃
like-minded *adj.* 具有相似意向或目的的

An **ethical** obligation to retain rights on certain pieces may **validate relinquishing** accessibility. In spite of the current intellectual property rights issues, perhaps both leaps in technology and the ideas of Kenneth Hamma and his **like-minded** colleagues will inspire a growing acceptance and ultimate embrace of digitization.

Exercises

I. **Read the passage and decide whether the following statements are true or false. Write T for True and F for False in the brackets.**

1. The advantages of digital libraries are widely recognized by people from all walks of life. ( )
2. Compared with traditional libraries, digital libraries cost more because they need more maintenance staff. ( )
3. People can search information provided by digital libraries without the limitation of place and time. ( )
4. By getting rid of stains and discoloration, digitization can improve the quality of images of objects. ( )
5. Some people argue that digital libraries violate the copyright law. ( )
6. Complex intellectual property matters may become involved since digital material is owned by only one library. ( )
7. Libraries' copying digital resources should comply with some specific guidelines defined in the copyright law of the Fair Use Provisions. ( )
8. Art institutions become generous to share their art resources with the ease of digital reproduction. ( )
9. Some supporters of intellectual property believe that controlling ownership of public domain artwork is beneficial to art institutions. ( )
10. Many people relate the Impressionist works copied on stationery sets and umbrellas to the violation of intellectual property rights. ( )

II. **Read the passage again, and complete the following items.**

1. The difference between traditional libraries and digital libraries in space: ___________
2. The innovation capacity of digital libraries in technology: ___________
3. A great merit to digital conversion: ___________
4. Multiple access means: ___________
5. The four factors that influence the fair use of digital resources under copyright law: ___________
6. Reasons why digital libraries are involved in copyright concerns: ___________
7. The consequence of institutions frequently seeking control over education usage: ___________
8. The possible trend of digitization of libraries: ___________

## Quotations from Great Scientists

*If I have a thousand ideas and only one turns out to be good, I am satisfied.*

*—Alfred Nobel*

假如我有上千个想法，而最终只有一个是有用的，我就心满意足了。

——阿尔弗雷德·诺贝尔

# Chapter 9 Medical Science 医学

## Section 1 Reading and Translation

### I What You Are Going to Read

Ever since cancer became a curse of human life, exploration on cancer cures has never stopped. As cancer is characterized by the development of abnormal cells where one or more genes are destroyed or altered, research is in progress to find ways of blocking, repairing or replacing abnormal genes in cancer cells and to find vaccines that would stimulate immune system to make antibodies against cancer cells.

In this article published in *Scientific American* (July, 2008), the author reviews the investigations made by himself and other researchers in the discovery of the function of heat shock proteins (HSPs) as cellular chaperones. Through their diverse interactions, these proteins pick up telltale "fingerprints" of each cell's contents, which has allowed them to evolve a critical role in immune responses to cancer or pathogens. Therapies that take advantage of these proteins as inhibitors and enhancers of their various natural functions are underway to provide immunotherapic cure for cancers.

### II About the Author

Pramod K. Srivastava is a professor of medicine and director of the Center for Immunotherapy of Cancer and Infectious Diseases at the University of Connecticut School of Medicine. In a series of discoveries beginning when he was a graduate student, Srivastava pioneered the study of the role of heat shock proteins in the immune system. Based on those insights, he co-founded the company Antigenics to develop cancer vaccines made with HSPs derived from individual patients' tumors. He remains a consultant to the company and continues to investigate the role of HSPs in immune responses.

### III Reading Passage

**Could Our Own Proteins Be Used to Help Us Fight Cancer?**

**人类自身的蛋白质能够帮助我们抗击癌症吗？**

By Pramod K. Srivastava

*Protective heat shock proteins[1] present in every cell have long been known to counteract stress. Newly recognized roles in cancer and immunity make them potential therapeutic allies.*

incubator *n.* 孵育器，孵化器
fruit fly 果蝇
geneticist *n.* 遗传学家
chromosome *n.* 染色体
puff up 变得蓬松，膨胀
loci *n.*（locus的复数）位置；【生】基因座

1 In 1962 someone at the Genetics Institute in Pavia[2], Italy, turned up the temperature in an **incubator** holding **fruit flies**. When Ferruccio Ritossa[3], then a young **geneticist**, examined the cells of these "heat shocked" flies, he noticed that their **chromosomes** had **puffed up** at discrete locations. The puffy appearance was a known sign that genes were being activated in those regions to give rise to their encoded proteins, so those sites of activity became known as the heat shock **loci**.

reproducible *adj.* 可重复的，可复制的
absorbing *adj.* 吸引人的，非常有趣的

2 The effect was **reproducible** but initially considered to be unique to the fruit fly. It took another 15 years before the proteins generated when these chromosome puffs appear were detected in mammals and other forms of life. In what is certainly among the most **absorbing** stories in contemporary biology, heat shock proteins (HSPs) have since been recognized as occupying a central role in all life—not just at the level of cells but of organisms and whole populations.

cellular *adj.* 细胞的
adversity *n.* 逆境；苦难
pathogen *n.* 病原体
vaccine *n.* 疫苗

3 Indeed, these ubiquitous molecules are among the most ancient survival mechanisms to have been conserved throughout evolution. They have even been shown to facilitate evolution itself. Produced in response to stressful conditions, including (but not limited to) heat, HSPs help individual cells to cope by keeping **cellular** processes working smoothly in the face of **adversity**. In the past decade scientists have realized that HSPs also play additional roles in higher organisms, such as humans. They are integral to our immune defenses against cancer and **pathogens** and might therefore prove valuable in developing a wide variety of new medicines and **vaccines**.

versatile *adj.* 万能的，通用的
chaperone *n.* 陪护
inhibit *v.* 阻止，约束

4 To understand how these **versatile** proteins can be harnessed therapeutically, it is helpful to look at the diverse ways they perform their core job, which is to act as "**chaperones**" for other proteins. Like the chaperoning of people, the work of HSPs has two objectives: to **inhibit** undesirable interactions and to promote desirable ones, so that a stable and productive bond forms between protein partners.

**Versatile Escorts**

5 Proteins inside a cell often have just one or a very few correct

"mates" with which they can interact effectively—for example, a **receptor** and its ligand[4], which behave like a lock and key, respectively. The ligand has little effect on other receptor types, and the receptor is typically activated only by its particular ligand or molecules very close to it in structure. In contrast, HSPs tend to associate with a wide range of "client" proteins, allowing the HSPs to perform a dizzying array of jobs. These can include helping newly formed **amino acid** chains to fold into their proper protein shapes, **dismantling** them after they have been damaged, **escorting** proteins to their intended mates and keeping them away from **interlopers**.

receptor *n.* 接受体
amino acid 氨基酸
dismantle *v.* 分解，拆除
escort *v.* 护送；为……护航
interloper *v.* 闯入者，干涉者

6 Specific examples can highlight just how critical these tasks are and can illustrate some of the ways that major HSP chaperones serve their clients. A protein's ability to carry out its intended functions depends not only on it getting to the right place at the right time, but on it having the correct shape. Newly formed chains of amino acids are subject to various forces that help them to take on the right conformation. Each amino acid, for instance, has a characteristic response to water in the cellular **cytoplasm**. **Hydrophobic** amino acids **abhor** water and try to get away from it by nestling inside the protein structure, whereas **hydrophilic** amino acids prefer to face outward. Such mechanisms are not always enough to ensure proper folding, though, so HSPs, such as HSP60, get involved.

cytoplasm *n.* 细胞质
hydrophobic *adj.* 排水性的，不溶于水的
abhor *v.* 厌恶；拒绝
hydrophilic *adj.* 亲水的，溶于水的

7 Arthur L. Horwich[5] of Yale University has provided much of the current understanding of the HSP60 chaperone, which resembles a cage composed of multiple HSP60 molecules. Its inner rim is highly hydrophobic and therefore attracts the exposed hydrophobic amino acids of an unfolded protein to bind to it. Once such a chain is drawn into this cage, it encounters a hydrophilic interior, which the hydrophobic amino acids want to avoid at all costs, so the trapped molecule is forced to change shape.[6] This process may not happen **in one go**, and the cage may release and recapture the protein multiple times before the protein acquires a correctly folded conformation. Thus, the HSP60 protein is known as a foldase[7]. Conversely, the HSP100 protein is an unfoldase. It, too, is a **multisubunit** ring, which, in cooperation with HSP70, can disassemble damaged proteins or undesirable protein **aggregates** or can even cause a fully folded protein to unfold.

in one go 一鼓作气
multisubunit *n.* 多亚单元
aggregate *n.* 聚集体，集成体

substrate *n.* 培养基，酶作用物
cleft *n.* 缝隙，裂口
clamp *v.* 夹住，夹紧

8 In contrast to the cagelike chaperones, most HSPs do not enclose their **substrates** but rather grab them by the "elbows" to help them along. HSP70, for example, binds directly to short stretches of amino acid sequences, also known as peptides[8]. The molecule has a peptide-binding **cleft** that is open when HSP70 is bound to the cellular energy source ATP[9], but when ATP is absent, a lidlike structure on HSP70 **clamps** down on the bound peptide and traps the larger protein chain in place. The ability of HSP70 to grab a variety of different peptides allows the molecule to play chaperone in many fundamental cellular processes, such as helping new amino acid chains to assume a mature conformation, facilitating the assembly of complex proteins and protecting proteins from falling apart in high temperatures.

deprivation *n.* 剥夺；丧失
dehydration *n.* 脱水
degrade *v.* 降级；退化
churn out *v.* 艰难地产生

9 Although heat shock proteins are active in cells in normal circumstances, it is easy to see how their help would be even more valuable to a cell in a difficult situation. Under emergency conditions, such as extreme heat or cold, oxygen **deprivation**, **dehydration** or starvation, a cell would be struggling just to survive. Critical proteins might be **degraded** by the harsh environment, even as the cell would try to **churn out** replacements. In these circumstances, heat shock proteins would mitigate the stress by rescuing essential proteins, dismantling and recycling damaged ones, and generally keeping cell operations running as smoothly as possible. Hence, when a cell is under high stress, one of its first responses will be to manufacture more of the HSPs themselves, as Ritossa first witnessed 46 years ago. This important role of HSPs has been well documented since its discovery. Beginning in the 1980s, however, a completely different function of HSPs—just as integral to survival for complex organisms—also began to be revealed.

## Antigenic Fingerprints

immunize *v.* 使免疫，赋予免疫性

10 As a graduate student in the early 1980s at the Center for Cellular and Molecular Biology in Hyderabad, India[10], I became interested in a phenomenon that had been observed since the 1940s but never explained. Many scientists had demonstrated that one can **immunize** rodents against their own cancers, just as people are routinely immunized against pathogens. Proteins from a pathogen are recognized by the mammalian immune system as foreign, however,

and that is why they act as **antigens**—provoking an immune response. A cancer, on the other hand, is made up of an individual's own cells, so the antigenic element remained a great mystery. I began trying to isolate these cancer-specific antigens.

antigen *n.* 抗原

11 During my graduate and postdoctoral work, I identified a protein, called gp96, which could indeed **elicit** immune resistance to tumors. This molecule, surprisingly, **turned out** to be a member of the HSP90 family—many HSP proteins come in several related forms—which occurs in normal tissues as well as cancer cells. Stephen J. Ullrich[11] and his colleagues at the National Institutes of Health[12] independently made a similar observation two years later. The gp96 molecules found in tumors and in normal tissues were identical in their amino acid sequences, so the cancer-derived gp96 was not cancer-specific. What, then, was the basis of its ability to immunize against cancers?

elicit *v.* 引起，得出
turn out 结果（是）；原来(是)；证明（是）

12 The answer began to emerge in 1990, when Heiichiro Udono, then a postdoctoral fellow in my laboratory at the Mount Sinai School of Medicine[13], and I were isolating HSP70 from tumors to test if it, too, elicited tumor immunity. We found that it could. The biggest surprise came, however, when we put the HSP70 through a final **purification** step called ATP-affinity chromatography[14], and the molecule's very **potent** tumor-immunizing activity disappeared!

purification *n.* 提纯，净化
potent *adj.* 有力的，有效的

13 We realized immediately that exposure of HSP70 to ATP was causing HSP70 to shed material, which we determined to be peptides. The work of several research groups in the **ensuing** years has revealed that HSP70 changes conformation when it binds to ATP, causing it to **let go of** any bound peptide. In fact, researchers learned that members of the HSP60, HSP70 and HSP90 families all regularly carry around peptides generated within cells. And when HSP70 or HSP90 are taken from cancers or from virus- or **tuberculosis**-infected cells, in nearly all instances they bear peptides derived from cancer-specific antigens, **viral** antigens or tuberculosis antigens. Thus, the HSP-associated peptides represent the "antigenic fingerprint" of the cells or tissues from which they come.

ensuing *adj.* 接着发生的；连续发生的
let go of 放开；释放
tuberculosis *n.* 结核病，肺结核
viral *adj.* 病毒的

14 This characteristic ability of certain chaperones to retain peptides representative of their cell of origin has given HSPs an

essential role in one of the most fundamental processes of the immune system—recognition of cancerous and virus-infected cells. T lymphocytes[15] recognize antigens on such cells through an elaborate process known as antigen presentation[16]. Essentially all antigens made inside cells are degraded into peptides that then associate with HSPs of the HSP60, HSP70 and HSP90 families in a sequence of events that is still unclear. The peptides are eventually loaded onto a special class of proteins, known as the major histocompatibility complex I (MHC I) proteins[17], displayed on the surface of most mammalian cells. The T cells recognize these MHC I-peptide complexes and destroy any that signifies the cell is diseased.

hypothesize *v.* 假设，假定

15 The chaperoning of peptides by HSPs is essential for their eventual loading onto MHC I molecules; when the HSPs are chemically silenced, the MHC I molecules remain empty of peptides and cannot be recognized by the T cells. This role of the HSP-chaperoned peptides in antigen presentation by MHC molecules was **hypothesized** by my colleagues and me in 1994 and shown to be true through our work and others.

intracellular *adj.* 细胞内的
foe *n.* 危害物；敌人

16 It is this antigen-chaperoning property of the peptide-binding HSPs that is the basis of the ability of HSPs derived from tumors or pathogen-infected cells to immunize against those same tumors or **intracellular** pathogens. But the HSP-peptide complexes also have another critical part in the T cells' recognition of friend and **foe** antigens—through their interactions with different types of immune cells known as antigen-presenting cells[18].

## Sounding the Alarm

sentinel *n.* 哨兵，卫士
home in on （靠信号、雷达等）导向目标追踪

17 **Sentinels** of the immune system, antigen-presenting cells occur in perhaps every tissue of the body, where they can "sample" their surroundings for any antigens that might be nearby. They present whatever they encounter to the T cells that will eventually **home in on** and attempt to destroy cancerous or infected cells.

18 It turns out that antigen-presenting cells carry receptors on their surface for the peptide-binding chaperones. The first such receptor was identified by Robert J. Binder, then a graduate student in my laboratory and currently an assistant professor at the University of Pittsburgh, as CD91. When the cells encounter an HSP-peptide complex, they

internalize it through the CD91 doorway and present the HSP-chaperoned peptides to the T cells, which can then multiply and fight off the cancer or pathogen. Broadly speaking, this mechanism is the reason HSPs isolated from a cancer are able to immunize against that cancer, whereas the HSPs isolated from normal tissues do not do so.

19 Beyond delivering a description of the invader to the immune system, HSPs seem to sound an alarm as well. Sreyashi Basu of the University of Connecticut School of Medicine and I have shown in laboratory studies that just exposing antigen-presenting cells to HSP70 and HSP90 family members causes the cells to undergo a number of changes, including initiation of signals that cause **inflammation**, which is part of a strong immune defense. Although HSPs normally do their work inside cells, scientists have known for some time that when mammalian cells are under stress, selected HSPs are released from the cells or displayed on the cell surface in small but significant quantities. Thus, the ability of HSPs to activate antigen-presenting cells by their mere presence suggests that an anomalous appearance of HSPs outside cells may be a mechanism to alert the immune system of danger.

inflammation *n.* 炎症，发炎

20 My work toward using HSP-peptide complexes purified from cancers to elicit tumor rejection is based on this immunizing function and on my belief that each patient's tumor is antigenically unique. I have developed a process for **extracting** HSP-bound peptides from the individual patient and then reintroducing them in purified form as a vaccine that would stimulate the immune system to attack cells bearing those specific tumor-associated antigens. This approach has been tested in the US and Europe in a series of early human (phases I and II) trials for several cancers. More advanced tests of **efficacy** (**randomized** phase III trials) in the US, Europe, Australia and Russia have just been concluded in patients with **melanoma** and **renal** cancer. Those latest studies showed that patients with melanoma who received sufficient doses of HSP-peptide-complex vaccine and whose disease was limited to the skin, **lymph nodes** and lungs lived significantly longer than patients who received other standard treatments, including **chemotherapy**. In the trial on renal cancers, the vaccine extended the recurrence-free survival[19] time in some groups of patients by more than a year and a half.

extract *v.* 萃取，提炼
efficacy *n.* 功效；效验
randomized *adj.* 随机化的，任意的
melanoma *n.* 黑素瘤
renal *adj.* 肾脏的，肾的
lymph node 淋巴结
chemotherapy *n.* 化学疗法

infectious *adj.* 传染性的，传染的
genital *adj.* 生殖的；生殖器的
herpes *n.* 疱疹

21 The results were enough for the Russian government to approve the treatment, making it the first cancer vaccine to enter actual clinical use. An application for approval in Europe will be filed shortly, and an application to the US Food and Drug Administration[20] is awaiting more data on the patients' long-term outcomes. Meanwhile this approach seems as if it should be just as applicable for treatment of serious **infectious** diseases, including **genital herpes**, tuberculosis and others. Clinical trials investigating those applications are at various stages.

## Wide Influence

amplify *v.* 扩大，增强
mutation *n.* 【遗】突变，变种
deleterious *adj.* 有害的，有毒的
buffer *v.* 缓冲
foster *v.* 鼓励，促进
compromise *v.* 危及；损害
variant *adj.* 不同的，变异的
potentiate *v.* 促进或加强（生化或生理行为、效力）

22 **Amplifying** the natural effect of HSPs on the immune system by using them in vaccines is not the only way to employ these versatile proteins therapeutically. Work by Suzanne L. Rutherford[21] of the University of Washington and Susan L. Lindquist[22] of the Whitehead Institute for Biomedical Research in Cambridge, Mass.[23], has provided a stunning example of how effectively HSPs perform their core job of mitigating stressful conditions inside cells. They have shown that when HSP90 functioning was suppressed in fruit flies, a large number of preexisting genetic **mutations** were unmasked, indicating that potentially **deleterious** effects were being **buffered** by HSP90. Rutherford and Lindquist have argued that widespread genetic variation that would otherwise affect the functioning of organisms exists in nature but is usually not manifested because HSP90 essentially hides the variation—an effect that **fosters** the quiet accumulation of genetic changes. When the buffering function is **compromised**, for example, by extreme temperature, **variant** traits emerge and then natural selection can act on them. Thus, HSP90, by fostering genetic variation, **potentiates** evolution.

collaborator *n.* 合作者
fungi *n.* （fungus的复数）真菌类
inhibitor *n.* 抑制剂
antibiotics *n.* 抗生素
viable *adj.* 能生存的
malignancy *n.* 【医】恶性；恶性肿瘤

23 Lindquist and her **collaborators** have provided further evidence of a role for HSP90 in the rapid evolution of novel traits, such as resistance to specific drugs in diverse species of **fungi**. As a result, she has suggested that species-specific **inhibitors** of HSP90 may be used as a new generation of **antibiotics**. Similarly, HSPs are believed to provide buffering against the accumulating mutations that should make cancer cells less and less **viable** but instead seem to drive their **malignancy**. Because HSP90 affects a wider variety of intracellular

signaling pathways than any other HSP does, loss of its function should make cancer cells more sensitive to stress and therefore more easily killed by chemotherapy. Hence, **pharmacological** inhibitors of increasing specificity for HSP90 are being tested in cancer patients, in combination with chemotherapy.

pharmacological *adj.* 药理学的

## Notes

1. **heat shock proteins:** (HSPs) 热应急蛋白，又称热休克蛋白，是广泛存在于从细菌到哺乳动物各种生物中的一类蛋白质。当有机体暴露于高温时，就会由热激发合成此种蛋白，以保护有机体自身。按照蛋白的大小，热应急蛋白共分为五类，分别为HSP100，HSP90，HSP70，HSP60 以及小分子热应急蛋白 (small heat shock proteins/sHSPs)。
2. **Pavia:** 帕维亚，意大利西北部城市，位于米兰南面。
3. **Ferruccio Ritossa:** 费鲁乔·里托萨（1936–），意大利遗传学家，于1959年获得博洛尼亚大学农学学士，后曾在帕维亚、那不勒斯、美国等地研究、教授遗传学。
4. **ligand:** 配合基，配位体。配位化合物中与另一化学个体连接的离子、分子或分子团。
5. **Arthur L. Horwich:** 亚瑟·豪威奇（1951– ），美国生物学家，耶鲁大学医学院遗传学与儿科教授。
6. **Once such a chain is drawn into this cage, it encounters a hydrophilic interior, which the hydrophobic amino acids want to avoid at all costs, so the trapped molecule is forced to change shape:** 这样的蛋白质链一旦被吸入笼子，就会碰到亲水性内壁，而这正是排水性氨基酸极力躲避的，其结果是陷入笼中的分子被迫改变形状。本句中which the hydrophobic amino acids want to avoid at all costs采用的是拟人化的表达方法，短语at all costs原意为“不惜任何代价”，在这里强调排水性氨基酸对亲水性内壁的排斥。
7. **foldase:** 折叠酶，蛋白质二硫键异构酶。它能识别和水解非正确配对的二硫键，使它们在正确的半胱氨酸残基位置重新形成二硫键，从而改变二硫键的连接位置。
8. **peptides:** 肽，缩氨酸。一种自然或人工的合成物，包括两个或两个以上氨基酸，通过一个氨基酸的酸基和另一个的羧基结合而成。
9. ATP: (adenosine-triphosphate) 三磷酸腺苷，一种由腺苷衍生的核苷酸，通过水解形成二磷酸腺苷，可为细胞的各类生化过程提供大量能量，包括肌肉收缩及糖分释放。
10. **Hyderabad, India:** 海德拉巴，印度中南部城市，位于孟买东南偏东，是金融中心和交通枢纽。
11. **Stephen J. Ullrich:** 史蒂芬·阿尔里奇，美国德克萨斯大学免疫学系教授。1979年毕业于美国乔治敦大学微生物专业，获博士学位。2001年曾在美国国家卫生研究院从事研究工作。
12. **the National Institutes of Health:** （美国）国家卫生研究院，美国主要的医学与行为学研究机构，初创于 1887年，拥有27个研究所及研究中心，任务是探索生命本质和行为学方面的基础知识，并充分运用这些知识延长人类寿命，预防、诊断和治疗各种疾病和残障。

13. **the Mount Sinai Hospital:** （美国）西奈山医学院，成立于1852年，位于纽约曼哈顿区，素以成果丰硕的临床与基础科学研究著称于世。

14. **ATP-affinity chromatography:** 腺三磷亲和层析。层析是根据蛋白质的形态、大小和电荷的不同而设计的物理分离蛋白质的方法，使用广泛。有些生物分子的特定结构部位能够同其他分子相互识别并结合，生物分子间的这种结合能力称为亲和力。亲和层析就是根据这样的原理设计的蛋白质分离纯化方法。

15. **T lymphocytes:** T淋巴细胞。淋巴细胞指的是淋巴组织中近乎透明的细胞，如淋巴结、脾、胸腺和扁桃体中淋巴细胞的数目相当于正常成人体内白细胞的22%到28%。它们具有免疫功能，主要有B细胞和T细胞两种类型。

16. **antigen presentation:** 抗原递呈，是辅佐细胞向辅助性T细胞展示抗原和MHC II类分子的复合物，并使之与TCR结合的过程。这个过程是几乎所有淋巴细胞活化的必需步骤。

17. **the major histocompatibility complex I (MHC I) proteins:** 主要组织相容性复合体I蛋白质。主要组织相容性复合体（MHC）由一群紧密连锁的基因群组成，定位于动物或人类某对染色体的特定区域，呈高度多态性，其编码的分子表达于不同细胞表面，参与抗原递呈，制约细胞间相互识别及诱导免疫应答。

18. **antigen presenting cell:** 抗原呈递细胞，是免疫应答起始阶段的重要辅佐细胞，有多种类型。其中巨噬细胞分布最广，是处理抗原的主要细胞。

19. **reoccurrence-free survival:** 无复发生存，指的是癌症病人在一定的治疗期内病情没有复发的存活状态。

20. **the US Food and Drug Administration:** 美国食品与药品管理局，是美国历史最悠久的保护消费者的联邦机构之一，隶属于美国卫生与公共服务部。它不仅负责依法审批和监督食品和药品的生产，而且负责有关医疗设备、化妆品、手机和微波炉之类放射性产品以及宠物和动物饲料及药品的审批和监督。

21. **Suzanne L. Rutherford:** 苏珊娜·卢瑟福，美国弗莱德·哈金森癌症研究中心助理研究员，华盛顿大学分子与细胞生物学助理教授。

22. **Susan L. Lindquist:** 苏珊·林奎斯特，麻省理工学院生物学教授，怀特黑德研究所研究员。

23. **the Whitehead Institute for Biomedical Research in Cambridge, Mass.:** 位于美国马萨诸塞州坎布里奇市的怀特黑德生物医学研究所，建立于1982年，是一家顶尖的、非营利性的研究机构，始终站在生物医学领域最前沿，并与麻省理工学院合作办学。

Exercises

## I. Answer the following questions.

1. What are the objectives of the work of HSPs as "chaperones" for other proteins?
2. Why is the HSP100 protein known as an unfoldase?
3. How do heat shock proteins help a cell in a difficult situation?
4. Why are HSPs isolated from a cancer able to immunize against that cancer?
5. What did the author notice while he was testing the efficacy of HSP-peptide complexes in cancer immunotherapy?

## II. The following statements are incomplete. Search the missing information in the passage and fill in the blanks.

1. Those regions where genes were being activated to give rise to their encoded proteins are known as ___________.
2. In the past 10 years scientists have realized that HSPs also play additional roles in ___________, such as human beings.
3. A protein's ability to carry out its intended functions depends on it getting to the right place at the right time and also having the ___________.
4. HSP70 binds directly to short stretches of ___________, also known as peptides.
5. The mammalian immune system recognized proteins from a pathogen as foreign; this is why these proteins act as antigens—provoking ___________.
6. When the HSP70 was put through a final purification step called ATP-affinity chromatography, and the molecule's very potent ___________ was gone!
7. The ability of HSPs derived from tumors or pathogen-infected cells to immunize against those same tumors or intracellular pathogens is based on ___________ of the peptide-binding HSPs.
8. Pharmacological inhibitors of increasing specificity for HSP90 are being tested in ___________, in combination with ___________.

## III. Identify the implied meanings of the underlined parts of the following sentences according to the context of the passage, and translate the sentences into Chinese.

1. When Ferruccio Ritossa, then a young geneticist, examined the cells of *these* "heat shocked" flies, he noticed that their chromosomes had puffed up at discrete locations.
2. To understand how these versatile proteins can be harnessed therapeutically, it is helpful to look at the diverse ways they perform their core job, which is to act as "chaperones" for other proteins.
3. In contrast, HSPs tend to associate with a wide range of "client" proteins, allowing the HSPs to perform a dizzying array of jobs.

4. In contrast to the cagelike chaperones, most HSPs do not enclose their substrates but rather grab them by the "elbows" to help them along.
5. Thus, the HSP-associated peptides represent the "antigenic fingerprint" of the cells or tissues from which they come.
6. The chaperoning of peptides by HSPs is essential for their eventual loading onto MHC I molecules; when the HSPs are chemically silenced, the MHC I molecules remain empty of peptides and cannot be recognized by the T cells.
7. But the HSP-peptide complexes also have another critical part in the T cells' recognition of friend and foe antigens—through their interactions with different types of immune cells known as antigen-presenting cells.
8. They present whatever they encounter to the T cells that will eventually home in on and attempt to destroy cancerous or infected cells.
9. In the trial on renal cancers, the vaccine extended the recurrence-free survival time in some groups of patients by more than a year and a half.
10. Similarly, HSPs are believed to provide buffering against the accumulating mutations that should make cancer cells less and less viable but instead seem to drive their malignancy.

## Translation Techniques (1)

### Translation of "so...that" (so...that句型的翻译方法)

You may be quite familiar with the sentence pattern of "so...that" in English. It is used to indicate the result or the extent to which something has reached, and the basic meaning is "如此……以至". However, accuracy of technical translation and usual ways of Chinese expressions often require changes of expressions on the basis of original versions. The following comparison will show this point clearly:

An atom is so small that we cannot see it.

Translation 1: 原子是如此的小，以至我们不能看见它。

Translation 2: 原子小得我们无法看到。

The second translation is obviously more concise and sounds more natural.

Here are more examples:

1. *So* fast does the electric current pass through conductors *that* it is difficult for us to imagine its speed.

   电流通过导体的速度快得难以想象。

2. Advances in office communications are occurring *so* rapidly *that* what is on the design boards this morning is fact this afternoon and becomes obsolescent tomorrow.

   办公室通信的进展极快，快到今天早上还处在设计之中，下午已成为现实，明天就要过时。

3. Receiver noise temperatures were *so* high *that* the other noise components seemed insignificant.

   接收机的噪声温度极高，相比之下其他噪声成分显得无足轻重了。

4. All these properties are *so* strange *that* most scientists are reluctant to accept the notion of a fifth force.

   由于所有这些性质都十分陌生，因此大多数科学家都不愿接受"第五种力"的概念。

5. Under such conditions the circuit losses at the singing frequency are *so* low *that* oscillation will continue, even after cessation of its original impulse.

   这种情况下，蜂鸣频率上的电路损耗很小，甚至在原始脉冲停止之后，振荡还会继续下去。

Sometimes the relation established by "so...that" is implied in translation without using such linking words as ……得，……到，因此，etc. For example:

1. The scale of this system is *so* large *that* optical fibers 400km long are required.

   该系统规模很大，需要400公里长的光缆。

2. This problem is *so* formidable *that* for systems with a highest frequency above 12MHz to 25MHz, other manufacturers use a double amplifier arrangement, with a separate amplifier for each transmission direction.

   这个问题相当棘手，其他厂家对于12兆赫至25兆赫以上的高频率系统，都采用两个放大器，即每一传输方向用一个单独的放大器。

## IV. Translate the following sentences into Chinese. Pay attention to how the sentence pattern "so...that" should be translated.

1. Fast cooling the hot steel in water gives very hard steels, *so* hard *that* in certain cases they are used for cutting softer steels.
2. The action is *so* fast *that* the dots blend to produce a moving scene of the action that is going on in the television studio.
3. The temperature in the sun is *so* high *that* nothing can exist in solid state.
4. However, since there are many fiber transmission modes, the noise can be assumed to

be averaged and so made *so* small *that* it can be ignored.

5. Reflectors are located *so* closely to each other *that* echoes overlap, resulting in the single continuous echo signal train.
6. The computations are *so* complicated *that* it would take one human computer years to work them out.
7. The layout design for the circuit components installed in the waveguide circuit is *so* difficult *that* exact computation is nearly impossible.
8. The product contains *so* little impurity *that* it is impossible to measure it by an ordinary method.

## Translation Techniques (2)

### Translation of Compounds in Technical Usage
### （科技英语中复合词的翻译）

Compounds are widely used in technical English to name things and discovery, to define phenomena, and to describe processes in scientific studies. They are nouns and adjectives composed of one or more words and many of them can be translated according to their literal meanings. For example, *acid rain* (酸雨), *air filter* (空气过滤器), *carbon fiber* (碳纤维), *download* (下载), *internal combustion engine* (内燃机), *light-year* (光年), *nuclear explosion* (核爆炸), *rock salt* (岩盐), etc.

However, some compounds need more attention because their constructions are different from the Chinese equivalents, and therefore it is necessary to make certain changes when translation is done. The translation techniques in this respect can be put into the following three categories: making up the missing words, reversing the word order, and free translation based on the original meaning.

**1. Making up the missing words**

In some compounds, part of the meaning is "hidden", which is quite normal in English, but when they are translated into Chinese that part should be made explicit by using extra words. For example, *magnetic levitation* (or *maglev*) refers to the kind of train which floats above the track, using magnetism, and therefore it travels at a high speed. Although there is no word "train" in the term, the word should be added in translation to make the meaning complete and comprehensible: 磁（力）悬浮火车.

More examples, *cut-and-cover*随挖随填的明挖施工法；*earth-filled dam*土石坝；

*erase head*消磁头，抹音头；*eye surgery*眼科手术；*fly trap*捕蝇草；*ground station* 卫星地面站.

**2. Reversing the word order**

Sometimes the order of the words should be reversed when English compounds are translated into Chinese so that the exact meanings can be conveyed. For example, *copper sulphate*硫酸铜；*fiberglass*玻璃纤维；*hydrocarbon*碳氢化合物；*hydrogen sulphide*硫化氢；*input*输入；*viewfinder*（照相机的）取景器.

**3. Free translation**

Some compounds are translated on the basis of their meanings because literal translation may confuse the readers. For example, a *greenhouse* is a building with sides and roof of glass used for growing plants that need protection from the cold weather. It is reasonable to call this kind of building "greenhouse" for it can keep the plants green even if it is winter, but if the word is translated directly as 绿房子, the Chinese readers would be confused. Instead, *greenhouse* should be translated as 温室、暖房. Another example is *Milky Way*（银河）. This type of compounds account for a larger proportion compared with the above two categories.

More examples: *athlete's foot*足癣；*bypass surgery*（心脏）搭桥手术；*bone meal* 骨粉（用作肥料或加在动物的饲料中）；*civil engineering*土木工程（学）；*cottage industry*家庭小工业；*freezing point*冰点；*reinforced concrete*钢筋混凝土.

**V. Use your dictionary and find out how the following terms are translated into Chinese. Discuss the translation technique and put the translated terms into the three categories.**

(1) fresh water; (2) general anaesthetic; (3) good conductor; (4) groundwater;
(5) gunpowder; (6) half-life; (7) hay fever; (8) heating;
(9) high explosive; (10) identical twin; (11) jawless fish; (12) lead oxide;
(13) letter press; (14) light-sensitive; (15) liquid hydrogen; (16) mouthpart;
(17) nerve gas; (18) pacemaker; (19) periodic table; (20) pig-iron;
(21) playback; (22) rainfall; (23) rapid-transit system; (24) red giant;
(25) redwood; (26) ROM (Read Only Memory); (27) sand-blast;
(28) satellite dish; (29) seed bank; (30) self-fertilization; (31) sodium chloride;
(32) solar cell; (33) space shuttle; (34) starfish;
(35) surrounded-sound headphone; (36) trial and error; (37) voice box;
(38) warhead; (39) windpipe; (40) bush baby

| Making up the missing words | Reversing the word order | Free translation |
| --- | --- | --- |
| | | |

**VI. Translate the following sentences into Chinese. Pay attention to how the compounds should be translated properly.**

1. ...but when ATP is absent, a *lidlike* structure on HSP70 clamps down on the bound peptide and traps the larger *protein chain* in place.
2. Under emergency conditions, such as extreme heat or cold, *oxygen deprivation*, dehydration or starvation, a cell would be struggling just to survive.
3. The gp96 molecules found in tumors and in normal tissues were identical in their amino acid sequences, so the *cancer-derived* gp96 was not *cancer-specific*.
4. A postdoctoral fellow in my laboratory at the Mount Sinai School of Medicine, and I were isolating HSP70 from tumors to test if it, too, elicited *tumor immunity*.
5. And when HSP70 or HSP90 are taken from cancers or from *virus- or tuberculosis-infected* cells, in nearly all instances they bear peptides derived from *cancer-specific* antigens, viral antigens or tuberculosis antigens.

**VII. Translate the following passage into Chinese.**

US scientists are considering creating a complete catalog of genetic anomalies characteristic of cancer in a new effort to combat the deadly disease, the director of the National Human Genome Institute said. The institute plans to establish the DNA sequence in at least 12,500 samples of tumors, or create 250 genetic maps for each of the 50 most common types of cancer, in order to be able to compare these maps with those of healthy cells. "With the completion of the Human Genome Project in April 2003, it is now possible to identify the complete universe of genes involved in every type of cancer," Human Genome Institute Director Francis Collins said in a statement. "That is the intent of this bold new proposal for a Human Cancer Genome Project. Such an inventory will give researchers powerful new ways to prevent, diagnose and treat every major form of the disease."

While the project "is still in its conceptual stages," his scientists were looking forward to working in partnership with the National Cancer Institute to explore how this project can be implemented. Knowing the defects of the cancer cell "points you to the Achilles' heel of tumors," Eric Lander, director of the Broad Institute, an affiliate of Harvard University and the Massachusetts Institute of Technology, told *The Times*. Lander and Leland Hartwell, a Nobel Prize winner and president of the Fred Hutchinson Cancer Research Center located in Seattle, Washington, presented their cancer genome study project before a consultative committee of the National Cancer Institute in February.

**VIII. Translate the following passage into English.**

### 热应急蛋白

热应急蛋白是一组存在于一切生命形式、所有细胞中的蛋白质，细胞处于不同形式的环境压力（如遇热、遇冷或缺氧）下就会诱发这种蛋白质的产生。

在完全正常的条件下，细胞内也会产生热应急蛋白，这些蛋白质就像“陪护”一样，确保细胞内的蛋白质在合适的时间处于合适的位置、具有合适的形状。例如，它们帮助新生细胞或变形的细胞走上正轨，这对于细胞功能的发挥极其重要。热应急蛋白还将蛋白质从细胞的一个部分转移到另一个部分，或将衰老的蛋白质送入细胞的“垃圾处理厂”。据信它们在细胞表面蛋白质或肽的生成过程中也起着作用，帮助免疫系统辨认不健康的细胞。

几十年前人们就已经了解到动物可以通过“接种疫苗”抗癌，其原理是：肿瘤细胞可以被弱化并像疫苗一样注射到小鼠体内。此后，如果将相同的、已充分发育的肿瘤细胞注入小鼠，小鼠就会对这些肿瘤细胞产生排斥，因此不会患上癌症。然而，如果小鼠没有以上述方式接受免疫，肿瘤细胞就会“生根”，导致小鼠患上癌症。

虽然人们已经清楚地了解到动物可以通过免疫抗癌，但在很长一段时间内对于该过程的原理并不了解。大约25年前，一位名叫Pramod Srivastava的研究生做了一系列实验，他取出若干肿瘤细胞，将它们剖开，并将细胞的各部分分割成块。然后他把每一块当作“疫苗”，以观察哪一块保护小鼠不患癌症。经过多次实验，他发现起到保护作用的那一部分正是热应急蛋白。

## Section 2 Reading for Academic Purposes

*As a graduate student, you may have realized the importance of information collection for the success of your researches. Therefore, the question you ask yourself before you set about the task is most probably what kind of information is needed. This is closely related to the research methodology and methods you are going to adopt. The former is a more general conception referring to the analysis of the principles of methods, rules, and postulates employed by a discipline, and the development of methods to be applied within a discipline, while the latter would be more accurate. For example, "Since students were not available to complete the survey about academic success, we changed our methodology and gathered data from instructors instead." In this instance the methodology (gathering data via surveys, and the assumption that this produces accurate results) did not change, but the method (asking teachers instead of students) did.*

*Most sciences have their own specific methods, which are supported by methodologies (i.e. rationale that supports the method's validity). Although academic researches are methodologically diverse, there are generally two categories: quantitative approach and qualitative approach, depending on the nature of the research, and some researchers adopt the mixed-method approaches, which are the combination of the two.*

*Let's start from quantitative research and the methods involved, which are the focus of the following essay.*

## Understanding Quantitative Research
## 什么是定量研究?

Quantitative research is the systematic scientific investigation of properties and phenomena and their relationships. Quantitative research is widely used in both the natural sciences and social sciences, from physics and biology to sociology and journalism. It is also used as a way to research different aspects of education.

pertain to 属于，关于
empirical *adj.* 经验的，根据经验的

The objective of quantitative research is to develop and employ mathematical models, theories and hypotheses **pertaining to** natural phenomena. The process of measurement is central to quantitative research because it provides the fundamental connection between **empirical** observation and mathematical expression of quantitative relationships.

The term quantitative research is most often used in the social sciences in contrast to qualitative research.

### Overview and background

Quantitative research is generally approached using scientific methods, which include:

1. The generation of models, theories and hypotheses
2. The development of instruments and methods for measurement
3. Experimental control and **manipulation** of **variables**
4. Collection of empirical data

manipulation *n.* 操作，处理
variable *n.* 变量；变数

5. Modeling and analysis of data
6. Evaluation of results

Quantitative research is often an **iterative** process whereby evidence is evaluated, theories and hypotheses are refined, technical advances are made, and so on. Virtually all research in physics is quantitative whereas research in other scientific disciplines, such as taxonomy and **anatomy**, may involve a combination of quantitative and other analytic approaches and methods.

iterative *adj.* 重复的，反复的
anatomy *n.* 解剖学

In the social science particularly, quantitative research is often contrasted with qualitative research which is the examination, analysis and interpretation of observations for the purpose of discovering underlying meanings and patterns of relationships, including classifications of types of phenomena and **entities**, in a manner that does not involve mathematical models. Approaches to quantitative psychology were first modeled on quantitative approaches in the physical sciences by Gustav Fechner in his work on **psychophysics**, which built on the work of Ernst Heinrich Weber. Although a distinction is commonly drawn between qualitative and quantitative aspects of scientific investigation, it has been argued that the two go hand in hand. For example, based on analysis of the history of science, Thomas Kuhn (*The Function of Measurement in Modern Physical Science*, 1961, p. 162) concludes that "large amounts of qualitative work have usually been **prerequisite** to fruitful quantification in the physical sciences". Qualitative research is often used to gain a general sense of phenomena and to form theories that can be tested using further quantitative research. For instance, in the social sciences qualitative research methods are often used to gain better understanding of such things as intentionality (from the speech response of the **researchee**) and meaning (why did this person/group say something and what did it mean to them?).

entities *n.* 实体；本质存在
psychophysics *n.* 精神物理学
prerequisite *n.* 先决条件
researchee *n.* 研究对象

Although quantitative investigation of the world has existed since people first began to record events or objects that had been counted, the modern idea of quantitative processes have their roots in Auguste Comte's positivist framework.

## Statistics in quantitative research

statistical mechanics 统计力学
inferential *adj.* 推论性的
dietary *adj.* 饮食的，有关饮食的
respondent *n.* 应答者，回复者
tabulate *v.* 把……制成表格

Statistics is the most widely used branch of mathematics in quantitative research outside of the physical sciences, and also finds applications within the physical sciences, such as in **statistical mechanics**. Statistical methods are used extensively within fields such as economics, social sciences and biology. Quantitative research using statistical methods typically begins with the collection of data based on a theory or hypothesis, followed by the application of descriptive or **inferential** statistical methods. Causal relationships are studied by manipulating factors thought to influence the phenomena of interest while controlling other variables relevant to the experimental outcomes. In the field of health, for example, researchers might measure and study the relationship between **dietary** intake and measurable physiological effects such as weight loss, controlling for other key variables such as exercise. Quantitatively based opinion surveys are widely used in the media, with statistics such as the proportion of **respondents** in favor of a position commonly reported. In opinion surveys, respondents are asked a set of structured questions and their responses are **tabulated**. In the field of climate science, researchers compile and compare statistics such as temperature or atmospheric concentrations of carbon dioxide.

non-linear *adj.* 非线性的
factor analysis 因子分析
causation *n.* 原因，发生因果关系
spurious *adj.* 不真实的；伪造的

Empirical relationships and associations are also frequently studied by using some form of general linear model, **non-linear** model, or by using **factor analysis**. A fundamental principle in quantitative research is that correlation does not imply **causation**. This principle follows from the fact that it is always possible a **spurious** relationship exists for variables between which covariance is found in some degree. Associations may be examined between any combination of continuous and categorical variables using methods of statistics.

## Measurement in quantitative research

divergent *adj.* 有分歧的，偏离的

Views regarding the role of measurement in quantitative research are somewhat **divergent**. Measurement is often regarded as being only a means by which observations are expressed numerically in order to investigate causal relations or associations. However, it has been argued that measurement often plays a more important role in quantitative research. For example, Kuhn (1961) argued that results which appear anomalous in the context of accepted theory potentially lead to the **genesis** of a search for a new, natural phenomenon. He believed that such anomalies are most striking when encountered during the process of obtaining measurements, as reflected in the following observations regarding the function of measurement in science:

genesis *n.* 起源，生成

> *When measurement departs from theory, it is likely to yield mere numbers, and their very **neutrality** makes them particularly **sterile** as a source of remedial suggestions. But numbers register the departure from theory with an authority and **finesse** that no qualitative technique can duplicate, and that departure is often enough to start a search (Kuhn, 1961, p. 180).*

neutrality *n.* 中性，中立
sterile *adj.* 没有结果的
finesse *n.*策略；技巧

In classical physics, the theory and definitions which **underpin** measurement are generally **deterministic** in nature. In contrast, **probabilistic** measurement models known as the Rasch model and Item response theory models are generally employed in the social sciences. **Psychometrics** is the field of study concerned with the theory and technique for measuring social and psychological attributes and phenomena. This field is central to much quantitative research that is undertaken within the social sciences.

underpin *v.* 加强……基础；支持，巩固
deterministic *adj.* 确定性的
probabilistic *adj.* 可能性的
psychometrics *n.* 心理测验学

Quantitative research may involve the use of **proxies** as **stand-ins** for other quantities that cannot be directly measured. Tree-ring width, for example, is considered a reliable proxy of **ambient** environmental conditions such as the warmth of growing seasons or amount of rainfall. Although scientists cannot directly measure the temperature of past years, tree-ring width and other climate proxies have been used to provide a semi-quantitative record of average temperature in the Northern **Hemisphere** back to 1000 AD. When used in this way, the proxy record (tree-ring width, say) only reconstructs a certain amount of the variance of the original record. The proxy may be **calibrated** (for example, during the period of the instrumental record) to determine how much variation is captured, including whether both short- and long-term variation is revealed. In the case of tree-ring width, different species in different places may show more or less **sensitivity** to, say, rainfall or temperature: when reconstructing a temperature record there is considerable skill in selecting proxies that are well correlated with

proxy *n.* 代理
stand-in *n.* 代替品
ambient *adj.* 周边的
hemisphere *n.* 半球
calibrate *v.* 校准
sensitivity *n.* 敏感，灵敏度

the desired variable.

## Quantitative methods

Quantitative methods are research methods dealing with numbers and anything that is measurable. They are therefore to be distinguished from qualitative methods.

Counting and measuring are common forms of quantitative methods. The result of the research is a number, or a series of numbers. These are often presented in tables, graphs or other forms of statistics.

uncontroversial *adj.* 无可争议的
controversy *n.* 争议，论争
ideology *n.* 意识形态
scorn *n.* 轻蔑，不屑
obscure *v.* 使暗淡，使模糊

In most physical and biological sciences, the use of either quantitative or qualitative methods is **uncontroversial**, and each is used when appropriate. In the social sciences, particularly in sociology, social anthropology and psychology, the use of one or the other type of method has become a matter of **controversy** and even **ideology**, with particular schools of thought within each discipline favoring one type of method and pouring **scorn** on to the other. Advocates of quantitative methods argue that only by using such methods can the social sciences become truly scientific; advocates of qualitative methods argue that quantitative methods tend to **obscure** the reality of the social phenomena under study because they underestimate or neglect the non-measurable factors, which may be the most important.

eclectic *adj.* 折中的，折中主义的

The modern tendency (and in reality the majority tendency throughout the history of social sciences) is to use **eclectic** approaches. Quantitative methods might be used with a global qualitative frame. Qualitative methods might be used to understand the meaning of the numbers produced by quantitative methods. Using quantitative methods, it is possible to give precise and testable expression to qualitative ideas. This combination of quantitative and qualitative data gathering is often referred to as mixed-methods research.

## Exercises

**I. Read the passage and decide whether the following statements are true or false. Write T for True and F for False in the brackets.**

1. Quantitative research is widely used in the natural sciences while qualitative research used in the social sciences. (  )

2. Almost all research in physics is quantitative rather than qualitative. ( )
3. The purpose of quantitative research in the social sciences is to discover underlying meanings and patterns of relationships. ( )
4. Gustav Fechner first modeled approaches to quantitative psychology on those in the physical sciences. ( )
5. Auguste Comte's positivist framework initiated the modern idea of quantitative processes. ( )
6. Qualitatively based opinion surveys are widely used in the media. ( )
7. It is a basic principle in quantitative research that correlation does not necessarily mean causation. ( )
8. Kuhn maintained that measurement is very important in quantitative research. ( )
9. Tree-ring width is considered an unreliable proxy of ambient environmental conditions. ( )
10. Those who favor qualitative method in social sciences may deem certain non-measurable factors as crucial to their research. ( )

**II. Read the passage again, and complete the following items.**

1. The general definition of quantitative research: __________
2. The objective of quantitative research: __________
3. The reason why the process of measurements is central to quantitative research: __________
4. The relationship between qualitative and quantitative research in the physical sciences according to Thomas Kuhn: __________
5. The most widely used branch of mathematics in quantitative research outside of the physical sciences: __________
6. A fundamental principle in quantitative research is: __________
7. Psychometrics can be defined as: __________
8. Despite the controversy over the use of quantitative or qualitative methods in the social sciences, the modern tendency is: __________

**Quotations from Great Scientists**

*In the fields of observation chance favors only the prepared mind.*

*—Louis Pasteur*

在科学观察领域，机会只会光顾那些有准备的人。

——路易·巴斯德

# Chapter 10 Particle Physics 粒子物理学

## Section 1 Reading and Translation

### I What You Are Going to Read

The Greeks thought the atom was the smallest thing in the universe; the atom was indivisible. Scientists have since smashed that assumption to pieces—literally. In the past century physicists have discovered hundreds of particles more minute than an atom (subatomic particles). Those that cannot be further broken down are called fundamental, or elementary, particles. For example, matter is composed of molecules that are made up of atoms that are made up of protons, neutrons, and electrons. While protons and neutrons can be broken down into fundamental particles known as quarks and gluons, electrons are themselves fundamental—at least for now. It's always possible that as physicists deepen their understanding of the universe and wield more powerful technology they will discover an even tinier unit underlying the universe.

The following paper published in *Scientific American* (February 2008) will introduce you to the coming revolutions in particle physics.

### II About the Author

Chris Quigg is a senior scientist at Fermi National Accelerator Laboratory, where for 10 years he led the Theoretical Physics Department. He is the author of a celebrated textbook on the so-called gauge theories that underlie the Standard Model, as well as the former editor of the *Annual Review of Nuclear and Particle Science*. Quigg's research on electroweak symmetry breaking and supercollider physics highlighted the importance of the terascale. When not blazing the trail to the deepest workings of nature, he can be found hiking on one of France's Sentiers de Grande Randonne.

### III Reading Passage

**The Coming Revolutions in Particle Physics**

**粒子物理学的革命**

By Chris Quigg

*The current standard model[1] of particle physics[2] begins to unravel when probed much beyond the range of current particle accelerators. So no matter what the Large Hadron Collider[3] finds, it is going to take physics into new territory.*

1 When physicists are forced to give a single-word answer to the question of why we are building the Large Hadron Collider (LHC), we usually reply "Higgs". The Higgs particle[4]—the last remaining undiscovered piece of our current theory of matter—is the **marquee attraction**. But the full story is much more interesting. The new collider provides the greatest leap in capability of any instrument in the history of particle physics. We do not know what it will find, but the discoveries we make and the new puzzles we encounter are certain to change the face of particle physics and to echo through neighboring sciences.

marquee attraction 特别吸引人的东西

2 In this new world, we expect to learn what distinguishes two of the forces of nature—**electromagnetism** and the weak interactions—with broad implications for our conception of the everyday world. We will gain a new understanding of simple and profound questions: Why are there atoms? Why chemistry? What makes stable structures possible?

electromagnetism *n.* 电磁学；电磁性

3 The search for the Higgs particle is a **pivotal** step, but only the first step. Beyond it lie phenomena that may clarify why gravity is so much weaker than the other forces of nature and that could reveal what the unknown dark matter that fills the universe is.[5] Even deeper lies the prospect of insights into the different forms of matter, the unity of outwardly distinct particle categories and the nature of spacetime. The questions in play all seem linked to one another and to the knot of problems that motivated the prediction of the Higgs particle to begin with.[6] The LHC will help us refine these questions and will set us on the road to answering them.

pivotal *adj.* 关键的，中枢的

## The Matter at Hand

4 What physicists call the "Standard Model" of particle physics, to indicate that it is still a work in progress, can explain much about the known world. The main elements of the Standard Model fell into place during the **heady** days of the 1970s and 1980s, when waves of landmark experimental discoveries engaged emerging theoretical ideas in productive conversation. Many particle physicists look on

heady *adj.* 振奋人心的；令人兴奋的

ferment *n.* 纷扰；不安
purview *n.* 范围；界限
brew *v.* 酝酿；策划

the past 15 years as an era of consolidation in contrast to the **ferment** of earlier decades. Yet even as the Standard Model has gained ever more experimental support, a growing list of phenomena lies outside its **purview**, and new theoretical ideas have expanded our conception of what a richer and more comprehensive worldview might look like. Taken together, the continuing progress in experiment and theory point to a very lively decade ahead. Perhaps we will look back and see that revolution had been **brewing** all along.

quark *n.* 【核】夸克
neutron *n.* 【核】中子
akin *adj.* [常作表语] 同类的，相似的
metaphorical *adj.* 隐喻性的；比喻性的

5 Our current conception of matter comprises two main particle categories, **quarks** and leptons[7], together with three of the four known fundamental forces, electromagnetism and the strong and weak interactions. Gravity is, for the moment, left to the side. Quarks, which make up protons and **neutrons**, generate and feel all three forces. Leptons, the best known of which is the electron, are immune to the strong force. What distinguishes these two categories is a property **akin** to electric charge, called color[8]. (This name is **metaphorical**; it has nothing to do with ordinary colors.) Quarks have color, and leptons do not.

symmetrical *adj.* 对称的
sphere *n.* 球面

6 The guiding principle of the Standard Model is that its equations are **symmetrical**. Just as a **sphere** looks the same whatever your viewing angle is, the equations remain unchanged even when you change the perspective from which they are defined. Moreover, they remain unchanged even when the perspective shifts by different amounts at different points in space and time.

geometric *adj.* 几何的，几何学的
beget *v.* 产生；引起

7 Ensuring the symmetry of a **geometric** object places very tight constraints on its shape. A sphere with a bump no longer looks the same from every angle. Likewise, the symmetry of the equations places very tight constraints on them. These symmetries **beget** forces that are carried by special particles called bosons[9].

dictum *n.* 名言；宣言
dictate *v.* 规定；控制；支配

8 In this way, the Standard Model inverts Louis Sullivan's[10] architectural **dictum**: instead of "form follows function", function follows form. That is, the form of the theory, expressed in the symmetry of the equations that define it, **dictates** the function—the interactions among particles—that the theory describes. For instance, the strong nuclear force follows from the requirement that the equations describing quarks must be the same no matter how

one chooses to define quark colors (and even if this convention is set independently at each point in space and time). The strong force is carried by eight particles known as gluons[11]. The other two forces, electromagnetism and the weak nuclear force, fall under the **rubric** of the "**electroweak**" forces and are based on a different symmetry. The electroweak forces are carried by a **quartet** of particles: the photon, Z boson, W+ boson and W– boson.

rubric *n.* 成规，成例
electroweak *n.* 弱电
quartet *n.* 四件一套，四个一组

## Breaking the Mirror

9 The theory of the electroweak forces was formulated by Sheldon Glashow, Steven Weinberg and Abdus Salam[12], who won the 1979 Nobel Prize in Physics for their efforts. The weak force, which is involved in radioactive beta **decay**, does not act on all the quarks and leptons. Each of these particles comes in mirror-image varieties, termed left-handed and right-handed, and the beta-decay force acts only on the left-handed ones—a striking fact still unexplained 50 years after its discovery.[13] The family symmetry among the left-handed particles helps to define the electroweak theory.

decay *n.* 【原】(放射性物质的)衰变，蜕变

10 In the initial stages of its construction, the theory had two essential shortcomings. First, it foresaw four long-range force particles—referred to as gauge bosons[14]—whereas nature has but one: the photon[15]. The other three have a short range, less than about 10–17 meters, less than 1 percent of the proton's **radius**. According to Heisenberg's uncertainty principle[16], this limited range implies that the force particles must have a mass approaching 100 billion electron volts (GeV). The second shortcoming is that the family symmetry does not permit masses for the quarks and leptons, yet these particles do have mass.

radius *n.* 【物】辐射线

11 The way out of this unsatisfactory situation is to recognize that a symmetry of the laws of nature need not be reflected in the outcome of those laws. Physicists say that the symmetry is "broken". The needed theoretical apparatus was worked out in the mid-1960s by physicists Peter Higgs, Robert Brout, François Englert[17] and others. The inspiration came from a seemingly unrelated phenomenon: **superconductivity**, in which certain materials carry electric current with zero resistance at low temperatures. Although the laws of electromagnetism themselves are symmetrical, the behavior of

superconductivity *n.* 【物】超导电性

electromagnetism within the superconducting material is not. A photon gains mass within a superconductor, thereby limiting the intrusion of magnetic fields into the material.

prototype *n.* 原型

12 As it turns out, this phenomenon is a perfect **prototype** for the electroweak theory. If space is filled with a type of "superconductor" that affects the weak interaction rather than electromagnetism, it gives mass to the W and Z bosons and limits the range of the weak interactions. This superconductor consists of particles called Higgs bosons. The quarks and leptons also acquire their mass through their interactions with the Higgs boson. By obtaining mass in this way, instead of possessing it intrinsically, these particles remain consistent with the symmetry requirements of the weak force.

constituent *n.* 成分，要素

13 The modern electroweak theory (with the Higgs) accounts very precisely for a broad range of experimental results. Indeed, the paradigm of quark and lepton **constituents** interacting by means of gauge bosons completely revised our conception of matter and pointed to the possibility that the strong, weak and electromagnetic interactions meld into one when the particles are given very high energies. The electroweak theory is a stunning conceptual achievement, but it is still incomplete. It shows how the quarks and leptons might acquire masses but does not predict what those masses should be. The electroweak theory is similarly indefinite in regard to the mass of the Higgs boson itself: the existence of the particle is essential, but the theory does not predict its mass. Many of the outstanding problems of particle physics and cosmology are linked to the question of exactly how the electroweak symmetry is broken.

## Where the Standard Model Tells Its Tale

Ill. *n.* (=Illinois)（美国）伊利诺伊州

14 Encouraged by a string of promising observations in the 1970s, theorists began to take the Standard Model seriously enough to begin to probe its limits. Toward the end of 1976 Benjamin W. Lee[18] of Fermi National Accelerator Laboratory[19] in Batavia, **Ill.**, Harry B. Thacker[20], now at the University of Virginia, and I devised a thought experiment to investigate how the electroweak forces would behave at very high energies. We imagined collisions among pairs of W, Z and Higgs bosons. The exercise might seem slightly fanciful because, at the time of our work, not one of these particles had been observed.

But physicists have an obligation to test any theory by considering its implications as if all its elements were real.

15 What we noticed was a subtle **interplay** among the forces generated by these particles. Extended to very high energies, our calculations made sense only if the mass of the Higgs boson were not too large—the equivalent of less than one trillion electron volts, or 1 TeV. If the Higgs is lighter than 1 TeV, weak interactions remain feeble and the theory works reliably at all energies. If the Higgs is heavier than 1 TeV, the weak interactions strengthen near that energy scale and all manner of **exotic particle** processes ensue. Finding a condition of this kind is interesting because the electroweak theory does not directly predict the Higgs mass. This mass threshold means, among other things, that something new—either a Higgs boson or other novel phenomena—is to be found when the LHC turns the thought experiment into a real one.

interplay *n.* 相互影响
exotic particle【核】奇异粒子

16 Experiments may already have observed the behind-the-scenes influence of the Higgs. This effect is another consequence of the uncertainty principle, which implies that particles such as the Higgs can exist for moments too fleeting to be observed directly but long enough to leave a subtle mark on particle processes. The Large Electron Positron Collider[21] at CERN[22], the previous inhabitant of the tunnel now used by the LHC, detected the work of such an unseen hand. Comparison of precise measurements with theory strongly hints that the Higgs exists and has a mass less than about 192 GeV.

17 For the Higgs to weigh less than 1 TeV, as required, poses an interesting riddle. In quantum theory, quantities such as mass are not set once and for all but are modified by quantum effects. Just as the Higgs can exert a behind-the-scenes influence on other particles, other particles can do the same to the Higgs. Those particles come in a range of energies, and their net effect depends on where precisely the Standard Model gives way to a deeper theory. If the model holds all the way to 1015 GeV, where the strong and electroweak interactions appear to unify, particles with truly **titanic** energies act on the Higgs and give it a comparably high mass. Why, then, does the Higgs appear to have a mass of no more than 1 TeV?

titanic *adj.* 巨大的，极大的

18 This tension is known as the **hierarchy** problem. One resolution

hierarchy *n.* 层次；层级

precarious *adj*. 无法保证的；不确定的；未证实的
contend *v*. 争论；竞争
cancellation *n*. 取消；撤销
mandate *v*. 命令；授权

would be a **precarious** balance of additions and subtractions of large numbers, standing for the **contending** contributions of different particles. Physicists have learned to be suspicious of immensely precise **cancellations** that are not **mandated** by deeper principles. Accordingly, in common with many of my colleagues, I think it highly likely that both the Higgs boson and other new phenomena will be found with the LHC.

## Supertechnifragilisticexpialidocious[23]

superpartner *n*. 超对称粒子，超伴子（超对称假说中的一种粒子）

19 Theorists have explored many ways in which new phenomena could resolve the hierarchy problem. A leading contender known as supersymmetry[24] supposes that every particle has an as yet unseen **superpartner** that differs in spin. If nature were exactly supersymmetric, the masses of particles and superpartners would be identical, and their influences on the Higgs would cancel each other out exactly. In that case, though, physicists would have seen the superpartners by now. We have not, so if supersymmetry exists, it must be a broken symmetry. The net influence on the Higgs could still be acceptably small if superpartner masses were less than about 1 TeV, which would put them within the LHC's reach.

technicolor *n*.【核】人工色（理论）
allude *v*. 暗指；提及（*to*）
veritable *adj*. 真正的；确实的
menagerie *n*. 各种各样的一组人或物

20 Another option, called **technicolor**, supposes that the Higgs boson is not truly a fundamental particle but is built out of as yet unobserved constituents. (The term "technicolor" **alludes** to a generalization of the color charge that defines the strong force.) If so, the Higgs is not fundamental. Collisions at energies around 1 TeV (the energy associated with the force that binds together the Higgs) would allow us to look within it and thus reveal its composite nature. Like supersymmetry, technicolor implies that the LHC will set free a **veritable menagerie** of exotic particles.

21 A third, highly provocative idea is that the hierarchy problem will go away on closer examination, because space has additional dimensions beyond the three that we move around in. Extra dimensions might modify how the forces vary in strength with energy and eventually meld together. Then the melding—and the onset of new physics—might not happen at 1012 TeV but at a much lower energy related to the size of the extra dimensions, perhaps only a few TeV. If so, the LHC could offer a peek into those extra dimensions.

22 One more piece of evidence points to new phenomena on the TeV scale. The dark matter that makes up the bulk of the material content of the universe appears to be a novel type of particle. If this particle interacts with the strength of the weak force, then the Big Bang would have produced it in the **requisite** numbers as long as its mass lies between approximately 100 GeV and 1 TeV. Whatever resolves the hierarchy problem will probably suggest a candidate for the dark matter particle.

requisite *adj.* 必不可少的，必备的

## Revolutions on the Horizon

23 Opening the TeV scale to exploration means entering a new world of experimental physics. Making a thorough exploration of this world—where we will **come to terms with** electroweak symmetry breaking, the hierarchy problem and dark matter—is the top priority for accelerator experiments. The goals are well motivated and matched by our experimental tools, with the LHC succeeding the current workhorse, Fermilab's **Tevatron** collider. The answers will not only be satisfying for particle physics, they will deepen our understanding of the everyday world.

come to terms with 使自己顺从（或接受）某事物
tevatron *n.* 一万亿电子伏加速器

24 But these expectations, high as they are, are still not the end of the story. The LHC could well find clues to the full unification of forces or indications that the particle masses follow a rational pattern. Any proposed interpretation of new particles will have consequences for rare decays of the particles we already know. It is very likely that lifting the electroweak veil will bring these problems into clearer relief, change the way we think about them and inspire future experimental **thrusts**.

thrust *n.* 要点，要旨；目标

25 Cecil Powell[25] won the 1950 Nobel Prize in Physics for discovering particles called pions[26]—proposed in 1935 by physicist Hideki Yukawa[27] to account for nuclear forces—by exposing highly sensitive photographic **emulsions** to cosmic rays on a high mountain. He later **reminisced**: "When the emulsions were recovered and developed in Bristol, it was immediately apparent that a whole new world had been revealed... It was as if, suddenly, we had broken into a walled orchard, where protected trees had flourished and all kinds of exotic fruits had ripened in great **profusion**." That is just how I imagine our first look at the TeV scale.

emulsion *n.* 感光乳剂
reminisce *v.* 回忆
profusion *n.* 丰富，大量

Notes

1. **standard model:** 标准模型，一套描述强作用力、弱作用力及电磁力这三种基本力以及组成所有物质的基本粒子的理论。
2. **particle physics:** 粒子物理学，又称高能物理学，研究比原子核更深层次的微观世界中物质的结构、性质，以及这些物质在高能量下相互转化及其原因和规律的物理学分支。
3. **Large Hadron Collider:** 大型强子对撞机（LHC），是一台粒子加速器，位于瑞士日内瓦的欧洲粒子物理实验室。
4. **the Higgs particle:** 希格斯粒子，又称希格斯玻色子（Higgs Boson），粒子物理学标准模型预言的一种自旋为零的玻色子，至今尚未在实验中观察到。它也是标准模型中唯一一种未被发现的粒子。
5. **Beyond it lie phenomena that may clarify why gravity is so much weaker than the other forces of nature and that could reveal what the unknown dark matter that fills the universe is:** 除此以外，还要研究能够说明为什么引力比其他自然力弱得多的现象，研究有可能揭示宇宙中无处不在的暗物质到底是什么，人们对此仍不了解。由于本句的主语phenomena带有一个比较长的定语从句，为了使句子保持平衡，作者使用了倒装句，把介词短语beyond it和谓语动词lie置于主语的前面，使用倒装句还可以更好地与上一句连接。
6. **The questions in play all seem linked to one another and to the knot of problems that motivated the prediction of the Higgs particle to begin with:** 这方面的问题似乎全都相互关联，同时又与一系列难题有关，正是这些难题最初促使人们对希格斯粒子进行预测。短语in play原指"(足球、板球等）按规则处于可比赛的状态中"，在本句中用来修饰the questions，说明这些问题属于上文提到的研究范畴。
7. **lepton:** 轻子。一种参与微弱互反应的基本粒子，包括电子、介子以及与它们相关的中子。
8. **color:** 这里指物理学中的"颜色"，用以确定夸克在强烈反应中作用的量的属性。
9. **boson:** 玻色子。指物理学中具有零或整数自旋状态、并且任意数目的相同粒子占据相同量子状态的统计学规则的粒子，如光子、π介子或α粒子。
10. **Louis Sullivan:** 路易斯·沙里文，芝加哥著名建筑师，现代主义建筑之父，有"摩天楼之父"之称。
11. **gluon:** 胶子。假设的无质量的中性基本粒子之一，被认为能传递一种把夸克结合在一起的强大的相互作用力。
12. **Sheldon Glashow, Steven Weinberg, Abdus Salam:** 谢尔登·格拉肖博士（美国），斯蒂芬·温伯格博士（美国），阿布杜斯·萨拉姆（巴基斯坦），他们对基本粒子间弱相互作用和电磁作用的统一理论做出贡献，并预言弱中性流的存在，1979年共同获得诺贝尔物理学奖。
13. **Each of these particles comes in mirror-image varieties, termed left-handed and right-handed, and the beta-decay force acts only on the left-handed ones—a striking fact still unexplained 50 years after its discovery:** 这些粒子每一个都以镜像的方式存在，被称作"左手型"与"右手型"，β衰变力只对左手型粒子产生影响，50年前人们就发现了这种现象，但至今仍无法解释。本句中的短语act on意为"对……起作用"。
14. **gauge boson:** 规范玻色子，传递基本相互作用的媒介粒子，它们的自旋都为整数，属于玻色子，在粒子物理学的标准模型内都是基本粒子。

15. **photon:** 光子，光量子。电磁能的量子，一般认为是零质量、无电荷和不定长寿命的离散性粒子。
16. **uncertainty principle:** 测不准原理（亦称“不定性原理”或“不确定原理”），由量子力学创始人海森堡（Heisenberg）提出。该原理揭示了微观粒子运动的基本规律，即粒子在客观上不能同时具有确定的坐标位置及相应的动量。
17. **Peter Higgs, Robert Brout, François Englert:** 彼得·希格斯，英国著名物理家，希格斯玻色子（Higgs boson）之父；罗伯特·布鲁特，比利时著名物理学家；弗朗索瓦·艾格勒特，比利时著名物理学家。他们共同获得1997年欧洲物理学会“高能与粒子物理学奖”和2004年度“沃尔夫物理奖”。
18. **Benjamin W. Lee:** 本杰明·李（or Ben Lee, 1935–1977），著名韩裔美国理论物理学家。
19. **Fermi National Accelerator Laboratory:** (=Fermilab or FNAL)（美国）费米国家加速器实验室，简称费米国家实验室或费米实验室。成立于1967年，隶属于美国能源部，主要研究领域为高能物理学和粒子物理学。
20. **Harry B. Thacker:** 哈里·撒克，弗吉尼亚大学教授，美国著名物理学家。
21. **Large Electron Positron Collider:** （LEP）大型正负电子对撞机。
22. **CERN:** 此为法语缩略词，英译为 European Organization for Nuclear Research 欧洲核子研究委员会，成立于1954年，是世界最大的粒子物理研究中心。CERN建有世界上最大的正负电子对撞机和超级质子同步加速器。
23. **supertechnifragilisticexpialidocious:** 该词是作者根据supercalifragilisticexpialidocious 一词杜撰的词，原词源自电影《欢乐满人间》（*Mary Poppins*）的插曲，为儿童用语或口语，意为“奇妙的”，“难以置信的”。本文作者将其中supercali几个字母改为supertechni，指下文中提到的supersymmetry，technicolor等通过高科技手段发现的物理现象层出不穷，令人难以置信。
24. **supersymmetry:** 超对称性，联系费密子与玻子、万有引力与亚原子作用力的假想对称性。
25. **Cecil Powell:** 塞西尔·鲍威尔（1903–1969），英国布利斯托大学教授，著名物理学家，1950年获得诺贝尔物理学奖，以表彰他发现研究核过程的光学方法，并用这一方法作出有关介子的发现。
26. **pions:** (=pi-meson）π介子。在各种介子中，π介子是最轻且最重要的介子。关于自由空间中π介子的结构与性质、核介质内π介子的性质、π-核子相互作用与π-核相互作用等问题始终受到人们的关注。
27. **Hideki Yukawa:** 汤川秀树（1907–1981），日本著名理论物理学家。1935年，汤川秀树在物理学界普遍不接受新粒子的情况下大胆提出一种新的核力场理论，认为存在强相互作用的π介子，介子理论的提出推动了核物理研究的发展。1949年他获得了诺贝尔物理学奖。

Exercises

## I. Answer the following questions.

1. Why did scientists build the Large Hadron Collider?
2. What can distinguish the two main particle categories, quarks and leptons?
3. What is the guiding principle of the Standard Model?
4. What is the difference between the Standard Model and Louis Sullivan's architectural dictum?
5. How does photon limit the intrusion of magnetic fields into the material?

## II. The following statements are incomplete. Search the missing information in the passage and fill in the blanks.

1. In this new world, we expect to learn what distinguishes two of the forces of nature (__________ and __________) with broad implications for our conception of the everyday world.
2. Our current conception of matter comprises two main particle categories, __________ and __________, together with three of the four known fundamental forces, __________ and the strong and weak __________.
3. In contrast to the chaos of earlier decades, many particle physicists regard the past 15 years as a period of __________.
4. Although many experiments support __________, many phenomena are still outside its purview.
5. The __________ of the Standard Model remain unchanged just like a sphere looks the same whatever the perspective is.
6. The phenomenon that certain materials carry electric current with zero resistance at low temperatures is called __________.
7. In quantum theory, quantities of mass are not set one time and for all time but are modified by __________.
8. The Higgs mass threshold means that something new is to be found when the LHC turns the __________ into a real one.

## III. Identify the implied meanings of the underlined parts of the following sentences according to the context of the passage, and translate the sentences into Chinese.

1. The Higgs particle—the last remaining undiscovered piece of our current theory of matter—is the marquee attraction.
2. We do not know what it will find, but the discoveries we make and the new puzzles we encounter are certain to change the face of particle physics and to echo through

neighboring sciences.

3. The LHC will help us refine these questions and will set us on the road to answering them.
4. The main elements of the Standard Model fell into place during the heady days of the 1970s and 1980s, when waves of landmark experimental discoveries engaged emerging theoretical ideas in productive conversation.
5. Perhaps we will look back and see that revolution had been brewing all along.
6. Leptons, the best known of which is the electron, are immune to the strong force.
7. The other two forces, electromagnetism and the weak nuclear force, fall under the rubric of the "electroweak" forces and are based on a different symmetry.
8. The Large Electron Positron Collider at CERN, the previous inhabitant of the tunnel now used by the LHC, detected the work of such an unseen hand.
9. But these expectations, high as they are, are still not the end of the story.
10. It is very likely that lifting the electroweak veil will bring these problems into clearer relief, change the way we think about them and inspire future experimental thrusts.

## Translation Techniques (1)

### Division（拆译法）

Division is a technique in translation, meaning the necessary splitting of a long and complex English sentence into shorter sentences in Chinese.

In order to explain the complicated phenomena thoroughly and precisely, scientists and professionals often adopt long and complex sentences in their English writing, which contain a certain number of phrases and clauses connected logically by linking words. In contrast, the sentences in Chinese tend to be shorter, and therefore it is necessary, or inevitable, to make divisions and rearrangements when the English sentences are translated into Chinese.

Obviously, whether the division technique is used appropriately or not depends on whether the translator comprehends the original thoroughly and correctly. Meanwhile, careful analysis of the logical relations between different parts of long and complex sentences and the consideration of the contexts are needed so that the Chinese translation is accurate and smooth.

Last but not the least, successful translation of long and complex sentences is also based on the combination of the techniques discussed in previous chapters of this book, such as amplification, omission, conversion, and extension, etc.

***Example 1***

Confidence that windtunnels are expected to continue to contribute such advances is evidenced by the construction of the new large subsonic tunnel capabilities in Europe and in America, the construction of the cryogenic National Transonic Facility in America, and the planning for the European Transonic Windtunnel.

*Analysis:*

(1) The main clause: Confidence...is evidenced by...（信念被……证实）

(2) The noun clause that explains the word “confidence”: that windtunnels are expected to continue to contribute such advances（风洞将继续为这些进展做出贡献）

(3) The three phrases following “by”:

the construction of the new large subsonic tunnel capabilities in Europe and in America（新的大型亚音速风洞在欧美的建造）

the construction of the cryogenic National Transonic Facility in America（低温国家跨声速设备在美国的建造）

the planning for the European Transonic Windtunnel（在欧洲建造跨声速风洞的计划）

*Rearrangement of the sentence in translation:*

欧美建造新的大型亚音速风洞，美国建造低温国家跨声速设备，以及计划建造欧洲跨声速风洞等事实都证实了人们的这样一种信念：风洞将继续为这些进展做出贡献。

***Example 2***

Closely associated with heat-transfer measurement is the problem of detecting the state of the boundary layer: it is possible to go some way toward the solution of this problem by the use of static and pitot comb pressure measurements, but these are limited in scope and application, and the development of more elegant methods is one of the primary tracks now in hand.

*Analysis:*

(1) Inverted structure: Closely associated with...is...（与传热测量问题密切相关的是测定附面层的状态）

(2) “It is...to...” structure: it is possible to go some way towards the solution of this problem（可能促进这个问题的解决）

(3) Prepositional phrase: by the use of static and pitot comb pressure measurements（利用总压排管和静压排管测量）

(4) Parallel structure:

these are...（这类压力测量的范围和应用受到限制）

and the development of...（研究出更为精巧的方法成为目前的主要任务）

*Rearrangement of the sentence in translation:*

与传热测量问题密切相关的是测定附面层的状态：利用总压排管和静压排管测量，可能使这个问题的解决前进一步。但是，由于这类压力测量的范围和应用都受到限制，因此研究出更为精巧的方法就成为目前的主要任务。

## IV. Translate the following sentences into Chinese. Pay attention to how the translation technique of division should be used.

1. Yet even as the Standard Model has gained ever more experimental support, a growing list of phenomena lies outside its purview, and new theoretical ideas have expanded our conception of what a richer and more comprehensive worldview might look like.
2. The inspiration came from a seemingly unrelated phenomenon: superconductivity, in which certain materials carry electric current with zero resistance at low temperatures.
3. Indeed, the paradigm of quark and lepton constituents interacting by means of gauge bosons completely revised our conception of matter and pointed to the possibility that the strong, weak and electromagnetic interactions meld into one when the particles are given very high energies.
4. This effect is another consequence of the uncertainty principle, which implies that particles such as the Higgs can exist for moments too fleeting to be observed directly but long enough to leave a subtle mark on particle processes.
5. If the model holds all the way to 1015 GeV, where the strong and electroweak interactions appear to unify, particles with truly titanic energies act on the Higgs and give it a comparably high mass.
6. If nature were exactly supersymmetric, the masses of particles and superpartners would be identical, and their influences on the Higgs would cancel each other out exactly.
7. If this particle interacts with the strength of the weak force, then the Big Bang would have produced it in the requisite numbers as long as its mass lies between approximately 100 GeV and 1 TeV.
8. Cecil Powell won the 1950 Nobel Prize in Physics for discovering particles called pions—proposed in 1935 by physicist Hideki Yukawa to account for nuclear forces—by exposing highly sensitive photographic emulsions to cosmic rays on a high mountain.

## Translation Techniques (2)

### Translation of Figurative Extensions（比喻的翻译）

Words are the basic units of a language which are indispensable for expressing the meanings. Although colorful language is not used often in technical writings owing to the seriousness of the subjects discussed in scientific circle, it does not necessarily mean that the technical language is always dry and direct. Take the following sentence for example: "The beauty of lasers is that they can do machining without ever physically touching the material." The word "beauty" is used figuratively to show the merit of laser. However, when we translate the sentence into Chinese, we should avoid the word-for-word approach as the expression美丽 is obviously inappropriate in this context, and therefore it is necessary to extend the meaning of the word so that readers can understand what the word intends to express. The sentence may be translated as 激光的妙处就在于它能进行机械加工而不必与加工的材料进行物理上的接触.

The following are more examples:

1. Materials science—once the least *sexy* technology—is bursting with new, practical discoveries, led by superconducting ceramics that may revolutionize electronics.

   过去材料科学是最无吸引力的技术，而今天崭新而实用的发现正在涌现，居首位的是可使电子学彻底变革的超导陶瓷。

2. Consumers may *shy away from* genetically engineered food as weird and even dangerous, so scientists are developing special testing plans for these new and unknown breeds.

   消费者可能会嫌弃遗传工程食品，总觉得它们古怪，甚至危险，因此科学家正在制订专门计划，对这些新奇的未知品种进行检验。

3. Taylor argues that microwaves are a *better bet* for enhancement, for unlike X rays, microwaves can penetrate the atmosphere, reaching the earth's surface from space.

   泰勒论证说，微波可能更适合于强化作用，因为微波不像X射线，它能穿透大气层，从空中到达地球表面。

4. The major problem in manufacture is the control of contamination and *foreign materials.*

   制造过程中的一个主要问题是如何控制污染和杂质。

Sometimes it is also possible that figurative expressions are used in Chinese translation when appropriate. For example: "In actual classroom situations, computers are gradually overcoming such obstacles as the shortage of courseware and prohibitive school system budgets—and such less-obvious hindrances as resistance by teachers who *fear for their jobs*."（在课堂实际教学中，计算机的使用正在逐步克服教学软件短缺、学校资金有限等障碍，并逐步冲破那些不太明显的障碍，如怕丢掉饭碗的教师的抵制等。）

**V. Discuss the translation technique and the ways of applying the technique to the translation of the following sentences. Complete each of the Chinese translations.**

1. The pace of discovery has been exhilarating, but for the scientists involved success *has been tempered by deep frustration*.（尽管这项发现进展之快令人兴奋，但对从事这项研究的科学家来说，在成功的喜悦中却 __________。）
2. The periodontal ligament around the root *acts as a suspension mechanism*, absorbing the mechanical shock of chewing.（牙根周围的牙周韧带 __________，承受着咀嚼时的机械性冲击。）
3. For legitimate emergencies, the body also *keeps on hand* a supply of platelets, disk-shaped bodies whose *job* is to facilitate clotting.（为了应付真正的紧急情况，人体内还存有 __________ 血小板，血小板形似圆盘，其 __________ 是促使血液凝固。）
4. With even more recent systems, using a combination of radar and other instruments on the aircraft and on the ground, the pilot can now land completely *blind* in perfect safety.（如果使用新式系统，将飞机和地面上的雷达和其他仪器结合起来，驾驶员现在已能十分安全地 __________ 着陆了。）
5. Though still in their *infancy*, these programs represent the industry's first steps toward software that will one day be able to diagnose a patient's illness and suggest treatments, most experts agree.（大多数专家认为，尽管这些程序仍处于 __________，却代表着工业界朝软件迈出的第一步，这些软件总有一天能为病人诊断疾病并提出治疗方案。）
6. Half-mile asteroids are *a dime a dozen* in the solar system, and they run into the planet once every 100,000 years, on average.（半英里大小的小行星 __________，它们平均每10万年碰撞地球一次。）
7. No mere *armchair theorist*, but an experimentalist of consummate skill, his optical researches would have assured him of immortal fame if he had done nothing else.（他并非只是一个 __________，而是一位技艺精湛的实验家，即使他别的什么也没干过，仅是光学方面的研究就足以使他名垂千古。）

8. Computers in education are neither proceeding at *the breakneck pace* that was originally projected, nor are they headed for the debacle that skeptics predicted.（计算机在教学上的应用，既不像原先规划的那样__________，也不像怀疑论者所预言的那样垮了下来。）

**VI. Translate the following sentences into Chinese. Pay attention to how the figurative expressions should be translated.**

1. Many particle physicists look on the past 15 years as an era of consolidation in contrast to the *ferment* of earlier decades.
2. Just as the Higgs can exert a *behind-the-scenes* influence on other particles, other particles can do the same to the Higgs.
3. Those particles come in a range of energies, and their net effect depends on where precisely the Standard Model *gives way to* a deeper theory.
4. The goals are well motivated and matched by our experimental tools, with the LHC succeeding the current *workhorse*, Fermilab's Tevatron collider.
5. While the robot is initially quite expensive to buy, it quickly *pays for itself* because at certain jobs it can work faster and better than a human, thus boosting productivity and lowering production costs.
6. Scientists have long acknowledged the existence of a "finagle factor"—a tendency by many scientists to *give a helpful nudge* to the data to produce desired results.
7. Even first-class surgery will *go downhill* if the inspection is not addressed and the patient does not practice strict oral hygiene.
8. Several facts about the proposed plane are already known although it is still barely *on the drawing* board.

**VII. Translate the following passage into Chinese.**

Particle physicists internationally agree on the most important goals of particle physics research in the near and intermediate future. The overarching goal, which is pursued in several distinct ways, is to find and understand what physics may lie beyond the standard model. There are several powerful experimental reasons to expect new physics, including dark matter and neutrino mass. Most importantly, though, there may be unexpected and unpredicted surprises which will give us the most opportunity to learn about nature.

Much of the efforts to find this new physics are focused on new collider experiments. A relatively near term goal is the completion of the Large Hadron Collider (LHC) in 2008 which will continue the search for the Higgs boson, supersymmetric particles, and other new physics. An intermediate goal is the construction of the International Linear Collider (ILC) which will complement the LHC by allowing more precise measurements of the properties of newly found particles. A decision for the technology of the ILC was taken in August 2004, but the site has still to be agreed upon.

Additionally, there are important non-collider experiments which also attempt to find and understand physics beyond the standard model. One important non-collider effort is the determination of the neutrino masses since these masses may arise from neutrinos mixing with very heavy particles. In addition, cosmological observations provide many useful constraints on the dark matter, although it may be impossible to determine the exact nature of the dark matter without the colliders.

**VIII. Translate the following passage into English.**

### 粒子物理学

粒子物理学是物理学的一个分支，其研究内容是物质和辐射的基本成分及其相互影响。该学科也称作高能物理学，因为许多基本粒子在自然界的正常情况下不会出现，却可以在其他粒子的能量碰撞过程中产生或观察到（例如在粒子加速器中）。

现代粒子物理学重点研究亚原子粒子，包括原子成分（如电子、质子和中子），由辐射和离散过程产生的粒子（如光子、中微子、μ介子），以及范围广泛的奇异粒子。

严格地说，“粒子”这个术语不够恰当，因为粒子物理学的力学原理是受到量子机械学制约的，因此它们表现出波粒子的二元性，即在某些实验条件下表现出和粒子一样的运动方式，而在其他条件下则表现出波状运动方式。根据粒子物理学家的传统观点，“基本粒子”指的是电子和光子之类的物质，因为这些“粒子”也具有波状运动方式。

目前人们观察到的所有粒子及其相互作用几乎都可以用一种叫做“标准模型”的量子场理论加以描述。标准模型有40种基本粒子，结合在一起形成复合粒子，这些复合粒子包括数百种自20世纪60年代以来发现的其他种类的粒子。虽然标准模型与迄今为止几乎所有的测试相符，但是多数粒子物理学家认为该模型对自然的描述是不全面的，他们相信还有待于发现一种更加基本的理论。近年来，在对中微子质量的测量已经发现偏离标准模型的第一批数据。

## Section 2 Reading for Academic Purposes

*In scientific research, especially in social science, qualitative research is often contrasted with quantitative research. Qualitative research is a field of inquiry that crosscuts disciplines and subject matters, and qualitative researchers aim to gather an in-depth understanding of human behavior and the reasons that govern human behavior. Qualitative research relies on reasons behind various aspects of behavior. To put it more simply, it investigates the why and how of decision making, not just what, where, and when. Hence, the need is for smaller but focused samples rather than large random samples, by which qualitative research categorizes data into patterns as the primary basis for organizing and reporting results. Qualitative researchers typically rely on four methods for gathering information:*

*participation in the setting, direct observation, in depth interviews, and analysis of documents and materials.*

*The following essay presents a more detailed description of the origins and methods of this approach.*

## Understanding Qualitative Research

## 什么是定性研究？

numerical *adj.* 用数字表示的；数值的
quantifiable *adj.* 可以计量的

Qualitative research is one of the two major approaches to research methodology in social sciences. Qualitative research involves an in-depth understanding of human behavior and the reasons that govern human behavior. Unlike quantitative research, qualitative research relies on reasons behind various aspects of behavior. Simply put, it investigates the "why" and "how" of decision making, as compared to "what", "where", and "when" of quantitative research. Hence, the need is for smaller but focused samples rather than large random samples, which qualitative research categorizes data into patterns as the primary basis for organizing and reporting results. Unlike quantitative research, which relies exclusively on the analysis of **numerical** or **quantifiable** data, data for qualitative research comes in many media—including text, sound, still and moving images.

phenomenology *n.* 现象学
grounded theory 扎根理论；基础理论
ethnography *n.* 人种学；民族志

Qualitative methods include the case study, **phenomenology**, **grounded theory**, and **ethnography**, among others.

### History

marginalize *v.* 使……边缘化
fieldwork *n.* 实地调查；野外考察
spate *n.* 大量；许多

Qualitative research approaches began to gain recognition in the 1970s. The phrase "qualitative research" was until then **marginalized** as a discipline of anthropology or sociology, and terms like ethnography, **fieldwork**, participant observation and Chicago school (sociology) were used instead. During the 1970s and 1980s qualitative research began to be used in other disciplines, and became a dominant—or at least significant—type of research in the fields of women's studies, disability studies, education studies, social work studies, information studies, management studies, nursing service studies, human service studies, psychology, communication studies, and other. In the late 1980s and 1990s after a **spate** of criticisms from the quantitative side, paralleling a slowdown in traditional media spending for the decade, new methods of qualitative research evolved, to address the perceived problems with reliability and imprecise modes of data analysis.

One way of differentiating qualitative research from quantitative research is that largely qualitative research is exploratory, while quantitative research hopes to be conclusive. However it may be argued that each reflect a particular **discourse**, neither being definitively more conclusive or "true" than the other. Quantitative data is measurable, while qualitative data cannot be put into a context that can be **graphed** or displayed as a mathematical term.

discourse *n.* 论述；话语
graph *v.* 用图表表示；把……绘入图表

## Case study

Rather than using large samples and following a rigid **protocol** to examine a limited number of variables, case study methods involve an in-depth, **longitudinal** examination of a single instance or event: a case. They provide a systematic way of looking at events, collecting data, analyzing information, and reporting the results. As a result the researcher may gain a sharpened understanding of why the instance happened as it did, and what might become important to look at more extensively in future research. Case studies lend themselves to both generating and testing hypotheses.

protocol *n.* 协议
longitudinal *adj.* 纵观的

As a distinct approach to research, use of the case study originated only in the early 20th century. *The Oxford English Dictionary* traces the phrase *case study* or *case-study* back as far as 1934, after the establishment of the concept of a *case history* in medicine.

The use of case studies for the creation of new theory in social sciences has been further developed by the sociologists Barney Glaster and Anselm Strauss who presented their research method, Grounded theory, in 1967.

The popularity of case studies in testing hypotheses has developed only in recent decades. One of the areas in which case studies have been gaining popularity is education and in particular educational evaluation.

Case studies have also been used as a teaching method and as part of professional development, especially in business and legal critical incidents 关键事件(法)

education. The problem-based learning (PBL) movement is such an example. When used in (non-business) education and professional development, case studies are often referred to as **critical incidents**.

## Phenomenology

Phenomenology is the study of structures of consciousness as experienced from the first-person point of view. The central structure of an experience is its intentionality, its being directed toward something, as it is an experience of or about some object. An experience is directed toward an object by virtue of its content or meaning (which represents the object) together with appropriate enabling conditions.

ontology *n.* 本体论
epistemology *n.* 认识论
ethics *n.* 道德规范
guise *n.* 伪装；外观
come into one's own 进入繁盛期，盛行起来
qualia *n.* 主观感受

Phenomenology as a discipline is distinct from but related to other key disciplines in philosophy, such as **ontology**, **epistemology**, logic, and **ethics**. Phenomenology has been practiced in various **guises** for centuries, but it **came into its own** in the early 20th century in the works of Husserl, Heidegger, Sartre, Merleau-Ponty and others. Phenomenological issues of intentionality, consciousness, **qualia**, and first-person perspective have been prominent in recent philosophy of mind.

## Grounded theory

Grounded theory (GT) is a systematic qualitative research methodology in the social sciences emphasizing generation of theory from data in the process of conducting research.

extract *v.* 摘录；选取

It is a research method that operates almost in a reverse fashion to traditional research and at first may appear to be in contradiction of the scientific method. Rather than beginning by researching and developing a hypothesis, a variety of data collection methods are the first step. From the data collected from this first step, the key points are marked with a series of codes, which are **extracted** from the text. The codes are grouped into similar concepts, in order to make them more workable. From these concepts categories are formed, which are the basis for the creation of a theory, or a reverse engineered hypothesis. This contradicts the traditional model of research, where the researcher chooses a theoretical framework, and only then applies this model to the studied phenomenon.

## Ethnography

Ethnography is a **genre** of writing that uses fieldwork to provide a descriptive study of human societies. Ethnography presents the results of a **holistic** research method founded on the idea that a system's properties cannot necessarily be accurately understood independently of each other. The genre has both formal and historical connections to travel writing and **colonial** office reports. Several academic traditions, in particular the **constructivist** and **relativist** paradigms, employ ethnographic research as a crucial research method. Many cultural anthropologists consider ethnography the essence of the discipline.

genre *n.* 体裁，类型
holistic *adj.* 整体的；注重整体的；整体论的
colonial *adj.* 殖民地的
constructivist *n.* 建构主义者
relativist *n.* 相对论者；相对主义者

Exercises

**I. Read the passage and decide whether the following statements are true or false. Write T for True and F for False in the brackets.**

1. Qualitative research, which is different from quantitative research, depends on reasons underlying various aspects of behavior. ( )
2. Data for quantitative research comes from many media, such as text, sound, still and moving images. ( )
3. Qualitative research approaches was relegated to a lower social status and were not accepted until the 1970s. ( )
4. Qualitative data is measurable and can be graphed and displayed as a mathematical term. ( )
5. Case study methods exam a single instance or event in a thorough and longitudinal way. ( )
6. Case studies do not generate and test hypotheses, but focus on the implications for the future research. ( )
7. Critical incidents are case studies used in non-business education and professional development. ( )
8. In recent philosophy of mind, phenomenological issues of unintentionality, unconsciousness and objectivity are prominent. ( )
9. Grounded theory was presented by two sociologists, Barney Glaster and Anselm Strauss in 1967 and was a qualitative research method.( )
10. Ethnography depends on first-hand observation to provide a descriptive study for human societies and concerned with wholes rather than analysis or separation into parts. ( )

## II. Read the passage again, and complete the following items.

1. What qualitative research involves is: __________
2. The general difference between qualitative research and quantitative research: __________
3. The fields in which qualitative research became a dominant or significant type of research during the 1970s and 1980s: __________
4. The things case study does systematically include: __________
5. One of the areas in which case studies have been gaining popularity: __________
6. The study of phenomenology focuses on: __________
7. The other key disciplines in philosophy related to phenomenology include: __________
8. The emphasis of grounded theory: __________

### Quotations from Great Scientists

*I have learned to use the word "impossible" with the greatest caution.*

—*Wernher von Braun*

我学会了极为慎重地使用"不可能"这个词。

——沃纳·冯·布劳恩

# Chapter 11

## Space Exploration
## 太空探索

## Section 1 Reading and Translation

### I What You Are Going to Read

At around 125 AD, a Greek satirist named Lucian wrote a book on space flight called *True Histories*. The book was full of tall, unbelievable tales and travelogues on visits to the sun and the moon. Today, the book could easily be discarded as the fantasy of a people of a bygone era. But it was significant in the sense that it kindled the curiosities of the people of the day and stimulated interest in outer space and space travel.

It is man's dream, technology, and understanding of science that forms the basis of all forms of space exploration. From ancient times well into the 20th century, the only technologically feasible method to explore space was astronomy—the studying of the millions of stars and neighboring planets, which invade night sky, as they have done for billions of years. The mysterious movements of the planets and the ebbing of stars across the sky had originally found explanations in religion, but as man's understanding of the science of astronomy increased, natural laws, and not dogma, took form. And, as a solid foundation was laid with ground-based astronomy, man walked resolutely into the Space Age, upon the advent of the modern rocket. Given this stepping stone of the liquid fueled rocket, man was able to enter the cosmic "ocean". Since then, space exploration has developed from unmanned probes to manned probes, and on July 20, 1969, the human race accomplished its single greatest technological achievement of all time when human first set foot on another celestial body—the moon. As you can see from the following essay published in *Scientific American* (September, 2007), forty years after that historic moment, American scientists are planning to go back to the moon—and beyond.

### II About the Authors

Charles Dingell, William A. Johns and Julie Kramer White manage engineering and technology operations for the NASA/Lockheed Martin *Orion* project. Dingell serves as the project's technical director for NASA. During a quarter of a century at the agency, he held leadership positions supporting the space shuttle, crew return vehicle and orbital space plane programs. Johns is chief engineer and technical director of the *Orion* crew exploration vehicle (CEV) for Lockheed Martin, the lead contractor on the project. After starting at Martin Marietta in 1980, he worked on successive versions of the Centaur upper stage and later led the development of the Atlas V evolved expendable launch vehicle. Kramer White is NASA's chief engineer for the *Orion* CEV project. She has more than 20 years of technical management experience on the space shuttle, space station and crew return vehicle.

## III Reading Passage

### To the Moon and Beyond
### 飞向月球，飞向更远的宇宙空间

By Charles Dingell, William A. Johns and Julie Kramer White

*Humans are returning to the moon. This time the plan is to stay a while.*

luminous *adj.* 发光的，明亮的
crescent *n.* 新月
well-timed *adj.* 时机正好的，正合时宜的
boost *n.* 推进，增进
translunar *adj.* 越过月球的；月球外侧的
injection *n.*（人造卫星、宇宙飞船等的）射入轨道
ignition *n* 点火，点燃
erupt *v.* 喷出，爆发
nozzle *n.* 管口，喷嘴
astern *adv.* 在船尾，向船尾
stack *n.* 堆，一堆
module *n.* 登月舱，指令舱
celestial *adj.* 天上的
outpost *n.* 前哨

1 The moon, a **luminous** disk in the inky sky, appears suddenly above the broad **crescent** of Earth's horizon. The four astronauts in the *Orion* crew exploration vehicle have witnessed several such spectacular moonrises since their spacecraft reached orbit some 300 kilometers above the vast expanse of our home planet. But now, with a **well-timed** rocket **boost**, the pilot is ready to accelerate their vessel toward the distant target ahead. "**Translunar injection** burn in 10 seconds..." comes the call over the headset. "Five, four, three, two, one, mark... **ignition**..." White-hot flames **erupt** from a rocket **nozzle** far **astern**, and the entire ship—a **stack** of functional **modules**—vibrates as the crew starts the voyage to our nearest **celestial** neighbor, a still mysterious place that humans have not visited in nearly half a century. The year is 2020, and Americans are returning to the moon. This time, however, the goal is not just to come and go but to establish an **outpost** for a new generation of space explorers.

contractor *n.* 订约人，承包人
mount *v.* 设置，安放

2 The *Orion* vehicle[1] is a key component of the Constellation program[2], NASA's ambitious, multibillion-dollar effort to build a space transportation system that can not only bring humans to the moon and back but also resupply the International Space Station (ISS) and eventually place people on the planet Mars. Since the program was established in mid-2006, engineers and researchers at NASA, as well as at Lockheed Martin[3], *Orion*'s prime **contractor**, have been working to develop the rocket launchers, crew and service modules, upper stages and landing systems necessary for the US to **mount** a robust and affordable human spaceflight effort after its current launch

**workhorse**, the space shuttle, retires in 2010.

workhorse *n.* 载重量大的车（船、机器）

3 To minimize development risks and costs, NASA planners based the Constellation program on many of the **tried-and-true** technical principles and **know-how** established during the *Apollo* program, an engineering feat that put men safely on the moon in the late 1960s and early 1970s. At the same time, NASA engineers are redesigning many systems and components using updated technology.

tried-and-true *adj.* 经考验证明是好的；实践证明可取的
know-how *n.* 实际知识；技能；诀窍

4 *Orion* starts with much the same general functionality as the *Apollo* spacecraft[4], and its crew capsule has a similar shape, but the resemblance is only skin-deep. *Orion* will, for example, accommodate larger crews than *Apollo* did. Four people will ride in a pressurized cabin with a volume of approximately 20 cubic meters for lunar missions (six will ride for visits to the space station starting around 2015), compared with *Apollo*'s three astronauts (plus equipment) in a cramped volume of about 10 cubic meters.

5 The latest structural designs, electronics, and computing and communications technologies will help project designers expand the new spacecraft's operational flexibility beyond that of *Apollo*. *Orion*, for instance, will be able to **dock** with other craft automatically and to **loiter** in lunar orbit for six months with no one onboard. Engineers are widening safety margins as well. In the event of an emergency during launch, for example, a powerful escape rocket will quickly remove the crew from danger, a benefit space shuttle astronauts do not enjoy. But to give you a better feel for what the program involves, let us start on the ground, before the *Orion* crew leaves Earth. From there, we will trace the progress of a **prototypical** lunar mission and the technologies planned to accomplish each stage.

dock *v.*（宇宙飞行器）在外层空间对接
loiter *v.* 徘徊；闲荡
prototypical *adj.* 原型的

## Up, Up and Away

6 Towering 110 meters above the salt marshes of Florida's Kennedy Space Center, the two-stage Ares V cargo launch vehicle stands **poised** to blast off. The uncrewed vehicle, which contains a cluster of five powerful rocket engines, has almost the height and **girth** of the massive Saturn V rocket of *Apollo* fame. Derived from the space shuttle's external tank, Ares V's central booster tank delivers liquid-oxygen-hydrogen **propellants** to the vehicle's RS-68 engines—each a modified version of the ones currently used in the Delta IV

poised *adj.* 做好准备的；摆好姿势不动的
girth *n.* 围长
propellant *n.* 火箭燃料；推进剂

strap-on *n.* 捆绑
flank *v.* 位于……侧面
thrust *n.* 推力；牵引力
loft *v.* 把……向高处发射
propulsion *n.* 推进；推进力

military and commercial launcher. Two "**strap-on**", solid-fuel rocket boosters adapted from the space shuttle's system **flank** Ares V's central cylinder. They add the extra **thrust** that the launcher will need to **loft** the buglike lunar lander and the "Earth departure stage"—a **propulsion** module that contains a liquid-oxygen-hydrogen-fueled J-2X engine (a descendant of NASA's *Apollo*-era Saturn V J-2 motor, built by Pratt & Whitney Rocketdyne[5]) that will enable *Orion* to escape Earth's gravity and travel to the moon.

mound *n.* 一堆；许多
billowing *adj.* 巨浪似的，汹涌的
gantry *n.*（固定火箭等的）连接架
launchpad *n.*（火箭等的）反射台
jettison *v.* 抛弃
sheath *n.* 护套，外壳
rendezvous *v.*（宇宙飞船等）在指定地点会合，集结

7 Abruptly, a flash exits the tail of the Ares V, and **mounds** of **billowing** smoke clouds soon envelop the booster, **gantry** and **launchpad**. After a momentary pause, a tremendous roar echoes across the spaceport, sending birds fleeing in all directions. Slowly at first, the big rocket ascends atop an ever expanding column of gray-white exhaust. Accelerating steadily, the vehicle blazes a smoky trail across the sky and disappears into the heavens.[6] Minutes later, amid the silence of near-Earth space, Ares V **jettisons** its strap-on boosters, which fall into the sea, where they will be recovered. It then sheds the protective cargo **sheath** that covers its nose, revealing the lunar landing module. Circling the globe at an altitude of about 300 kilometers, the robot spacecraft now awaits the next step in the lunar excursion plan: **rendezvous** with *Orion*.

perch *v.*（在较高或较险的位置）坐落
imminent *adj.* 即将来临的，逼近的
conical *adj.* 圆锥的，圆锥形的
capsule *n.* 太空舱
fairing *n.*【空】整流罩；减阻装置
aerodynamic *adj.* 空气动力学的

8 That same day the four moon-bound astronauts **perch** 98 meters above another Kennedy launchpad, anticipating **imminent** liftoff. Just below their **conical** *Orion* crew **capsule** is a drum-shaped service module that contains the spacecraft's on-orbit propulsion engine and much of its life-support system. Protective **fairings** envelop both to shield them from the strong **aerodynamic** forces and harsh conditions they will encounter during ascent. The crew capsule and the service module sit atop NASA's two-stage Ares I crew launch vehicle. Slimmer than its big brother, the "Stick", as it is known by some, comprises another modified solid shuttle booster (constructed by Alliant Techsystems[7]) topped by a second stage that is powered by a single J-2X motor. A spacecraft adapter serves as the structural and electrical interface between the *Orion* spacecraft and Ares I.

9 Capping the tall stack is an escape tower that is primed to rocket the occupants away from danger in the event of a failure. As the 1986

*Challenger* accident[8] proved, space shuttle crews have little chance of survival if their ship sustains a major technical problem during launch and early ascent. *Orion*'s launch-abort system (LAS)[9], in contrast, can for a few seconds impart a thrust that is equivalent to about 15 times its own mass and that of the detached crew module. The rocket tower is set to rapidly remove the astronauts from harm's way during a mission abort while still on the launchpad or during ascent. Should a serious **glitch** occur on the ground, the separated system would reach an altitude of about 1,200 meters to allow for parachute deployment and a downrange, or horizontal, distance of about 1,000 meters to clear the launchpad. Mission planners estimate that the LAS, together with *Orion*'s advanced guidance and control system, would be able to return the crew safely 999 out of 1,000 times it is needed.

glitch *n.* 故障，突然的失灵

10 But any such thoughts recede rapidly as the **exhilaration** of the **impending** launch mounts. As the countdown nears zero, commander and pilot intently eye the flight instruments on the flat-screen displays of *Orion*'s "glass **cockpit**", adapted from a safety-redundant version of the **avionics** system used by advanced airliners such as the newly introduced Boeing 787 Dreamliner. The cockpit, with its computerized, fully electric "fly by wire" controls, energy-conserving electrical equipment and few mechanical switches, would be nearly unrecognizable to an *Apollo*-era astronaut.

exhilaration *n.* 高兴，愉悦
impend *v.* 即将发生
cockpit *n.* 驾驶员座舱
avionics *n.* 航空电子学；（用于航空、导弹和宇宙航行方面的）电子设备，控制系统

11 A shudder **ripples** up through the entire structure, followed by a thunderous **rumble**. The Stick starts to move skyward. Gaining speed with every second, it rises rapidly, pressing the astronauts into their seats.

ripple *v.* 波动，飘动
rumble *n.* 隆隆声

12 Almost two and a half minutes into the flight, the solid rocket booster is driving Ares I upward at a speed of Mach 6. At a height of about 61,000 meters, the first stage separates and falls back to Earth on parachutes so that it may be recovered and later recycled. Meanwhile the J-2X second-stage rocket motor ignites, sending the *Orion* crew module, the service module and the LAS through the last reaches of the atmosphere. Their usefulness ended now that the craft has exited the atmosphere, the aerodynamic **shrouds** break away to maximize ascent performance by shedding weight. By this time the vessel has gained enough velocity to reduce the risk of an emergency

shroud *n.* 覆盖物

abort, so the LAS and its protective fairing also separate and fall away. The second-stage engine cuts off as the crew capsule and the service module near an altitude of about 100 kilometers.

## Rendezvous in Earth Orbit

bay *n.* 分隔舱
polymer-composite *n.* 高分子合成材料
expendable *adj.* 可消耗的；可牺牲的

13 The service module engine then ignites, completing the job of inserting *Orion* into orbit and initiating the maneuvers it needs to rendezvous with the Earth departure stage and the lunar lander. *Orion*'s main engine is adapted from the flight-proved space shuttle orbital maneuvering engine, upgraded for greater propulsion thrust and efficiency. The service module contains power generation and storage systems, radiators that expel surplus heat into space, all necessary fluids and a science equipment **bay**. To maximize space in the crew vehicle, the service module also carries some of the avionics system, as well as part of the environmental control and life-support subsystems. A lightweight **polymer-composite** honeycomb reinforced with aluminum forms its structure; simple manufacturing methods should help keep down the cost of this **expendable** item.

14 One of the more notable differences between *Orion* and *Apollo* is the addition to the service module of umbrella-shaped solar arrays that unfold when needed in orbit. Because the *Apollo* spacecraft was designed for moon missions measured in days, it carried hydrogen fuel cells that could generate electrical power only for relatively short periods. *Orion*, in contrast, must be able to produce electricity for at least six months.

align *v.* 对准；校直

15 Gradually, *Orion* catches up to the lunar lander and departure stage that Ares V had earlier placed into low Earth orbit. When the two craft finally rendezvous, the crew performs (or monitors) the final maneuvers and keeps an eye on the automated "soft capture" system as it **aligns** the pair and then smoothly docks them. Force-feedback and electromechanical components sense loads, automatically capturing the mating rings of the vehicles and actively damping out any contact forces. Ship and crew are now nearly ready to head for the moon.

lithium *n.* 【化】锂
alloy *n.* 合金
titanium *n.* 【化】钛

16 The crew module is the only element of *Orion* that will make the entire trip, and it may be reused for up to 10 flights. A lightweight aluminum-**lithium alloy** with **titanium** reinforcements makes up

most of the capsule structure. The exterior of the crew vehicle is lined with a thermal protection system, which, in addition to protecting its living quarters from the **searing** heat of reentry, also incorporates a tough, impact-resistant layer that shields it against high-velocity **micrometeoroids** or other **debris** that may strike its outer surface.

sear *v.* 烤焦，使枯萎
micrometeoroid *n.* 【天】微流星体
debris *n.* 碎片，残骸

17 The crew module's reaction-control maneuvering system uses gaseous oxygen and methane propellants, a technology that builds on the progress engineers made during NASA's X-33 single-stage-to-orbit vehicle program, which was canceled in 2001.[10] One advantage of the oxygen-methane propulsion system is that its fuel will be nontoxic (unlike its predecessors that used **hypergolic** propellants), which will help ensure the safety of the flight and ground crews after they return to Earth.

hypergolic *adj.* （火箭燃料）自燃的，自发火的

18 When all is ready, the Earth departure stage rocket engine ignites to propel the spacecraft toward the moon. Engineers are configuring *Orion* to support both "lunar sortie missions", in which crew members spend four to seven days on the moon's surface to demonstrate the *Orion* system's ability to transport and land humans on Earth's satellite, and "lunar outpost missions", in which a **semicontinuous** human presence would be established there. Because the maximum duration of a crew's stay on the lunar surface is 210 days (determined by the available supplies of oxygen, water and other **consumables**), *Orion*'s continuous operation capability must exceed that period. The biggest design driver for *Orion* lunar missions is the amount of propellant required to meet these objectives.

semicontinuous *adj.* 半接续的
consumable *n.* 消费品

19 After a four-day trip **outbound**, the crew enters into lunar orbit, having dumped the Earth departure stage along the way. The four astronauts climb into the lander, leaving the crew capsule and service module to wait for them in orbit. As with the *Apollo* lunar excursion module, the lunar lander consists of two components. One is the descent stage, which has legs to support the craft on the surface as well as most of the crew's consumables and scientific equipment. The other part is the ascent stage that houses the crew. After landing and exploring the surface, the foursome blasts off the moon's surface and later docks with the crew and service modules in orbit. The ascent stage of the lander is discarded into outer space, and *Orion* rockets

outbound *adj.* 向外去的，向外开的

back to Earth.

## Return to the Home Planet

exacerbate *v.* 使加深；使加剧

20 As the *Orion* astronauts close in on the blue planet, they may have to prepare for a reentry and landing quite unlike those of *Apollo*. Like the *Gemini*[11] and *Mercury* spacecraft[12] before it, *Apollo* splashed down in the ocean after it had plunged through the atmosphere. But because water landings would require costly fleets of recovery ships and expose a reusable spacecraft to saltwater corrosion, NASA planners may decide that *Orion* should touch down on land, as the Russian *Soyuz* spacecraft[13] does. *Orion*'s greater size, weight and lift, however, **exacerbate** the engineering challenge. The "land landing" mode is also important to minimize life-cycle costs. If the agency instead opts to land in the ocean, *Orion* will be fitted with much the same capabilities as *Apollo*.

trajectory *n.* 【物】（射体）轨道，弹道

21 Unfortunately, setting down on American soil after a lunar mission presents a fundamental problem. For nearly half of the lunar month, orbital conditions would place any landing site in the Southern Hemisphere, away from the planned locations in the western continental US. Although the time of departure from lunar orbit can vary the longitude of the reentry point, its latitude is fixed by the declination (angular distance from the equator) of the moon relative to Earth at lunar departure.[14] Thus, to reach landing sites in the western US or waters near the continental US during unfavorable periods of the lunar month, *Orion* will stretch its landing point into the Northern Hemisphere by employing aerodynamic lift produced as it descends into Earth's outer atmosphere. A **trajectory** of this type, in which a spacecraft bounces across the upper atmosphere like a stone skipping across a pond, is sometimes known as a skip reentry.

fine-tune *v.* 微调

22 Having spent the four-day return journey from the moon **fine-tuning** *Orion*'s flight path for the first crewed skip-reentry maneuver ever attempted, anticipation builds among the astronauts as the blue-white visage of our home planet grows ever larger in their view screen. They are soon occupied, however, by reorienting the ship so that the service module can be jettisoned, a necessary operation that exposes the protective heat shield on the crew module's underside. Later, after using *Orion*'s redundant navigation system and flight computers to

check that the spacecraft's attitude is positioned properly for reentry and that its trajectory is following the correct, shallow-angle route, the crew prepares for the **onset** of **deceleration** forces as *Orion* encounters the atmosphere.

onset *n.* 开始
deceleration *n.* 减速

23 The skip-reentry process starts out slowly. At first, the crew begins to notice weak g-forces[15] caused by the resistance of the thin, high-altitude air. The g-forces, which push the crew members against their seats, grow steadily in strength as bits of glowing heat-shield material and streams of **ionized** gas **streak** past the windows. Shortly after *Orion* starts to scrape against the upper reaches of the atmosphere, the spacecraft rebounds briefly to a higher altitude. After the skip, the capsule dives deeply into the air on a path toward the landing site.

ionize *v.* 使电离
streak *v.* 疾驰，飞跑

24 The tragic loss of the *Columbia* space shuttle[16] and crew in 2003 demonstrated that the thermal protection system of a returning vehicle is critical. Atmospheric reentry generates tremendous heating on the undersurface of spacecraft (a couple of thousand degrees Celsius) caused by the friction of the air rushing by at **hypersonic** speeds. Because *Orion*'s reentry velocity from a moon mission (which is on the order of 11 kilometers a second) will be 41 percent faster than a shuttle's descent speed from low Earth orbit, the heat load will be several times greater. The fact that the *Orion* crew module is larger than that of *Apollo* compounds the challenge.

hypersonic *adj.* 高超音速的（指超过音速5倍以上的）

25 The leading candidate for *Orion*'s base heat shield is a material called PICA (phenolic impregnated carbon ablator)[17]. PICA is a **matrix** of carbon fibers embedded in a phenolic **resin**. At high temperatures, the outer surface of the PICA layer **ablates**, or burns away, to carry off much of that extreme heat. The ablator's surface **pyrolyzes** when heated, leaving a heat-resistant layer of **charred** material. PICA's low thermal conductivity also blocks heat transfer to the crew module. PICA was used in 2006, when it protected the *Stardust* spacecraft[18] (which carried a sample from Comet Wild 2) as it came back to Earth at 13,000 meters a second—the fastest controlled reentry ever. Being 40 times larger in area, *Orion*'s heat shield will need to be built in segments, thus adding new complexities.

matrix *n.* 母体；基体
resin *n.*（天然或合成的）树脂
ablate *v.* 融化掉；蒸发掉；腐蚀掉
pyrolyze *v.*【化】（使）热解，（使）裂解
char *v.* 烧焦

## Landing on Land

voluminous *adj.* 体积大的，大量的
canopy *n.* (降落伞的）伞盖
jar *v.* 震动，颠簸，摇晃
chute *n.*<口>降落伞（parachute的缩略形式）

26 Finally, three large parachutes—which closely resemble those used by *Apollo*—deploy to slow the vehicle's rate of descent. The reassuring sight of the **voluminous** red-and-white **canopies** opening above tells the astronauts that their amazing trip is almost complete. Before long, *Orion* is **jarred** by the release of its large heat shield. Hanging below the big **chutes**, the crew module now descends at about eight meters a second.

jolt *n.* 摇晃

27 In the case of a "land landing", an airbag system inflates on the crew module's underside to absorb and attenuate the upcoming landing shock. With a solid **jolt**, the spacecraft at last sets down on dry land in the western American desert. *Orion* has returned home.

## Notes

1. **The *Orion* vehicle:** "猎户座"飞船。
2. **the Constellation program:** 星座计划，美国国家航空航天局正在筹备的一项太空探索计划。整个计划将包括一系列新型航天器、运载火箭以及相关硬件，将用于国际空间站补给运输以及登月等各种太空任务中。
3. **Lockheed Martin:**（美国）洛克希德·马丁公司。由原洛克希德公司和马丁·玛丽埃塔公司于1995年合并而成。该公司目前是美国第一大国防承包商，创建于1913年，1932年改称"洛克希德公司"，1995年改为现名。
4. ***Apollo* spacecraft:** 美国"阿波罗"号宇宙飞船。该飞船由指挥舱、服务舱和登月舱3部分组成，于1969年7月21日登上月球，首次实现了人类登上月球的梦想。此后，美国又相继6次发射"阿波罗"号飞船，其中5次成功，总共有12名航天员登上月球。
5. **Pratt & Whitney Rocketdyne:**（美国）普拉蒂及惠特尼火箭发动机公司。
6. **Accelerating steadily, the vehicle blazes a smoky trail across the sky and disappears into the heavens:** 火箭随后稳定加速，拖着烟尾划过天空，消失在茫茫天际。句中the heavens（注意：heaven一词为复数形式）意为the sky, as seen from the Earth（天空）。
7. **Alliant Techsystems:**（美国）阿连特技术系统公司，为美国及其盟国提供航空和防御技术，特别是传统武器、推进系统、合成结构等的公司。
8. **the 1986 *Challenger* accident:** "挑战者"号航天飞机事故。美国东部时间1986年1月28日，"挑战者"号航天飞机发射后，其右侧固体火箭助推器的O型环密封圈失效，致使航天飞机在发射后的第73秒解体，所有7名宇航员罹难。
9. **launch-abort system (LAS):** 发射中断系统，一种保护宇航员安全逃生的系统。如果航天

器在发射过程中发生爆炸或者故障，航天地面指挥中心的一套计算机系统将自动发射一枚火箭撞击航天器，使乘载宇航员的空间舱弹离航天器。接着，这个空间舱会垂直降落，一段时间后舱上的降落伞会自动打开，最后空间舱将降落在海面或陆地上。

10. **The crew module's reaction-control maneuvering system uses gaseous oxygen and methane propellants, a technology that builds on the progress engineers made during NASA's X-33 single-stage-to-orbit vehicle program, which was canceled in 2001:** 乘员舱的反馈控制机动系统采用的推进剂是气态氧和甲烷，这项技术的基础是在NASA开发X-33单级入轨飞行器的过程中奠定的，不过X-33的研制计划已在2001年取消。句中短语build on (sth.) 的含义是use sth. as a foundation for further progress（用……作为进一步发展的基础）。
11. ***Gemini* spacecraft:** 美国"双子座"号载人飞船。该飞船从1965年3月到1966年11月共进行了10次载人飞行。主要目的是在轨道上进行机动飞行、交会、对接和航天员试作舱外活动等。
12. ***Mercury* spacecraft:** "水星" 号飞船，美国第一代载人飞船。该飞船总共进行了25次飞行试验，其中6次是载人飞行试验。
13. **Russian *Soyuz* spacecraft:** 俄"联盟"号飞船，俄罗斯使用时间最长的载人飞船系列。"联盟"号具有轨道机动、交会和对接能力，可为空间站接送航天员，又能在对接后与空间站一起飞行。1975年7月15日发射的"联盟-19"号飞船与美国的"阿波罗-18"号飞船在轨道上对接成功，实现了世界上首次太空国际联合飞行。
14. **Although the time of departure from lunar orbit can vary the longitude of the reentry point, its latitude is fixed by the declination (angular distance from the equator) of the moon relative to Earth at lunar departure:** 尽管调整脱离绕月轨道的时间可以改变再入点的经度，但再入点的纬度却是由飞船脱离绕月轨道时月球所处的赤纬（即月球与地球赤道的夹角）决定的，无法改变。reentry point指飞船再次进入地球大气层时的确切位置。
15. **g-forces:** （=G-force; g-load）G力，万有引力常数。在航空中G力定义为，飞行器在海平面飞行时的升力和受到地球引力向下吸引的力量达到平衡，当飞行器改变惯性（如加减速或进行非直线动作）时，则会产生正或负G力。
16. ***Columbia* space shuttle:** "哥伦比亚"号航天飞机，美国最早的航天飞机。2003年2月，该航天飞机在重返大气层时解体，机上7名宇航员全部遇难。
17. **PICA (phenolic impregnated carbon ablator):** 酚醛树脂浸渍碳烧蚀材料。
18. ***Stardust* spacecraft:** "星尘"号飞船，于1999年2月发射升空。2004年1月，该飞船近距离飞过"维尔特二号"（Comet Wild 2）彗星时，飞船上的尘埃采集器成功捕获到彗星物质粒子，并首次将彗星样本带回地球。

Exercises

## I. Answer the following questions.

1. What is the Constellation program about?
2. What is the *Orion* system mainly made up of?
3. What systems does the service module contain?
4. How does *Orion* work to land its crew members onto the moon after a four-day trip?
5. By what means will *Orion* descend into the Earth's outer atmosphere and land on the Earth?

## II. The following statements are incomplete. Search the missing information in the passage and fill in the blanks.

1. The Constellation program will adopt many ________ established during the *Apollo* program for the sake of fewer development risks and costs.
2. *Orion* will accommodate ________ astronauts while *Apollo* held only ________ occupants plus equipment in a narrow space.
3. If an emergency occurs during the launch of *Orion*, ________ will quickly help the astronauts in it out of danger, which space shuttle astronauts do not enjoy.
4. It is ________ that will propel *Orion* to escape the Earth's gravity and travel to the moon.
5. The ________ which are added to the service module of *Orion*, is able to supply enough electrical power for at least six months.
6. On *Orion*, the only thing which will experience the entire journey is ________ and it can undergo recycled use of ________ flights.
7. *Orion* is designed by engineers to support two great missions: ________ and ________.
8. According to the passage, the tragic loss of *Columbia* space shuttle and crew in 2003 seems to have something to do with the trouble in the ________.

## III. Identify the implied meanings of the underlined parts of the following sentences according to the context of the passage, and translate the sentences into Chinese.

1. *Orion* starts with much the same general functionality as the *Apollo* spacecraft, and its crew capsule has a similar shape, but the resemblance is only skin-deep.
2. Engineers are widening safety margins as well.
3. Towering 110 meters above the salt marshes of *Florida*'s Kennedy Space Center, the two-stage Ares V cargo launch vehicle stands poised to blast off.
4. That same day the four moon-bound astronauts perch 98 meters above another Kennedy launchpad, anticipating imminent liftoff.
5. Protective fairings envelop both to shield them from the strong aerodynamic forces and harsh conditions they will encounter during ascent.

6. Capping the tall stack is an escape tower that is primed to rocket the occupants away from danger in the event of a failure.
7. *Orion*'s launch-abort system (LAS), in contrast, can for a few seconds impart a thrust that is equivalent to about 15 times its own mass and that of the detached crew module.
8. Force-feedback and electromechanical components sense loads, automatically capturing the mating rings of the vehicles and actively damping out any contact forces.
9. The exterior of the crew vehicle is lined with a thermal protection system, which, in addition to protecting its living quarters from the searing heat of reentry, also incorporates a tough, impact-resistant layer that shields it against high-velocity micrometeoroids or other debris that may strike its outer surface.
10. The fact that the *Orion* crew module is larger than that of *Apollo* compounds the challenge.

## Translation Techniques (1)

### Translation of Numbers（数字的译法）

Numbers frequently appear in scientific and technical papers when comparisons are made between two objects or the changes of quantities are indicated. If the expressions are similar in using the numbers in English and Chinese, there will be no problems. For example, in comparison, the Chinese translation of the sentence "At its closest, the asteroid was about 450,000 miles away, roughly twice the distance between the earth and the moon." is最接近的时候，那颗小行星距离地球大约45万英里，差不多等于地球和月球之间距离的2倍；and the sentence "The oil production rises at the rate of 28 percent a year" can be translated into原油产量以每年28%的速率增长.

However, due to differences of the two languages, there are some expressions which cannot be translated into the target language directly, and some changes have to be made. These expressions mainly involve comparisons of quantities (including size, height, width, and volume, etc.), as well as the increase and decrease in quantities.

**1. Comparison of quantities**

There are several ways to express comparisons of quantities in English, and special attention should be paid to such expressions as "*n* times as…as", "*n* times more than", "*n* times over", "by a factor of *n*", which should be translated as是（相当于）……的*n*倍or比……大（长、宽、多……）$n-1$倍. For example,

(1) As it emerges, each beam is focused on about the width of three human hairs, yet is *1,000 trillion times brighter than* the sunlight that falls on the earth.

当光束射出时，每束光聚焦于三根头发粗细的一点上。然而，亮度却是照射到地球上的太阳光的1,000兆倍。(NOT: 比……亮1,000兆倍)

(2) The magnetic power of such materials is about *10 times stronger than* that of ordinary magnets and they have been used in microwave communication, magnetic therapy and precision instruments.

这些材料的磁力是普通磁铁的10倍左右，并已应用于微波通讯、磁疗和精密仪表等领域。(NOT: 比……大10倍左右)

(3) This partner, it seems, has a mass *10 or 20 times greater than* the sun—yet we can't see it.

这颗伴星的质量似乎有太阳的10至20倍那么大，但我们还是看不到它。(NOT: 比……大10至20倍)

If the expression is "*n* times less (fewer) than", the fraction (分数) should be used in Chinese translation even if there is the word "time" in the English version: (比……少(n－1) /n or 是……的1/n). For example,

Line AB *is five times* shorter than Line CD. 线段AB是线段CD的1/5。(OR: 线段A比线段B短4/5) (NOT: 线段A比线段B短5倍)

**2. Increase in quantities**

Similar to comparisons of quantities, students tend to make mistakes in the translation of sentences involving the increase in quantities. The expressions of"increase *n* times", "increase by *n* times", "increase *n*-fold", "increase by a factor of *n*" and "double" should be translated as增加到（原来的）*n*倍or增加了*n*－1倍。For example,

(1) Using the method, one machine part had its working life *increased by 14 times*.

利用这种方法，机器零件的工作寿命延长了13倍（延长到原来的14倍）。(NOT: 延长了14倍)

(2) The energy available in a wind stream is proportional to the cube of its speed, which means that *doubling* the wind speed *increases* the available energy *by a factor of eight*.

风中能够利用的能量同风速的立方成正比，就是说风速提高1倍，可用能量将增加7倍。(NOT: 提高2倍；增加8倍)

### 3. Decrease in quantities

In case of decrease in quantities, similar to the expression "*n* times less (fewer) than" discussed above, even if the word "time" is used in English sentences, the Chinese prefer not to use the word 倍; instead fractions（分数）are often used. Therefore, the expressions such as "reduce by *n* times", "reduce *n* times", "reduce by a factor of *n*", "reduce *n* times as much (many) as", and "*n*-fold reduction" should be translated as "减少了(n − 1)/n" or "减少到（原来的）1/n"。For example,

(1) The natural disaster has caused the output of the crops to *decrease three times*. 自然灾害使农作物的产量减少了2/3（减少到原来的1/3)。(NOT: 减少了三倍)

(2) According to the report, the new type of cells would reduce the cost of producing solar power *by a factor of 10*.
根据该报告，这种新型电池可使太阳能发电的成本降低到原来的1/10。(NOT: 降低了10倍)

(3) The error probability of the equipment was reduced by 2.5 times through technical innovation.
通过技术革新，该设备的误差概率降低了3/5。(NOT: 降低了2.5倍)

## IV. Translate the following sentences. Pay attention to how numbers in the sentences should be translated.

*1. English-Chinese Translation*

(1) Mission planners estimate that the LAS, together with *Orion*'s advanced guidance and control system, would be able to return the crew safely 999 out of 1,000 times it is needed.

(2) Because *Orion*'s reentry velocity from a moon mission (which is on the order of 11 kilometers a second) will be 41 percent faster than a shuttle's descent speed from low Earth orbit, the heat load will be several times greater.

(3) Being 40 times larger in area, *Orion*'s heat shield will need to be built in segments, thus adding new complexities.

(4) Scientists said the expansion next year will permit the system to deal with 1,000 million pieces of information each second, four times more than now.

(5) There are examples where factors of 20 and greater savings were achieved using micro-programmed problem-oriented processors, as compared with a typical general-purpose

processor.

*2. Chinese-English Translation*

(1) 只要在空气系统里注入3%的二氧化碳，宇航员的呼吸速率就会增加1倍，同时产生听觉障碍。

(2) 很多太阳灶厨房靠抛物镜工作，这种镜子能使太阳光线的强度提高到原来的1,000倍。

(3) 美国每年有近79万人死于心脏病和动脉血管病，几乎是死于癌症人数的2倍，大约是死于车祸人数的15倍。

(4) 传统喷气式发动机使用由石油提炼出的碳氢燃料，但当速度约3倍于音速时，这些发电机就会变得过热。

(5) 在对飞航式导弹的布局进行性能评定时，应用这种一体化试验技术使风洞时间降到原来的1/5。

## Translation Techniques (2)

### Translation of the Titles of Science Articles
### （科技文章题目的译法）

Besides academic journals which are intended for professionals, there are also numerous science journals published in English throughout the world aiming at the general public, such as *Nature, New Scientist, Science Magazine, National Geographic*, just to mention a few, and many others, though not science journals, cover important scientific discoveries and events. In translation of these articles and reports, one part that needs careful consideration is how their titles can be put into Chinese properly. This is not only because the title of an article is normally the first part the readers see, but also because the language used by the title has its own unique features.

In order to catch readers' attention and arouse their interest, titles of the articles and reports in the magazines often adopt short or incomplete sentences that only contain key words. Similarly, translated titles should also be concise and eye-catching, and when necessary some adjustments should be made to reflect the exact meaning of original versions.

The commonly adopted techniques discussed in this book are also applicable in the translation of titles of science articles and reports. For example, the translation of the title *First Light from Distant Planets Seen by Telescope* involves the following adjustments: the inversion of the passive voice into the active voice, the addition of the words科学家to indicate the doers, and the transformation of the adjective "first" into the adverb首次. Thus, the Chinese translation of the title is科学家首次通过望远镜看到来自遥远行星的光线. It is noticeable that the original title is a phrase, but when translated into Chinese it is necessary to change it into a complete sentence so that the meaning can be expressed clearly, and at the same time it is idiomatic.

Sometimes the change of tenses is also necessary. For example, in the title *Astronomers find oldest, most distant planet*, the verb is in the form of the present tense, but the action indicated by the verb actually took place at a certain moment in the past, and this should be reflected by adding the word了to the verb in the Chinese translation: 天文学家发现了最古老、最遥远的行星.

Mostly, English titles of science articles and reports are brief and concise with only concrete words, which can save space and at the same time save time of readers. However, when translated into Chinese they often need some words to be added to logically and clearly indicate original meanings. For example, the proper translation of the title *Trees, Less Global Warming, Right? — Not Exactly* is 多种树就能扭转全球变暖的趋势吗？答案是：不一定. Similarly, omission of words in translation is also possible if necessary as the following example shows: *Antibiotics: Handle Them with Care*（抗生素：慎用！）.

**V. Discuss the translation technique and the ways of applying the technique to the translation of the following titles. Choose the appropriate translation from each pair.**

1. Study Suggests Primates and Dinosaurs Shared the Earth
   a. 研究表明灵长目动物与恐龙曾同时在地球上生存
   b. 研究表明灵长目动物与恐龙曾经分享过地球
2. Study Finds Smog Raises Death Rate
   a. 研究发现烟雾提高了死亡率
   b. 研究发现烟雾污染造成更多的人死亡
3. Cell's "Quality Control" Mechanism Discovered
   a. 科学家已发现细胞的"质量监控"机制
   b. 细胞的"质量监控"机制被发现
4. Putting a Virtual Doctor in the Ambulance
   a. 把一个虚拟医生放在救护车里
   b. 救护车中的虚拟医生
5. Phone Therapy Seen as Anti-depression Aid

a. 被看作是有助于抗忧郁的电话疗法

b. 电话疗法——抗忧郁的好助手

## VI. Translate the following titles of science articles into Chinese. Pay attention to what changes should be made so that translated titles are accurate and impressive.

1. The Future of Mars: Plans for NASA's Next Decade of Red Planet Probing
2. More whole grains, exercise urged in new food pyramid
3. Aging: Blame the Brain for Hearing Loss
4. Scientists make bacteria behave like computers
5. Global Warming Unstoppable for 100 Years, Study Says
6. Supercomputer to simulate nuke explosion
7. Life in a Bubble: Mathematicians Explain How Insects Breathe under Water
8. Digestive Specialists Freeze Out Esophagus Cancer with New Therapy
9. Nanoparticles + Light = Dead Tumor Cells
10. New Materials for Microwave Cookware that Heats Faster with Less Energy
11. Mapping Out Future of Intelligent Robots
12. Scientists Race to Stay One Step Ahead of the Drug-taking and Genetic Manipulation That Threatens Sport

## VII. Translate the following passage into Chinese.

NASA is formulating a Mars exploration plan for the next decade, receiving advice from all quarters, from outside academic circles to internal NASA working groups, as well as from the White House. And if all goes according to plan, Mars will speak for itself, giving up surface and subsurface secrets as ever-more capable spacecraft—like the two rovers currently en route—survey that mysterious world. How to respond to the expected fast-paced rush of new discoveries, enough so that outgoing missions can take advantage of just-in findings, is a challenge.

One scenario has been advocated by the Committee on Planetary and Lunar Exploration (COMPLEX)—a study arm of the National Research Council. It calls for a legion of robotic return-sample craft that are needed to truly unravel the history of Mars and reveal whether life existed in the planets past or is present today.

Beginning in 2011, the COMPLEX study states, the first in a series of perhaps 10 automated Mars return-sample missions would be launched to dot the red planet. This robotic search and seize campaign of hauling back to Earth Mars specimens might stretch out for three or four decades, to as much as a century.

Rocketing back to Earth samples of Martian soil and rock via automated means has been on NASA's to-do list for decades. But it is a difficult task, with cost being the lead impediment. Over the years, NASA as well as industry teams have sketched out details for such an undertaking. So too have European Space Agency engineers, as well as independent groups in France and Russia. Those grappling with ways to land, pick up, and then safely haul back the goods to Earth from Mars—all by robotic hardware—wind up with an acute case of Space Age sticker shock.

**VIII. Translate the following passage into English.**

### 空间站

美国通过"阿波罗"计划对月球进行了地球物理探测之后，又继续利用"太空实验室"进行人类太空探索。"太空实验室"是环绕地球运动的空间站，曾作为3名宇航员的工作站和生活基地。首先由火箭助推器将主密封舱送入太空，随后宇航员乘坐与"阿波罗"号飞船类型相同的飞行器到达主密封舱，并与之对接。"太空实验室"的工作寿命为8个月，在此期间三批宇航员（每批3人）分别在此停留了1个月、2个月、3个月。第一批宇航员于1972年5月到达"太空实验室"。

"太空实验室"的科研任务有时侧重于对太阳进行的天体物理学研究，有时侧重于对地球自然资源的研究，此外，宇航员对长期处于失重状态的反应进行评估。太阳观测室装有8台高清晰度望远镜，每台望远镜研究太阳光谱的不同部分（例如可见光、紫外线、X光、红外光等）。对太阳耀斑的研究受到了特别关注。利用空间站对地球的研究包括地球资源的遥感观测，使用的是基于可见光和红外光的技术，叫做多光谱扫描。由此获得数据有助于科学家预测农作物和木材的产量，寻找具有生产潜力的土地，监察虫灾，绘制沙漠分布图，测量冰、雪覆盖层，寻找矿藏，追踪水生与野生动物的迁徙，以及探查空气与水污染的分布情况等。此外，通过雷达探测还获得了有关全球范围海洋表面的起伏状况和海水带电特性的信息。"太空实验室"于1979年7月脱离运行轨道，虽然经过努力，但仍有数个大块残骸落到地球表面。

此后只有俄罗斯的空间站继续为人类提供在地球轨道逗留的条件，宇航员在空间站的工作时间长达14个月以上。除了进行遥感探测和收集医学数据，宇航员还利用微重力环境生产电子和医学产品，这些产品在地球上是不可能制造出来的。国际空间站是美国、俄罗斯、日本、加拿大、巴西和欧洲宇航局的合作项目，为了准备该空间站的建设，来自阿富汗、奥地利、英国、保加利亚、法国、德国、日本、哈萨克斯坦、叙利亚和美国的宇航员与俄罗斯宇航员共同展开了工作。

## Section 2 Reading for Academic Purposes

*Technical communication is the process of conveying usable information through writing or speech about a specific domain to an intended audience. Information is usable if the intended audience is able to perform an action or make a decision based on its contents.*

*Technical communicators often work collaboratively to create products for various media, including paper, video, and the Internet. Technical communication covers wide areas such as online help, user manuals, technical manuals, specifications, process and procedure manuals, training, business papers and reports.*

*Technical communication is sometimes considered a professional task for which organizations either hire specialized employees or outsource their needs to communication firms. For example, a professional writer may work with a company to produce a user manual. Other times, technical communication is regarded as a responsibility that technical professionals employ on a daily basis as they work to convey technical information to co-workers and clients. For example, a computer scientist may need to provide software documentation to fellow programmers or clients.*

*In the early 1900s, technical communication was a burgeoning professional field, represented in academe by service courses taught primarily at engineering institutions. By the 1980s, however, it had become a significant professional and academic discipline in its own right, and now many colleges and universities include technical communication in their curricula.*

## Technical Communication Documents
## 技术资料

biotechnology *n.* 生物技术；生物工艺学

Technical communication is conducted in fields as diverse as computer hardware and software, chemistry, the aerospace industry, robotics, finance, consumer electronics, and **biotechnology**. Technical communication includes information development, technical documentation, or technical publications. In some organizations, technical writers may be called information developers, documentation specialist, documentation engineer, or technical content developers. Technical communication documents are developed to explain complex ideas to technical and nontechnical audiences. This could mean telling a programmer how to use a software library, or telling a consumer how to operate a television remote.

### Communicating with the audience

exposition *n.* 说明性讲话（或文章等）；讲解

A technical communication document conveys a particular piece of information to a particular audience for a particular purpose. It is often **exposition** about scientific subjects and technical subjects associated with sciences.

Technical communication translates complex technical concepts into simple language to enable a specific user or set of users to perform a specific task in a specific way. Effective communication requires the skills to produce quality content, language, format, and more. To present appropriate content, writers must understand the audience and purpose. Technical communicators play a vital role in bridging the gap between those who create new ideas or who are subject matter experts (such as scientists and engineers) and those who must understand how to implement their ideas.

Technical communication is most often associated with online help and user manuals; however, there are other forms of technical content created by technical writers, including: assembly instructions, developer guides, maintenance and repair procedures, installation guides, user guides, scientific reports, specifications, instruction manuels, technical papers, and **troubleshooting guides**, etc. The following is a brief introduction to the user guides and instruction manuals, the commonly used technical communication documents.

troubleshooting guide（机器等的）故障检修指南

**User guides**

A user guide, also commonly known as a manual, is a technical communication document intended to give assistance to people using a particular system. It is usually written by a technical writer, although user guides could be written by programmers, product or project managers, or other technical staff, particularly in smaller companies.

User guides are most commonly associated with electronic goods, computer hardware and software.

Most user guides contain both a written guide and the associated images. In the case of computer applications it is usual to include **screenshots** of how the program should look, and hardware manuals often include clear, simplified diagrams. The language is written to match up with the intended audience with jargon kept to a minimum or explained thoroughly.

screenshot *n.* 屏幕截图

The usual sections of a user manual often include:

- A **coverpage**;
- A titlepage and copyright page;
- A preface, containing details of related documents and information on how to best use the user guide;

coverpage *n.* 封面

- A contents page;
- A guide on how to use at least the main functions of the system;
- A troubleshooting section detailing possible errors or problems that may occur along with how to fix them;
- An FAQ (frequently asked questions);
- Where to find further help and contact details;
- A glossary and, for larger documents, an index.

### Car owner's manual

glove compartment 汽车仪表板上的小贮藏柜
defensive driving 防御性驾驶；安全驾驶

All new cars come with an owner's manual from the manufacturer. Most owners leave them in the **glove compartment** for easy reference. This can make their frequent absence in rental cars frustrating because it violates the driver's user expectations, as well as makes it difficult to use controls that aren't understood, which is not good because understanding control operation of an unfamiliar car is one of the first steps recommended in **defensive driving**. Owner's manuals usually cover three main areas—a description of the location and operation of all controls, a schedule and descriptions of maintenance required, both by the owner and by a mechanic, and specifications such as oil and fuel capacity and part numbers of light bulbs used.

liability *n.* 责任；义务
lawsuit *n.* 诉讼（案件）

Current car owner's manuals have become much bigger in part due to many safety warnings most likely designed to avoid product **liability lawsuits**, as well as from ever more complicated audio and navigational systems, which often have their own manual.

### Instruction manual (computer and video games)

storyline *n.* 故事情节

An instruction manual, in the context of computer and video games, is a booklet that instructs the player on how to play the game, gives descriptions of the controls and their effects, and shows a general outline of the concepts and goals of the game. It is also common for manuals to contain a brief summary of the **storyline** of the game, especially in games with complex storylines, such as role playing games. Manuals can be large, such as the *Civilization II* manual which runs hundreds of pages, or small, such as the single sheet of double-sided A5 paper included with *Half-Life 2*.

Computer games typically have larger manuals because some genres native to personal computers such as **simulators** or strategy games require a more in-depth explanation of the interface and game mechanics. Furthermore, instruction manuals for personal computer games tend to include installation instructions to assist a user in installing the game, but those instructions could also appear in a separate piece of paper or in a different leaflet. As some of these manuals are so large as to be **cumbersome** when searching for a specific section, some games include a quick reference card (usually a list of keyboard commands) on a separate sheet of paper or in the back cover of the manual.

simulator *n.* 模拟装置
cumbersome *adj.* 讨厌的，麻烦的，笨重的

Other manuals go much farther than being simple guides: some games based on historical or well developed **fictional** stories often include extensive information about the settings, like WWI combat **simulator** *Flying **Corps***, where every campaign was thoroughly described with historical information. In some genres, this led to the **aforementioned** large manuals traditional with computer games.

fictional *adj.* 虚构的
simulator *n.* 模拟装置
corps *n.* 军团，技术兵种，特殊兵种
aforementioned *adj.* 上述的，前述的

The trend in recent years is toward smaller manuals—sometimes just a single instruction sheet—for a number of reasons. **Console games** are no longer sold in large **cardboard** boxes; instead, since the early 2000s, DVD cases have been used (as today's major consoles use DVDs), which leave no room for a large manual. Printing is also expensive, and game publishers can save money by including a PDF of the manual on the disk (computer games) or **in-game help** (both computer and console games). This kind of manual may include the use of "spare" buttons which become available on controllers, screen resolution and an **obstacle course** to teach movement and target practice to teach shooting and introduce different types of weapons; in real-time strategy games, there are one or more missions that teach not only the issuing of basic commands (build, upgrade, move, harvest resources, attack) but also the uses of basic units.

console game 单机游戏
cardboard *n.* 纸板
in-game help 网络游戏帮助工具
obstacle course 超越障碍训练

## Exercises

**I. Read the passage and decide whether the following statements are true or false. Write T for True and F for False in the brackets.**

1. Technical communication can also be called information development, technical documentation, or technical publications. ( )
2. Technical communication interprets the complex technical concepts into simple language so that the user is able to accomplish a specific task in a specific way. ( )
3. Technical communication takes various forms of technical content such as assembly instructions, developer guides, maintenance and repair procedures, installation guides, user guides, etc. ( )
4. User guides are mostly attached to daily necessity goods. ( )
5. A user guide often provides solutions to detailed possible errors or problems that may occur when people use a particular system. ( )
6. People who rent cars seldom read the car owner's manuals in cars so that some of them may feel frustrated when they have trouble in driving an unfamiliar car. ( )
7. Instruction manuals have become much bigger owing to the addition of many safety warnings and sub-manuals of some more complicated systems. ( )
8. For searching convenience, there is a quick manual reference card on a specific paper or in the back cover of the manuals of computer games. ( )
9. Some instruction manuals for computer and video games function not only as simple guides, but also supply historical background or fictional stories for the games. ( )
10. Many game publishers of computer games tend to use electronic manuals rather than paper manuals for less cost. ( )

**II. Read the passage again, and complete the following items.**

1. Fields in which technical communication may be conducted: ___________
2. Synonyms of technical writers: ___________
3. The function of technical communication documents: ___________
4. The people who write user guides include: ___________
5. The components of a user manual: ___________
6. The general components of a car owner's manual: ___________
7. The definition of an instruction manual (computer and video games): ___________
8. The tendency of instruction manuals in recent years: ___________

### Quotations from Great Scientists

*Give me a lever long enough and a fulcrum on which to place it, and I shall move the world.*

*—Archimedes*

假如给我一支足够长的杠杆和用来支撑杠杆的支点，那么我就能撬动整个地球。

——阿基米德

# Chapter 12 Farming Techniques 农业技术

## Section 1 Reading and Translation

### I What You Are Going to Read

Since ancient times, farmers in many cultures have prepared the land for growing crops. They used plows to turn the soil in their fields. This is called tilling. Now, a method called no-till farming (sometimes called zero tillage) is gaining popularity all over the world on big and small farms.

Producing crops usually involves regular tilling to remove weeds, and lift up the remains of last season's crops and unwanted plants. The process brings air into the soil so dead plant material breaks down quickly to form natural fertilizer. However, conventional plow-based farming leaves soil vulnerable to erosion and reduce the amount of water in soil. Over the last several years, farmers have sought ways to protect soil by avoiding unneeded tilling. For example, soybeans, wheat, corn, and cotton are crops often farmed without tilling. In spite of the advantages of no-till farming, some problems, even risks, are hindering widespread adoption of this approach. You will learn more about no-till farming and its development from the following essay published in *Scientific American* (June, 2008).

### II About the Authors

David R. Huggins is a soil scientist with the USDA-Agricultural Research Service, Land Management and Water Conservation Research Unit in Pullman, Wash. He specializes in conservation cropping systems and their influence on the cycling and flow of soil carbon and nitrogen. John P. Reganold, Regents Professor of Soil Science at Washington State University at Pullman, specializes in sustainable agriculture.

### III Reading Passage

**No-Till: How Farmers Are Saving the Soil by Parking Their Plows**

**免耕法：耕作者为保护土壤而放下手中的犁**

By David R. Huggins and John P. Reganold

*The age-old practice of turning the soil before planting a new crop is a leading cause of farmland degradation. Many farmers are thus looking to make plowing a thing of the past.*

crumble *v.* 破碎，碎裂
porous *adj.* 疏松的；易渗水的

1 John Aeschliman turns over a shovelful of topsoil on his 4,000-acre farm in the Palouse region[1] of eastern Washington State. The black earth **crumbles** easily, revealing a **porous** structure and an abundance of organic matter that facilitate root growth. Loads of earthworms are visible, too—another healthy sign.

residue *n.* 剩余物，残留物
tillage *n.* 耕耘；耕地
take a toll 造成损失
no-till farming 免耕法

2 Thirty-four years ago only a few earthworms, if any, could be found in a spadeful of his soil. Back then, Aeschliman would plow the fields before each planting, burying the **residues** from the previous crop and readying the ground for the next one. The hilly Palouse region had been farmed that way for decades. But the **tillage** was **taking a toll** on the Palouse, and its famously fertile soil was eroding at an alarming rate. Convinced that there had to be a better way to work the land, Aeschliman decided to experiment in 1974 with an emerging method known as **no-till farming**.

manure *n.* 粪肥
aerate *v.* 通风
vulnerable *adj.* 易受伤害的
runoff *n.*（雨水、融雪等的）径流（量）
pesticide *n.* 杀虫剂
mulch *n.* 覆盖层
seeder *n.* 播种机，播种器

3 Most farmers worldwide plow their land in preparation for sowing crops. The practice of turning the soil before planting buries crop residues, animal **manure** and troublesome weeds and also **aerates** and warms the soil. But clearing and disturbing the soil in this way can also leave it **vulnerable** to erosion by wind and water. Tillage is a root cause of agricultural land degradation—one of the most serious environmental problems worldwide—which poses a threat to food production and rural livelihoods, particularly in poor and densely populated areas of the developing world. By the late 1970s in the Palouse, soil erosion had removed 100 percent of the topsoil from 10 percent of the cropland, along with another 25 to 75 percent of the topsoil from another 60 percent of that land. Furthermore, tillage can promote the **runoff** of sediment, fertilizers and **pesticides** into rivers, lakes and oceans. No-till farming, in contrast, seeks to minimize soil disruption. Practitioners leave crop residue on the fields after harvest, where it acts as a **mulch** to protect the soil from erosion and fosters soil productivity. To sow the seeds, farmers use specially designed **seeders** that penetrate through the residue to the undisturbed soil

below, where the seeds can **germinate** and surface as the new crop.

4 In its efforts to feed a growing world population, agriculture has expanded, resulting in a greater impact on the environment, human health and biodiversity. But given our current knowledge of the planet's capacity, we now realize that producing enough food is not enough—it must also be done sustainably. Farmers need to generate adequate crop yields of high quality, conserve natural resources for future generations, make enough money to live on, and be socially just to their workers and community. No-till farming is one system that has the potential to help realize this vision of a more sustainable agriculture. As with any new system, there are challenges and **trade-offs** with no-till. Nevertheless, growers in some parts of the world are increasingly abandoning their plows.

germinate *v.* 发芽；发育

trade-off *n.*（对不能同时兼顾的因素的）权衡

**Plowing Ahead**

5 People have used both no-till and tillage-based methods to produce food from the earth ever since they started growing their own crops around 10,000 years ago. In the transition from hunting and gathering to raising crops, our **Neolithic predecessors** planted garden plots near their dwellings and **foraged for** other foods in the wild. Some performed the earliest version of no-till by punching holes in the land with a stick, dropping seeds in each **divot** and then covering it with soil. Others scratched the ground with a stick, an **incipient** form of tillage, to place seeds under the surface. Thousands of farmers in developing countries still use these simple methods to sow their crops.

Neolithic *adj.* 新石器时代的
predecessor *n.* 祖先；祖辈
forage for 寻找，搜寻
divot *n.* 草皮层
incipient *adj.* 初始的；起初的

6 In time, working the soil mechanically became the standard for planting crops and controlling weeds, thanks to the **advent** of the plow, which permitted the labor of a few to sustain many. The first such tools were scratch plows, consisting of a frame holding a vertical wooden post that was dragged through the topsoil. Two people probably operated the earliest version of this device, one pulling the tool and the other guiding it. But the **domestication** of **draft animals**—such as oxen in Mesopotamia[2], perhaps as early as 6000 BC—replaced human power. The next major development occurred

advent *n.* 出现；来临
domestication *n.* 驯养；驯服
draft animal 牵引重物的动物

plowshare *n.* 犁头
wedge-shaped *adj.* 楔形的；V形的
moldboard *n.* 犁板

around 3500 BC, when the Egyptians and the Sumerians[3] created the **plowshare**—a **wedge-shaped** wooden implement tipped with an iron blade that could loosen the top layer of soil. By the 11th century, the Europeans were using an elaboration of this innovation that included a curved blade called a **moldboard** that turned the soil over once it was broken open.

7 Continuing advancements in plow design enabled the explosion of pioneer agriculture during the mid-1800s; farmers cultivated grass-dominated native prairies in eastern Europe, South Africa, Canada, Australia, New Zealand and the US, converting them to corn, wheat and other crops. One such region, the tall-grass prairie of the Midwestern US had resisted widespread farming because its thick, sticky sod was a barrier to cultivation. But in 1837 an Illinois blacksmith named John Deere[4] invented a smooth, steel moldboard plow that could break up the sod. Today this former grassland, which includes much of the famous Corn Belt[5], is home to one of the most agriculturally productive areas in the world.

scrutiny *n.* 详细审查
vulnerability *n.* 弱点
ravage *v.* 蹂躏；劫掠
spur *v.* 刺激；鞭策
folly *n.* 蠢事；耗费巨大而又无益的事
agronomist *n.* 农学家；农艺学家
tenable *adj.*（主张等）站得住脚的
herbicide *n.* 除草剂
atrazine *n.* 莠去津（一种除草剂）
paraquat *n.*【化】百草枯（一种除草剂）

8 Agricultural mechanization continued through the early 1900s with the development of many tools that helped farmers cultivate the earth ever more intensively, including tractors that could pull multiple plows at once. Tillage practices were about to undergo profound **scrutiny**, however. The Dust Bowl era[6] between 1931 and 1939 exposed the **vulnerability** of plow-based agriculture, as wind blew away precious topsoil from the drought-**ravaged** southern plains of the US, leaving behind failed crops and farms. Thus, the soil conservation movement was born, and agriculturalists began to explore reduced tillage methods that preserve crop residues as a protective ground cover. **Spurring** the movement was the controversial publication in 1943 of *Plowman's **Folly***, by **agronomist** Edward Faulkner, who challenged the necessity of the plow. Faulkner's radical proposition became more **tenable** with the development of **herbicides**—such as 2,4-D, **atrazine** and **paraquat**—after World War II, and research on modern methods of no-till agriculture began in earnest during the 1960s.

conceive *v.* 构想出（主意、计划等）；设想

9 Considering the pivotal role the plow has come to play in farming, **conceiving** a way to do without it has proved quite challenging, requiring the reinvention of virtually every aspect

of agricultural production. But specially designed seeders have been evolving since the 1960s to meet the unique mechanization requirements of no-till farming. These new seeders, along with chemical herbicides, are two of the main technologies that have at last enabled growers to effectively practice no-till on a commercial scale.

**Signing Up for No-Till**[7]

10 Farmers today prepare for planting in ways that disturb the soil to varying degrees. Tillage with a moldboard plow completely turns over the first 6 to 10 inches of soil, burying most of the residue. A **chisel plow**, meanwhile, only **fractures** the topsoil and preserves more surface residue. In contrast, no-till methods merely create in each planted row a **groove** just half an inch to three inches across into which seeds can be dropped, resulting in minimal overall soil disturbance. In the US, no-till agriculture fits under the broader US Department of Agriculture definition of conservation tillage. Conservation tillage includes any method that retains enough of the previous crop residues such that at least 30 percent of the soil surface is covered after planting. The protective effects of such residues are considerable. According to the USDA's *National Resources Inventory* data[8], soil erosion from water and wind on US cropland decreased 43 percent between 1982 and 2003, with much of this decline coming from the adoption of conservation tillage.

chisel plow 凿形犁
fracture *v.* 使破碎，使碎裂
groove *n.* 凹槽

11 Soil protection is not the only benefit of no-till. Leaving crop residues on the soil surface helps to increase water **infiltration** and limit runoff. Decreased runoff, in turn, can reduce pollution of nearby water sources with transported sediment, fertilizers and pesticides. The residues also promote water conservation by reducing evaporation. In instances where water availability limits crop production, greater water conservation can mean higher-yielding crops or new capabilities to grow alternative crops.

infiltration *n.* 渗透

12 The no-till approach also fosters the diversity of soil flora and fauna by providing soil organisms, such as earthworms, with food from the residues and by stabilizing their habitat. Together with associated increases in soil organic matter, these conditions encourage soils to develop a more stable internal structure, further improving the overall capacity to grow crops and to buffer them against stresses

caused by farming operations or environmental hazards. No-till can thus enable the more sustainable farming of moderately to steeply sloping lands that are at elevated risk of erosion and other problems.

13 Wildlife, too, gains from no-till, because standing crop residues and inevitable harvest losses of grain provide cover and food for upland game birds and other species.[9] In a study published in 1986, researchers in Iowa found 12 bird species nesting in no-till fields, compared with three species in tilled fields.

carbon sequestration 碳截存（二氧化碳的捕获、分离、存贮或再利用）
mitigation *n.* 减轻；缓解
photosynthesis *n.* 光合作用
sequester *v.* 使隔绝

14 Furthermore, reducing tillage increases soil **carbon sequestration**, compared with conventional moldboard plowing. One of agriculture's main greenhouse gas **mitigation** strategies is soil carbon sequestration, wherein crops remove carbon dioxide from the atmosphere during **photosynthesis**, and nonharvested residues and roots are converted to soil organic matter, which is 58 percent carbon. About half of the overall potential for US croplands to **sequester** soil carbon comes from conservation tillage, including no-till.

mundane *adj.* 世俗的；寻常的

15 In addition, no-till can offer economic advantages to farmers. The number of passes over a field needed to establish and harvest a crop with no-till typically decreases from seven or more to four or fewer. As such, it requires 50 to 80 percent less fuel and 30 to 50 percent less labor than tillage-based agriculture, significantly lowering production costs per acre. Although specialized no-till seeding equipment can be expensive, with some sophisticated seeders priced at more than $100,000, running and maintaining other tillage equipment is no longer necessary, lowering the total capital and operating costs of machinery required for crop establishment by up to 50 percent. With these savings in time and money, farmers can be more competitive at smaller scales, or they can expand and farm more acres, sometimes doubling farm size using the same equipment and labor. Furthermore, many farmers appreciate that the time they once devoted to rather **mundane** tillage tasks they can instead spend on more challenging aspects of farming, family life or recreation, thereby enhancing their overall quality of life.

## Betting the Farm[10]

16 No-till and other conservation tillage systems can work in a wide range of climates, soils and geographic areas. Continuous no-

till is also applicable to most crops, with the notable exceptions of wetland rice and root crops, such as potatoes. Yet in 2004, the most recent year for which data are available, farmers were practicing no-till on only 236 million acres worldwide—not even 7 percent of total global cropland.

17 Of the top five countries with the largest areas under no-till, the US ranks first, followed by Brazil, Argentina, Canada and Australia. About 85 percent of this no-till land lies in North and South America. In the US, roughly 41 percent of all planted cropland was farmed using conservation tillage systems in 2004, compared with 26 percent in 1990. Most of that growth came from expanded adoption of no-till, which more than tripled in that time, to the point where it was practiced on 22 percent of US farmland. This no doubt partly reflects the fact that US farmers are encouraged to meet the definition of conservation tillage to participate in government subsidy and other programs. In South America, adoption of no-till farming has been relatively rapid as a result of coordinated efforts by university agricultural-extension educators[11] and local farm communities to develop viable no-till cropping systems **tailored to** their particular needs.

tailor to 为某一特定目的或目标而制作、修改或调整

18 On the other hand, adoption rates are low in Europe, Africa and most parts of Asia. **Embracing** no-till has been especially difficult in developing countries in Africa and Asia, because farmers there often use the crop residues for fuel, animal feed and other purposes. Furthermore, the specialized seeders required for sowing crops and the herbicides needed for weed control may not be available or can be prohibitively expensive for growers in these parts of the world. Meanwhile, in Europe, an absence of government policies promoting no-till, along with elevated restrictions on pesticides (including herbicides), among other variables, leaves farmers with little incentive to adopt this approach.

embrace *v.* 乐意采纳

19 Changing from tillage-based farming to no-till is not easy. The difficulty of the transition, together with the common perception that no-till incurs a greater risk of crop failure or lower net returns than conventional agriculture, has seriously hindered more widespread adoption of this approach. Although farmers accept that agriculture is

fail-safe *adj.* 万无一失的
fungal *adj.* 真菌的

not a **fail-safe** profession, they will hesitate to adopt a new farming practice if the risk of failure is greater than in conventional practice. Because no-till is a radical departure from other farming practices, growers making the switch to no-till experience a steep learning curve.[12] In addition to the demands of different field practices, the conversion has profound impacts on farm soils and fields. Different pest species can arise with the shift from tillage-based agriculture to no-till, for instance. And the kinds of weeds and crop diseases can change. For example, the elevated moisture levels associated with no-till can promote soil-borne **fungal** diseases that tillage previously kept in check. Indeed, the discovery of new crop diseases has sometimes accompanied the shift to no-till.

vigilant *adj.* 警惕的

20 Some of the changes that follow from no-till can take years or even decades to unfold, and farmers need to remain **vigilant** and adaptable to new, sometimes unexpected, situations, such as those that arise from shifts in soil and residue conditions or fertilizer management. During this transition, there is a real risk of reduced yields and even failed crops. In the Palouse, for example, some farmers who attempted no-till in the 1980s are no longer in business. Consequently, farmers looking to switch to no-till should initially limit the converted acreage to 10 to 15 percent of their total farm.

underwrite *v.* 给……保险；签署（保险单）
lease *v.* 租用
agribusiness *n.* 农业综合企业
go a long way 大有帮助，大有作用

21 Farmers who are new to no-till techniques often visit successful operations and form local or regional support groups, where they share experience and discuss specific problems. But the advice they receive in areas with limited no-till adoption can be incomplete or contradictory, and gaps in knowledge, experience or technology can have potentially disastrous outcomes. If the perception that no-till is riskier than conventional techniques develops in a farming community, banks may not **underwrite** a no-till farmer's loan. Alternatively, growers who are **leasing** land may find that the owners are opposed to no-till because of fears that they will not get paid as much. Improving the quality of information exchange among farmers, universities, **agribusinesses** and government agencies will no doubt **go a long way** toward overcoming these obstacles.

22 Yet even in the hands of a seasoned no-till farmer, the system has drawbacks. No-till crop production on fine-textured, poorly drained soils can be particularly problematic, often resulting in

decreased yields. Yields of no-till corn, for instance, are often reduced by 5 to 10 percent on these kinds of soils, compared with yields with conventional tillage, particularly in northern regions. And because the crop residue blocks the sun's rays from warming the earth to the same degree as occurs with conventional tillage, soil temperatures are colder in the spring, which can slow seed **germination** and **curtail** the early growth of warm-season crops, such as corn, in northern latitudes.[13]

germination *n.* 萌芽；发育
curtail *v.* 减少；削减

23 In the first four to six years, no-till demands the use of extra nitrogen fertilizer to meet the **nutritional** requirements of some crops, too—up to 20 percent more than is used in conventional tillage systems—because increasing organic matter at the surface **immobilizes nutrients**, including nitrogen. And in the absence of tillage, farmers depend more heavily on herbicides to **keep** weeds **at bay**.

nutritional *adj.* 营养的
immobilize *v.* 使不动，使固定
nutrient *n.* 营养成分
keep at bay 不使……迫近，牵制

24 Herbicide-resistant weeds are already becoming more common on no-till farms. The continued practice of no-till is therefore highly dependent on the development of new herbicide formulations and other weed management options. Cost aside, greater reliance on **agrichemicals** may adversely affect nontarget species or **contaminate** air, water and soil.

agrichemical *n.* 农用化学品
contaminate *v.* 污染

## Integrating No-Till

25 No-till has the potential to deliver a host of benefits that are increasingly desirable in a world facing population growth, environmental degradation, rising energy costs and climate change, among other daunting challenges. But no-till is not a cure-all; such a thing does not exist in agriculture. Rather it is part of a larger, evolving vision of sustainable agriculture, in which a diversity of farming methods from no-till to organic—and combinations thereof—is considered healthy. We think that ultimately all farmers should integrate conservation tillage, and no-till if feasible, on their farms.

26 Future no-till farming will need to employ more diverse pest and weed management strategies, including biological, physical and chemical measures to lessen the threat of pesticide resistance. Practices from successful organic farming systems may be instructive in that regard. One such technique, **crop rotation**—in which farmers grow a series of different crops in the same space in **sequential** seasons—is

crop rotation【农】轮作制
sequential *adj.* 连续的；相继的

already helping no-till's war on pests and weeds by helping to break up the weed, pest and disease cycles that arise when one species is continuously grown.

ethanol *n.* 乙醇
monoculture *n.* 【农】单一栽培；单作
biofuel *n.* 生物燃料
bioenergy *n.* 生物能
perennial *n.* 【植】多年生植物
switchgrass *n.* 柳枝稷（生长于北美洲，用作牧场饲料及干草）

27 To that end, the capacity to grow a diverse selection of economically viable crops would advance no-till farming and make it more appealing to farmers. But the current emphasis on corn to produce **ethanol** in the Midwestern Corn Belt, for instance, is promoting **monoculture**—in which a single crop, such as corn, is grown over a wide area and replanted every year—and will likely make no-till farming more difficult in this region. Experts continue to debate the merits of growing fuel on farmland, but if we decide to proceed with **biofuel** crops, we will need to consider using no-till with crop rotation to produce them sustainably. Development of alternative crops for **bioenergy** production on marginal lands, including **perennials** such as **switchgrass**, could complement and promote no-till farming, as would perennial grain food crops currently under development.

sustainability *n.* 可持续性

28 Today, three decades after first attempting no-till on his Palouse farm, John Aeschliman uses the system on 100 percent of his land. His adoption of no-till has followed a gradual, cautious path that has helped minimize his risk of reduced yields and net returns. Consequently, he is one of many farmers, large and small, who is reaping the rewards of no-till farming and helping agriculture evolve toward **sustainability**.

## Notes

1. **Palouse region:** 帕洛斯地区，位于美国西北部，地跨东华盛顿州、爱达荷州中北部，向南延伸到俄勒冈州的东北部，是美国主要的小麦产区。
2. **Mesopotamia:** 美索不达米亚，亚洲西南部底格里斯河（Tigris）和幼发拉底河（Euphrates）两河流域间的古王国，今伊拉克所在地。这一地区孕育了众多的人类早期文明。
3. **Sumerian:** 苏美尔人，一个古代民族，很可能是非闪米特（non-Semitic）的起源，公元前四千年期间在苏美尔建立了一个城邦国家，这是已知最早的具有重大历史意义的文明之一。
4. **John Deere:** 约翰·迪尔（1804–1886），出生于美国佛蒙特州拉特兰市。1837年由于在佛蒙特的生意惨淡而来到伊利诺伊州重新开始干事业。足智多谋、勤奋刻苦的他迅速成为当地很受欢迎的铁匠，发明了能在草原湿地上耕地、具备自清功能的犁，并创建了美国约翰·迪尔（强鹿）公司。

5. **Corn Belt:** 玉米产区带，位于美国中西部，玉米是主要的经济作物。玉米产区带主要包括爱荷华州、印第安纳州、伊利诺伊州和俄亥俄州，美国大约50%的玉米产自这四个州。
6. **the Dust Bowl era:** 美国20世纪30年代发生的沙尘暴，又称为the dirty thirties，此时正值大萧条时期，美国西南部大平原地区发生了严重的沙尘暴，造成南部平原到加利福尼亚州的人口大迁移，影响了整个美国社会。这场严重的沙尘暴与长期的气候干旱和土地滥用有关。
7. **Signing Up for No-Till:** 实施免耕法好处多。短语sign up for原意为“(报名）参加（俱乐部、课程等)”，因此标题的含义为“实施免耕法会给农民带来种种好处”。
8. **the USDA's *National Resources Inventory* data:** (USDA=United States Department of Agriculture) 美国农业部《国家资源目录》提供的信息。
9. **Wildlife, too, gains from no-till, because standing crop residues and inevitable harvest losses of grain provide cover and food for upland game birds and other species:** 野生动物也会得益于免耕法，因为未倒伏的农作物残留部分和收割过程中不可避免地遗漏在田地里的谷物，会给生活在高地、可供人们捕猎的鸟类和其他种类的动物提供栖息地和食物。本句中的game作定语，用来修饰bird，意思是“可捕猎的”，尤其指人工繁殖和保护的猎物，如game reserve（猎物繁殖和保护区），game-warden（猎物繁殖与保护区的管理员或经理人)。
10. **Betting the Farm:** 免耕法有风险。该标题中的bet一词原意为“用（钱或物）打赌”，因此“用农场打赌”的含义是采用免耕法有可能导致减产等问题，转译为“有风险”。
11. **university agricultural-extension educators:** 大学农业函授教育工作者。这里extension指大学的附设部分（如夜校、函授班等)，可用作定语，例如：a course offered as part of an extension service为非全日制的学生（如夜校生、函授生）提供的课程。
12. **Because no-till is a radical departure from other farmingpractices, growers making the switch to no-till experience a steep learning curve:** 由于免耕法与其他耕作方式有很大不同，因此改用该法耕作的农民在学习过程中需要积累大量经验。learning curve意为“学习曲线”(又称为经验曲线)，该曲线描述的是人们反复不断地做同一件事情，由于知识的积累、经验的增加，会使后面做这件事情的时间或者成本按固定的比率减少。
13. **And because the crop residue blocks the sun's rays from warming the earth to the same degree as occurs with conventional tillage, soil temperatures are colder in the spring, which can slow seed germination and curtail the early growth of warm-season crops, such as corn, in northern latitudes:** 由于作物残留阻挡了阳光，土壤的温度达不到传统方法耕作时那么高，致使春天土壤温度较低，这种情况会延长种子的发芽时间，在北部地区还会阻碍暖季作物（如玉米）的早期生长。latitude（纬度）用作复数时，意为“地区（尤指从气候或温度上而言)”，因此本句中的northern latitudes是“北方”或“北部地区”的意思。

Exercises

## I. Answer the following questions.

1. What are the advantages and disadvantages of turning the soil before planting?
2. How did our Neolithic forefathers sow their crops?
3. What are the two main technologies that have enabled growers to effectively practice no-till on a commercial scale?
4. What are the advantages brought about by the no-till approach of farming?
5. What is the common perception that makes the change from tillage-based farming to no-till approach difficult?

## II. The following statements are incomplete. Search the missing information in the passage and fill in the blanks.

1. By the late 1970s in the Palouse, soil erosion had removed __________ percent of the topsoil from 10 percent of the cropland, along with another __________ percent of the topsoil from another 60 percent of that land.
2. People have used both no-till and tillage-based methods to produce food from the earth ever since they started growing their own crops around __________ years ago.
3. Adopting no-till farming, farmers use special __________ to penetrate through the residue to the undisturbed soil below, where the seeds can __________.
4. The Dust Bowl era between __________ and __________ exposed the vulnerability of plow-based agriculture, as wind blew away precious __________ from the drought-ravaged southern plains of the US, leaving behind failed crops and farms.
5. Different from tillage-based methods, no-till methods merely create in each planted row a groove just __________ to __________ across into which seeds can be dropped, resulting in minimal overall soil disturbance.
6. After the adoption of conservation tillage, soil erosion from water and wind on US cropland decreased __________ percent between 1982 and 2003.
7. No-till approach requires __________ percent less fuel and __________ percent less labor than tillage-based agriculture, and lowers the total capital and operating costs of machinery required for crop establishment by up to __________.
8. Future no-till farming will need to employ more diverse pest and weed management strategies, including __________, __________ and __________ measures to lessen the threat of pesticide resistance.

## III. Identify the implied meanings of the underlined parts of the following sentences according to the context of the passage, and translate the sentences into Chinese.

1. The hilly Palouse region had been farmed that way for decades. But the tillage was

taking a toll on the Palouse, and its famously fertile soil was eroding at an alarming rate.

2. As with any new system, there are challenges and trade-offs with no-till.
3. Continuing advancements in plow design enabled the explosion of pioneer agriculture during the mid-1800s.
4. Tillage practices were about to undergo profound scrutiny, however.
5. Together with associated increases in soil organic matter, these conditions encourage soils to develop a more stable internal structure, further improving the overall capacity to grow crops and to buffer them against stresses caused by farming operations or environmental hazards.
6. In South America, adoption of no-till farming has been relatively rapid as a result of coordinated efforts by university agricultural-extension educators and local farm communities to develop viable no-till cropping systems tailored to their particular needs.
7. Although farmers accept that agriculture is not a fail-safe profession, they will hesitate to adopt a new farming practice if the risk of failure is greater than in conventional practice.
8. And in the absence of tillage, farmers depend more heavily on herbicides to keep weeds at bay.
9. But no-till is not a cure-all; such a thing does not exist in agriculture.
10. His adoption of no-till has followed a gradual, cautious path that has helped minimize his risk of reduced yields and net returns.

## Translation Techniques (1)

### Translation of Prefixes（前缀的翻译方法）

Many technical words possess prefixes as part of the word formation, which are used to add new meanings to the words so their original meanings are expanded. For example, the prefix of *bio-* means "of living things" or "of life", and therefore when it is added to the word "medicine", the newly formed word is "biomedicine" and translated as生物医学. Another example is "thermochemistry" which is composed of the prefix *thermo-* (meaning "of heat") and "chemistry", and accordingly it is translated as 热化学.

Generally, the translation of technical terms which contain prefixes does not impose particular difficulties because of definite meanings of prefixes, but sometimes you may find that it is necessary to use different words in Chinese translation even if the same prefixes are used. Typical examples are prefixes *non-* and *anti-*. Although each has its basic meaning, when they are translated into Chinese there may be different expressions depending on the contexts in which they are used.

Let's have a look at the following examples.

1. *Non-* (used widely with nouns, adjectives and adverbs meaning "not"): noncombustible materials（不燃物质）；nonmetal（非金属）；non-radiative process（无辐射过程）；non-dissociated molecules（未离解的分子）；non-valent elements（零价元素）；non-shrink（抗缩）；nonskid tyre（防滑轮胎）；nonferrous metals（有色金属）；nonstop run（连续运转）；nonsoap detergent（合成洗涤剂）；non-return valve（止回阀）；noncondensing engine（排气蒸汽机）；non-delay（瞬时动作的）；non-watertight（透水的）

2. *Anti-* (used widely with nouns and adjectives meaning "opposed to", "against" and "preventing"):
antibody（抗体）；anticatalyst（反催化剂）；anticathode（对阴极）；anticlockwise（逆时针方向的）；anticorrosive（防腐蚀的）；antidote（解毒药）；antifebrile（退热药）；antifungal（抗真菌的）；antimatter（反物质）；antipersonnel（杀伤性的）；antiphlogistic（消炎的）；antispasmodic（镇痉药）

## IV. Translate the following sentences into Chinese. Pay attention to how the words with prefixes (the italicized words) should be translated.

1. One of agriculture's main greenhouse gas mitigation strategies is soil carbon sequestration, wherein crops remove carbon dioxide from the atmosphere during *photosynthesis*, and *nonharvested* residues and roots are converted to soil organic matter, which is 58 percent carbon.
2. Some of the changes that follow from no-till can take years or even decades to *unfold*, and farmers need to remain vigilant and adaptable to new, sometimes *unexpected*, situations.
3. If the perception that no-till is riskier than conventional techniques develops in a farming community, banks may not *underwrite* a no-till farmer's loan.
4. Improving the quality of information exchange among farmers, universities, *agribusinesses* and government agencies will no doubt go a long way toward overcoming these obstacles.
5. In the first four to six years, no-till demands the use of extra nitrogen fertilizer to meet the nutritional requirements of some crops, too...because increasing organic matter at the surface *immobilizes* nutrients, including nitrogen.

6. Cost aside, greater reliance on *agrichemicals* may adversely affect *nontarget* species or contaminate air, water and soil.
7. But the current emphasis on corn to produce ethanol in the Midwestern Corn Belt, for instance, is promoting *monoculture*...and will likely make no-till farming more difficult in this region.
8. Experts continue to debate the merits of growing fuel on farmland, but if we decide to proceed with *biofuel* crops, we will need to consider using no-till with crop rotation to produce them sustainably.

## Translation Techniques (2)

### The Use of Parentheses in Technical Translation
### （括号在科技英语翻译中的使用）

One of the problems faced by translators is how to transfer meanings of English sentences fully, faithfully and logically into Chinese sentences. Owing to differences between English and Chinese, it is often necessary to change the structures of sentences, but sometimes certain clauses or phrases seem not to fit into any part of the sentences properly and smoothly. To solve this problem, you may try the use of parentheses in your translation.

Take the following sentence as an example: "In these latter goods, which make up the bulk of goods on world trade, the addition of technological innovation has often been overlooked by Americans, but not by their competitors." You may compare the following three possibilities in translating the sentence.

**译文1** 美国人往往忽视在世界贸易中占大部分的这后一类商品的生产方面实行技术革新，而美国的竞争者却不然。

**译文2** 美国人往往忽视在这后一类商品的生产方面实行技术革新，这类商品在世界贸易中占大部分，而美国的竞争者却不然。

**译文3** 美国人往往忽视在后一类商品（这类商品在世界贸易中占大部分）的生产方面实行技术革新，而他们的竞争者却不然。

The comparison shows that the use of the parenthesis in the third sentence makes it not only concise in structure but also clear in meaning.

However, it should be noted that too many parentheses may destroy the smoothness of sentences, and whether parentheses should be used depends on the specific context. The following examples will show you that parentheses are used more often as supplementary elements or explanations.

(1) The $10 billion a year spent by the American government on defence R&D, *mainly in industrial contracts*, is widely regarded as a major subsidy to civilian research.

美国政府在国防研发方面每年要花费100亿美元（主要是通过签订工业合同的方式），人们普遍认为这是给民用科研事业的一大笔补贴。

(2) The two countries have consistently spent over 2 percent of GDP on R&D, until recently more than any other nation *except Russia*.

这两个国家的研发经费始终占其国民生产总值的2%以上，这个数目直到最近还高于其他任何国家（俄罗斯除外）。

(3) A major scientific advance such as the development of a comprehensive theory and knowledge of elementary particles, *the basic components of all matter and energy*, could profoundly change the way we live tomorrow.

有关基本粒子（即一切物质和能量的基本成分）综合理论与知识这一类的重大科学进步，能够改变人类未来的生活方式。

**V. Discuss the translation technique and the ways of applying the technique to the translation of the following sentences. Complete each of the Chinese translations.**

1. When photons, *which are energized particles of light*, strike certain specially prepared layers of semiconductor materials, their energy knocks electrons loose.（当光子________撞击某些经过特别处理的半导体材料层时，它们的能量使电子离散。）
2. Because the power was far too low—*1 percent of its capacity*—technicians removed all but a few of the control rods and disconnected the automatic-rod-control system.（由于功率太低__________，技术人员取出了几乎所有的控制棒，并断开了控制这些棒的自动系统。）
3. Since in a majority of cases AIDS is fatal—*at least, so far*—the death toll could be enormous.（从大多数病历来看，艾滋病是致命的__________，因此由艾滋病造成的死亡人数可能非常大。）
4. They claim that hypertension has numerous causes, including: too much salt; too much ennin, *a kidney-manufactured hormone that regulates blood pressure*; a combination of both; or a usually curable cause such as kidney or adrenal disease.（他们认为导致高血压的原因很多，其中包括食盐摄入量过多，高血压蛋白原酶__________过多，以及上述两种情况的综合，或者是由于某种通常可以治愈的疾病引起的，如肾病或肾上腺疾病。）
5. The first, *or top*, number is the systolic reading, the pressure of blood against artery walls when the heart beats—the period of highest tension.（第一个数字__________为收缩压，是心脏跳动时血液对动脉管壁的压力，即血管的最大伸张期。）

6. Even the least energetic gamma ray photons have an energy of 100,000 electronvolts, *about 100,000 times the energy of a visible photon*, and from there the energies spread upwards to at least a million million electronvolts.（即使是能量最低的伽马射线光子也有10万电子伏的能量__________，并由此向上一直增加到至少1万亿电子伏。）
7. This mixture of ions and electrons, *or plasma*, has its most significant effect on HF radio waves between altitudes of about 80 kilometers and 300 kilometers, where the concentration of free electrons is greatest.（电离层是由离子和电子构成的混合层__________，在高度大约80公里至300公里范围内对高频无线电波的影响最大，因为在这个范围内自由电子的浓度最高。）
8. An example of another type of anemometer in which the airstream displaces an indicator giving an instantaneous reading of velocity *or flowrate* is the rotameter.（另一类流速计的代表是转子流速计，这种仪器的工作原理是气流使指针偏转，显示瞬间速度值__________。）

## VI. Translate the following sentences into English. Pay attention to how the technique of parentheses should be used. Some of the sentences may not need parentheses.

1. Tillage is a root cause of agricultural land degradation—one of the most serious environmental problems worldwide—which poses a threat to food production and rural livelihoods, particularly in poor and densely populated areas of the developing world.
2. But given our current knowledge of the planet's capacity, we now realize that producing enough food is not enough—it must also be done sustainably.
3. Faulkner's radical proposition became more tenable with the development of herbicides—such as 2,4-D, atrazine and paraquat—after World War II, and research on modern methods of no-till agriculture began in earnest during the 1960s.
4. Yet in 2004, the most recent year for which data are available, farmers were practicing no-till on only 236 million acres worldwide—not even 7 percent of total global cropland.
5. One such technique, crop rotation—in which farmers grow a series of different crops in the same space in sequential seasons—is already helping no-till's war on pests and weeds by helping to break up the weed, pest and disease cycles that arise when one species is continuously grown.

## VII. Translate the following passage into Chinese.

The Dust Bowl era, or the dirty thirties, was a period of severe dust storms causing major ecological and agricultural damage to American and Canadian prairie lands from 1930 to 1936 (in some areas until 1940), caused by severe drought coupled with decades of extensive farming without crop rotation or other techniques to prevent erosion.

The Dust Bowl was a manmade disaster caused by deep plowing of the virgin topsoil of the Great Plains, which killed the natural grasses. Such grasses normally kept the soil in place and moisture trapped, even during periods of drought and high winds. During the drought of the 1930s, with the grasses destroyed, the soil dried, turned to dust, and blew away eastwards and southwards in large dark clouds. At times the clouds blackened the sky, reaching all the way to East Coast cities such as New York and Washington, D.C. Much of the soil ended up deposited in the Atlantic Ocean. The Dust Bowl affected 100,000,000 acres (400,000 $km^2$), centered on the panhandles of Texas, Oklahoma, New Mexico, Colorado, and Kansas.

The storms of the Dust Bowl were given names such as Black Blizzard and Black Roller because visibility was reduced to a few feet (around a meter). The Dust Bowl was an ecological and human disaster. It was caused by misuse of land and years of sustained drought. Millions of acres of farmland became useless, and hundreds of thousands of people were forced to leave their homes. Degradation of dry lands claimed peoples' cultural heritage and livelihoods. Hundreds of thousands of families from the Dust Bowl (often known as "Okies", since so many came from Oklahoma) traveled to California and other states, where they found conditions little better than those they had left. Owning no land, many traveled from farm to farm picking fruit and other crops at starvation wages.

**VIII. Translate the following passage into English.**

## 土壤保持

土壤保持是一整套管理策略，旨在防止土壤从地表流失，或由于过度耕作、盐碱化、酸化或其他化学污染而发生化学方面的改变。这些管理措施涉及多个学科，其中包括农艺学、水文学、土壤学、气象学、微生物学和环境化学。

在实行轮作制、种植覆盖作物和防风林方面做出的决定对于表土层保持稳定性的能力至关重要，这与侵蚀力和由营养流失引起的化学变化有关。简而言之，轮作制是在某一特定农田中进行农作物轮替耕作的通常作法，目的是避免由于单一作物生长反复吸收和沉积某化学成分而造成的营养流失。

覆盖作物的作用是保护土壤不受侵蚀、杂草群落和土壤水分过度蒸发的影响，除此之外，覆盖作物还可以起到关键的化学作用。例如，豆科植物通过耕犁翻入地下而增加土壤中的硝酸盐含量，其他植物则具有代谢土壤污染物和改善pH值的能力。以上技术还可以用于城市景观的建设，特别是在选择控制土壤侵蚀的覆盖植物和抑制杂草这两个方面。

防风林的建立方法是，在农田的上风易受风蚀的方向种植密度足够的成排的树木。选择长绿树种可以起到终年保护的作用，然而，只要树木在表土层裸露的季节仍长有树叶，落叶树也可达到同样的效果。树木、灌木和地被植物都可以种植在农田周围，通过阻挡地表水分流失来防止土壤侵蚀。

*As you conduct research, you will consult different sources of information. A professor may request primary, secondary, or tertiary sources. What does that mean?*

## Section 2 Reading for Academic Purposes

*Scientific literature comprises scientific publications that report original empirical and theoretical work in natural and social sciences. There are different types of literature depending on the readership and purposes for which it is put to use. Primary sources are original materials on which other research is based. They are usually the first formal appearance of results in physical, print or electronic format. They present original thinking, report a discovery, or share new information. Patents and technical reports, for minor research results and engineering and design work (including computer software) can also be considered primary literature. Secondary sources are interpretations and evaluations of primary sources. They are not evidence, but rather commentary on and discussion of evidence including articles in review journals (which provide a synthesis of research articles on a topic to highlight advances and new lines of research), and books for large projects, broad arguments, or compilations of articles. Tertiary sources are works that compile, analyze, and digest secondary sources, and general and specialized encyclopedias are familiar examples of tertiary sources.*

*Before you start reading the following essay about encyclopedia, the words of French philosopher and writer Denis Diderot (1731–1784) will give you an idea of the purpose of an encyclopedia:*

*"Indeed, the purpose of an encyclopedia is to collect knowledge disseminated around the globe; to set forth its general system to the men with whom we live, and transmit it to those who will come after us, so that the work of preceding centuries will not become useless to the centuries to come; and so that our offspring, becoming better instructed, will at the same time become more virtuous and happy, and that we should not die without having rendered a service to the human race."*

### Encyclopedia
### 百科全书

An **encyclopedia** (or encyclopædia) is a comprehensive written **compendium** that contains information on either all branches of knowledge or a particular branch of knowledge. Encyclopedias are divided into articles with one article on each subject covered. The articles on subjects in an encyclopedia are usually accessed

encyclopedia *n.* 百科全书
compendium (*pl.* compendia) *n.* 概要；简编

alphabetically *adv.* 按字母顺序地

**alphabetically** by article name and can be contained in one volume or many volumes, depending on the amount of material included.

The encyclopedia as we recognize it today was developed from the dictionary in the 18th century. A dictionary primarily focuses on words and their definitions, and typically provides limited information, analysis, or background for the word defined. While it may offer a definition, it may leave the reader still lacking in understanding the meaning or significance of a term, and how the term relates to a broader field of knowledge.

well-informed *adj.* 见识广博的；消息灵通的
content experts 熟谙某方面(书籍或课程）内容的专家

To address those needs, an encyclopedia treats each subject in more depth and conveys the most relevant accumulated knowledge on that subject or discipline, given the overall length of the particular work. An encyclopedia also often includes many maps and illustrations, as well as bibliography and statistics. Historically, both encyclopedias and dictionaries have been researched and written by well-educated, **well-informed content experts**.

Four major elements define an encyclopedia: its subject matter, its scope, its method of organization, and its method of production.

Encyclopedia Britannica 大英百科全书
gazetteer *n.* 地名辞典，地理学辞典或索引
Encyclopedia Judaica 犹太民族大全，犹太文物全书

1. Encyclopedias can be general, containing articles on topics in every field (the English-language ***Encyclopedia Britannica*** and *German Brockhaus* are well-known examples). General encyclopedias often contain guides on how to do a variety of things, as well as embedded dictionaries and **gazetteers**. There are also encyclopedias that cover a wide variety of topics but from a particular cultural, ethnic, or national perspective, such as the ***Encyclopedia Judaica.***

2. Works of encyclopedic scope aim to convey the important accumulated knowledge for their subject domain, such as an encyclopedia of medicine, philosophy, or law. Works vary in the breadth of material and the depth of discussion, depending on the target audience. (For example, the *Medical Encyclopedia* produced by A.D.A.M., Inc. for the US National Institutes of Health.)

hierarchical *adj.* 分级的

3. Some systematic method of organization is essential to making an encyclopedia usable as a work of reference. There have historically been two main methods of organizing printed encyclopedias: the alphabetical method (consisting of a number of separate articles, organized in alphabetical order), or organization by **hierarchical**

categories. The former method is today the most common by far, especially for general works. The **fluidity** of electronic media, however, allows new possibilities for multiple methods of organization of the same content. Further, electronic media offer previously unimaginable capabilities for search, indexing and cross reference. The **epigraph** from **Horace** on the title page of the 18th century *Encyclopédie* suggests the importance of the structure of an encyclopedia: "What grace may be added to commonplace matters by the power of order and connection."

fluidity *n.* 流动性
epigraph *n.* 碑文；铭文
Horace (Quintus Horatius Flaccus) 贺拉斯（公元前65至公元前8年），罗马诗人、讽刺家

4. As modern multimedia and the information age have evolved, they have had an ever-increasing effect on the collection, **verification, summation**, and presentation of information of all kinds. Projects such as *Everything2, Encarta, h2g2* and *Wikipedia* are examples of new forms of the encyclopedia as information retrieval becomes simpler.

verification *n.* 证实；核实
summation *n.* 合计；总计

Traditional encyclopedias are written by a number of employed text writers, usually people with an academic degree, and distributed as **proprietary** content. Encyclopedias are essentially **derivative** from what has gone before, and particularly in the 19th century, copyright **infringement** was common among encyclopedia editors. However, modern encyclopedias are not merely larger compendia, including all that came before them. To make space for modern topics, valuable material of historic use regularly had to be discarded, at least before the advent of digital encyclopedias. Moreover, the opinions and world views of a particular generation can be observed in the encyclopedic writing of the time. For these reasons, old encyclopedias are a useful source of historical information, especially for a record of changes in science and technology. As of 2007, old encyclopedias whose copyright has **expired**, such as the 1911 edition of *Britannica*, are also the only free content encyclopedias released in print form.

proprietary *adj.* 私有的；专利的
derivative *adj.* 被引出的；衍生的
infringement *n.* 侵权；侵害
expire *v.* 期满；终止

The concept of a new free encyclopedia began with the **Interpedia** proposal on **Usenet** in 1993, which outlined an Internet-based online encyclopedia to which anyone could submit content and that would be freely accessible. Early projects in this vein included *Everything2* and *Open Site*. In 1999, Richard Stallman proposed the GNUPedia, an online encyclopedia which, similar to the GNU

Interpedia *n.* (=Internet encyclopedia) 网上百科全书，因特网全书
Usenet *n.* 用户网（信息发送系统）

operating system, would be a "generic" resource. The concept was very similar to Interpedia, but more in line with Stallman's GNU philosophy.

copyleft *n.* 著作权（非盈利版权，是著作权的补充）
GNU Free Documentation License (=GNU FDL / GFDL) GNU自由文档许可证

It was not until Nupedia and later Wikipedia that a stable and thriving free encyclopedia project was able to be established on the Internet. The English Wikipedia became the world's largest encyclopedia in 2004 at the 300,000 article stage and by late 2005, Wikipedia had produced over 2 million articles in more than 80 languages with content licensed under the **copyleft GNU Free Documentation License**. As of July 2007, Wikipedia has over 2 million articles in English and well over 8 million combined in over 250 languages.

animation *n.* 动画片
hyperlink *n.* 超链接
yearbook *n.* 年鉴；年刊
supplemental *adj.* 补充的

The encyclopedia's hierarchical structure and evolving nature is particularly adaptable to a disk-based or online computer format, and all major printed multi-subject encyclopedias had moved to this method of delivery by the end of the 20th century. Disk-based (typically DVD-ROM or CD-ROM format) publications have the advantage of being cheaply produced and easily portable. Additionally, they can include media which are impossible to store in the printed format, such as **animations**, audio, and video. **Hyperlinking** between conceptually related items is also a significant benefit. Online encyclopedias, like Wikipedia, offer the additional advantage of being (potentially) dynamic: new information can be presented almost immediately, rather than waiting for the next release of a static format (as with a disk- or paper-based publication). Many printed encyclopedias traditionally published annual supplemental volumes ("**yearbooks**") to update events between editions, as a partial solution to the problem of staying up-to-date, but this of course required the reader to check both the main volumes and the **supplemental** volume(s). Some disk-based encyclopedias offer subscription-based access to online updates, which are then integrated with the content already on the user's hard disk in a manner not possible with a printed encyclopedia.

Information in a printed encyclopedia necessarily needs some form of hierarchical structure. Traditionally, the method employed is to present the information ordered alphabetically by the article title. However with the advent of dynamic electronic formats the need to impose a pre-determined structure is less necessary. Nonetheless, most electronic encyclopedias still offer a range of organizational strategies for the articles, such as by subject area or alphabetically.

CD-ROM and Internet-based encyclopedias also offer greater search abilities than printed versions. While the printed versions rely on indexes to assist in searching for topics, computer accessible versions allow searching through article text for keywords or phrases.

Exercises

**I. Read the passage and decide whether the following statements are true or false. Write T for True and F for False in the brackets.**

1. Encyclopedia is a written summary that includes information only on a particular branch of knowledge. ( )
2. According to the author, a dictionary can be regarded as an encyclopedia because they are of the same origin. ( )
3. An encyclopedia often contains maps, illustrations, references and collection of information shown in numbers. ( )
4. The compilers and researchers of encyclopedias and dictionaries are normally well-educated and well-informed content experts. ( )
5. Thanks to the fluidity of electronic media, it can provide new multiple methods of organization and searching according to indexing and cross reference besides alphabetical method. ( )
6. Even before the birth of Nupedia and Wikipedia, a stable and promising free online encyclopedia project had been established. ( )
7. Since the basic characteristic of an encyclopedia is its hierarchical structure and developing nature, it is adaptable to a disk-based or online computer format. ( )
8. Disk-based publications include media which are impossible to store in the printed format, such as animation, audio and video. ( )
9. People can subscribe supplements online to update their disk-based encyclopedia just like that of a printed encyclopedia. ( )
10. Traditionally, printed encyclopedias offer a range of organizational strategies for the articles such as by subject area or in alphabetical order. ( )

**II. Read the passage again, and complete the following items.**

1. The four major elements that define an encyclopedia: __________
2. The qualifications of the people that can be employed as encyclopedia text writers:

__________

3. The way to save space for modern topics before the arrival of digital encyclopedias: __________
4. The content of the Interpedia proposal on Usenet in 1993: __________
5. The world's largest encyclopedia on the Internet: __________
6. The advantage of disk-based publications: __________
7. The reason why online encyclopedia is dynamic: __________
8. The traditional way in which printed encyclopedia solved the problem of staying up-to-date: __________

## Quotations from Great Scientists

*Try to learn something about everything and everything about something.*

*—Thomas H. Huxley*

要努力做到既是对每个领域都有所了解的博学者，又是精通某个领域的专家。

——托马斯·赫胥黎

# Chapter 1

## Section 1 Reading and Translation

**I.**

1. Biodiversity has attracted so much attention these days mainly because of the number of species that are going extinct, either by natural causes, or because the space-hungry human beings are destroying their habitats.
2. The history of our knowledge of biodiversity is first and foremost a history of species collecting and collections.
3. The author shows his respect for the collectors of species throughout the history and praises them highly.
4. Remarkably little has been written about the craft of scientific collecting: it remains a field unfamiliar to the scientists.
5. The questions include: When and how were those inventories created and made robust? Who organized and paid for collecting expeditions, collected and prepared specimens in the field, compiled lists, built museums and herbaria, and kept vast collections in good physical and conceptual order?

**II.**

1. insects
2. 17,500
3. mid-18th
4. local inhabitants
5. practical skills; firsthand experience
6. accessibility; interest
7. transoceanic
8. biodiversity

**III.**

1. our fellow passengers on the global ark = the animals that coexist with the human beings on the earth
   译文：我们依靠为数很少的人对与我们共存于一个世界的生物进行搜集、描述、命名和分类。
2. romantic = interesting
   译文：安德森最后说："虽然有时其他生物学家把植物标本轻蔑地称为'生物分类学的干草'，其实这些枯燥无味的标本也有它们自己精彩的世界。"

3. a "black box" = an unknown field

   译文：科学收集工作，无论是从技术还是社会史的角度都很少见诸文字，该领域对我们来说一直是个"黑匣子"。

4. a fitful trickle = unsteady and slow development

   译文：自从20世纪中叶以来，人们发现新的脊椎动物的进程断断续续，速度也放缓了（尽管无脊椎动物的名单越来越长）。

5. it would appear that the Creator loves rodents as well as He does beetles = it seems rodents developed as fast as insects did

   译文：与此相反，采用林奈的方法对动物进行描述的人员几乎对啮齿目动物一无所知，在科学研究领域对其也没有充分的了解，这种情况一直延续到生物勘查的时代，这时人们才发现该物种数量的庞大，看起来造物主不仅喜欢甲虫类，而且对啮齿目动物也倍加宠爱。

6. faunas and floras that had seemed closed books were reopened = it had seemed that the research on faunas and floras had reached the end and produced no new results, but later it started again

   译文：某些地区是人们已经探索过的，但尚未进行过深入研究（如美国西部地区以及美国南部的大部分地区），对该地区动植物的研究似乎已经结束，但此后又重新开始，并大规模展开。

7. the flood of new knowledge = a large quantity of new knowledge

   译文：一种普遍的现象（也是无可置疑的事实）是，早期的现代博物学家受到了大量新知识的激励，这些新知识是欧洲贸易和征服在世界范围内扩张的副产品。

8. trade has followed flags = as soon as the new territories are opened, business people will enter, and open new markets there

   译文：哪里开辟了新的领土，哪里就会有贸易展开，对于博物学家和物种采集者来说也是如此。

9. became media events = became events that would draw a lot of attention from the public

   译文：由于人们对脊椎动物的调查已相当充分，因而此后只要发现新的种类就会引起轰动。

10. caught the eye = attracted the attention

    译文：正是早期那些富有戏剧性的探险吸引了人们的眼球，实际上这些探险的目的就是为了吸引诸如投资家、王子、出版商、读者和编年史的编者等的注意。

**IV.**

1. 在脊椎动物中，硬骨鱼类占的比重最大，共18,150种，此外还有63种无颌鱼类和843种软骨鱼类（如七鳃鳗和鲨鱼）。

2. 威尔逊估计在热带雨林中每年可能有17,500个物种（大部分是昆虫）灭绝，而且我们人类使物种灭绝的步伐比过去快了1,000 – 10,000倍。
3. 人们还担心，由于我们的无知，我们也许正在毁灭那些对于我们赖以生存的生态系统结构起着关键作用的物种。
4. 虽然人们一直在为动植物命名，但物种记录学是一个比较新的学科，起源于18世纪中期卡尔·冯·林奈的一项具有深刻意义的发明——（林奈）双名法。
5. 植物学家埃德加·安德森曾做过这样一个实验，他从植物标本盒任意取出一个用马尼拉麻制成的标本夹（后来发现盒中装的是一种西南部地区的草），他想看看采集这些标本的都是什么人。
6. 现代自然史是一门需要付出极大努力的学科，不仅如此，其研究者还要想办法克服野外环境带来的困难。
7. 正是由于这些标本采集者多种多样的经历，其身份不像实验室工作者那样固定。
8. 这些动物包括鸟类、肉食动物、灵长目动物和大型猎物。

**V.**

1. 体积；速度；价格　2. 再简单不过的任务；却成了难事　3. 这是令人
4. 增长　5. 能够过上　6. 现象　7. 查明　8. 其中；代表；代表

**VI.**

1. 物种收集人员往往轻装旅行，并依靠当地居民获得信息和支持，因而对于他们来说，物种收集与调查的过程变成了一种丰富多彩的社会经历。
2. 在野外实地调查的学科中，自然史的调查特别吸引人，原因是系统、科学的物种收集工作内容非常丰富。
3. 我们还可以清楚地看到，19世纪初，由于越洋汽轮得到越来越普遍的使用，以及欧洲帝国的扩张和移民定居（特别是在南半球富饶的热带地区），物种收集与命名工作蓬勃发展。
4. 我们认为从自然史的角度来看，19世纪末、20世纪初并不是一个重要的时期，然而事实上该时期是非常重要的。
5. 在物种考查年代，对物种进行科学的收集大多都是由人数不多的团队完成的（3人到6人不等），他们的目的不是搜集奇特的物品，也不是要向人们炫耀他们富于英雄色彩的探险经历和发现，而是要将成箱的标本带回去。

**VII.**

这段历史的另一个不足是，当时人们研究的主要是脊椎动物和某些植物，而昆虫和无脊椎动物几乎没有涉及。然而这并非是故意造成的缺憾，原因是在我所经历的那段时期，为了物种勘查进行的标本采集活动（特别是由展览馆进行的这类活动）主要集中在脊椎动物方面，因为当时的野外科学考察工作将采集脊椎动物制作展品作为重点。（昆虫、植物和软体动物在展览中不像脊椎动物那样引人注目。）此外，对于综合性物种勘查来说，无脊椎动物的数量之大使人们望而生畏，因此在对脊椎动物展开有组织的勘查后很长时间，无脊椎动物仍然是业余研究人员的研究对象。最近对无脊椎动物也展开了系统的勘查，但其方法与早期有很大不同。

与任何科学（或文化）活动一样，自然史调查也有其特定阶段和生命周期。它产生于特定的环境、文化和科学的土壤，自然发展，最后在自然多样性这一领域中被新的、不同的研究所取代。该研究在美国取得了不同寻常的发展，尽管不限于美国。我写此书的目的是告诉读者自然史调查活动在其鼎盛时期的情况，它之所以能够蓬勃发展的原因，以及在实际中它是怎样操作的。

**VIII.**

**Biodiversity**

Biodiversity is a term that describes the number of different species that live within a particular ecosystem. Scientists have variously estimated that there are from 3 to 30 million extant species, of which 2.5 million have been classified, including 900,000 insects, 41,000 vertebrates, and 250,000 plants; the remainder are invertebrates, fungi, algae, and microorganisms. Although other species remain to be discovered, many are becoming extinct through deforestation, pollution, and human settlement.

Much of this diversity is found in the world's tropical areas, particularly in the forest regions. A habitat in equilibrium has a balance between the number of species present and its resources. Diversity is affected by resources, productivity, and climate. The more pristine a diverse habitat, the better chance it has to survive a change or threat—either natural or human—because that change can be balanced by an adjustment elsewhere in the community; damaged habitats may be destroyed by breaking the food chain with removal of a single species. Thus, biological diversity helps prevent extinction of species and helps preserve the balance of nature. At the 1992 United Nations Conference on Environment and Development, more than 150 nations signed a treaty intended to protect the planet's biological diversity.

## Section 2 Reading for Academic Purposes

I.

1. T 2. F 3. T 4. F 5. F 6. F 7. F 8. T 9. F 10. F

II.

1. Scientific treatises of book length but otherwise variable format prepared by acknowledged experts on specialized topics for the benefit of others who have specialized in, or who wish to obtain a specialist's appreciation of, these topics.
2. The value of monographs lies in the coherence and comprehensiveness of the information and knowledge they contain, which is important to the specialized researchers to whom they are directed and, therefore, to the advancement of science and engineering generally.
3. The authors of monographs should have exceptional breadth and depth of knowledge, and must be able to collect, collate, analyze, integrate, and synthesize all relevant contributions to the archival literature of the scientific and engineering journals and to add original material as required.
4. Monographs generally are written by specialists for the benefit of other specialists. Textbooks are pedagogical works which, even if written on fairly narrow subjects, are designed to serve broader and more junior readerships than specialized research communities.
5. Conference papers commonly take the form of premature announcements of new scientific discoveries. Conference proceedings generally have a short shelf life.
6. The author, title and subtitle, date of publication, dust cover or blurb, content pages, bibliography and index, illustrations, preface and introduction.
7. The number of editions is an indication of the book's success.
8. It gives the reader a rapid overview of the contents and approach. It might also say what the book contains and for whom it is written.

## Chapter 2

### Section 1 Reading and Translation

I.

1. Over the past century, robot has become the familiar figure in popular culture through books such as Isaac Asimov's *I, Robot*, movies such as *Star Wars* and television shows such as *Star Trek*.
2. One reason for this gap is that it has been much harder than expected to enable computers and robots to sense their surrounding environment and to react quickly and accurately.
3. His goal was to help students understand how exciting and important computer

science can be, and he hoped to encourage a few of them to think about careers in technology.

4. One is tackling the problem of concurrency, and the other is the development of the technology DSS (decentralized software services) which helps to simplify the writing of distributed robotic applications.
5. The future robots will play an important role in providing physical assistance; in helping people with disabilities get around and extending the strength and endurance of soldiers, construction workers and medical professionals. Besides, robots will maintain dangerous industrial machines, handle hazardous materials and monitor remote oil pipelines. They will also enable health care workers to diagnose and treat patients who may be thousands of miles away, and they will be a central feature of security systems and search-and-rescue operations.

**II.**

1. the back-office operations
2. from university researchers to entrepreneurs, hobbyists and high school students
3. Czech playwright Karel Bapek
4. the potential of robotics
5. Microsoft BASIC
6. write a traditional, single-threaded program
7. create an affordable, open platform that allows robot developers to readily integrate hardware and software into their designs
8. not look like

**III.**

1. critical mass = a very important or crucial development
   译文：事实上，尽管对这个产业的未来充满热情和希望，但是没有人能明确地说出它什么时间或究竟是否有可能取得关键性的发展。
2. get their hands on = get hold of; have, possess
   译文：电子公司生产了会模仿人、狗或恐龙的机器玩具，玩家也渴望拥有最新版的乐高机器人系统。
3. best minds = most talented people
   译文：与此同时，世界尖端科技人员正试图解决机器人技术中最棘手的难题，诸如视觉识别、远程操控以及机器学习等。
4. intriguing parallel = interesting similarity
   译文：机器人与电脑产业之间还有另一个有趣的相同点：当今国际互联网络的前

身Arpanet，当初也是由DARPA赞助而催生的。

5. from square one = from the beginning
   译文：无论何时，任何人若想开发新型机器人，通常都得从头开始。
6. on our behalf = in place of us human beings
   译文：我相信，许多技术将为新一代的自动装置开启大门，包括分散式运算、声音与视觉识别，以及无线宽频连线等，电脑将能够代替人类完成现实世界里的各项工作。
7. devilishly tricky = extremely difficult, requiring skills
   译文：即使是区分一扇开着的门和窗户这般简单的事，对机器人来说都有可能是非常困难的。
8. quantum leap = great progress or advance
   译文：阅读了唐迪的报告，我似乎更加清楚这一点：机器人产业若想像30年前的个人电脑产业一样获得跳跃式的进步，就必须找到这项缺失的要素。
9. pave the way for = make development or progress easier
   译文：我相信这项进步会促使全新类型的机器人出现。最重要的是，这种机器人是可移动的，具有与台式个人电脑相连的无线设备，让电脑负责处理运算需求高的工作，比如视觉识别与远程操控。
10. work in concert = work together
    译文：由于这些设备能以网络彼此连接，可以想见，机器人将能集体合作，完成海底勘探或作物种植等工作。

**IV.**

1. 看到不同的潮流开始汇合在一起，我可以预见未来自动装置将进入我们日常生活的几乎每一个角落。
2. 列奥纳多·达·芬奇1495年画了一幅机械骑士的草图，图中的骑士可以坐起来，双臂和腿还可以移动，人们认为这是第一张具有人的特征的机器人设计图。
3. 现已证明要赋予机器人那些对人类来说再自然不过的能力（例如根据室内所放物品来调整自己位置的能力，对声音产生反应、理解话语的能力，以及抓握不同大小、不同质地和不同强度的物品的能力）是非常困难的。
4. 阻碍机器人发展的另一个因素是制作硬件所需的高昂成本，例如使机器人能够判断距离的感应器，以及使机器人能够在持物时把握力量和精确性的电动机和伺服器。
5. 我在每一所大学演讲后都有机会对该校计算机科学系最有趣的研究项目直接进行考察。
6. 虽然很多人都为个人电脑的发展做出了重要贡献，但微软的BASIC语言是软硬件创新最关键的推动力之一，掀起了一场个人电脑的革命。

7. 这种设计与宽带无线技术相结合，可以轻易地通过网上浏览器对机器人进行远程监控和调整。
8. 但是，在为老人提供体力帮助，甚至是陪伴老人等方面，机器人将很有可能发挥重要作用。

**V.**

| 1. 出现了 | 2. 染上了 | 3. 出现 | 4. 开发 |
| --- | --- | --- | --- |
| 5. 问世了 | 6. 发育得 | 7. 培育出 | 8. 患 |

**VI.**

1. a) 当然，上面这段可以说是对20世纪70年代中期计算机产业的描述，就在那时我和保罗·艾伦创立了微软公司。
   b) 该造船厂的发言人说，他们希望在两年之内使这艘新潜水艇下水。
2. a) 该产业针对市场需求的产品包括可以实施手术的机器人手臂，可在伊拉克和阿富汗拆卸路边炸弹的监控机器人，以及可以清扫地板的家用机器人。
   b) 已经证明这种新型计算机可以在太空飞行器上执行许多特殊任务。
3. a) 由于DSS允许软件系统的各个部分独立运转，因此如果计算机中的某个元件出现故障，我们可以把它关掉重启，甚至将其更换，而不必重新启动计算机。
   b) 去年由于缺雨造成小麦歉收。
4. a) 我们推出该产品的目的就是要创造一个价格合理的开放平台，让机器人开发人员能够轻而易举地将软、硬件与他们的设计结合起来。
   b) 集成电路是一种很小的电路，由安装在半导体材料上的许多微小元件组成。

**VII.**

家用机器人：由于机器人的价格下降，而与此同时其性能和计算能力却在提高，使其成为人们可以买得起的产品，并具备了足够的自动性，因此越来越多的机器人出现在家庭中做那些简单而人们不想做的事情，如吸尘、清洁地板、割草。虽然机器人在市场上出售已经有几年的时间，2006年是家用机器人销售量大涨的一年。截至2006年，iRobot公司销售了200多万台打扫房间用的机器人，这种机器人自动性较强，通常只需要一个指令就可以开始工作，接下去它们就会按照自己的方式工作。在这方面它们表现出良好的能动性，被看作智能机器人。

遥控机器人：由于危险大，距离远，或无法接近等原因人不能在现场的情况下，便可使用遥控机器人。遥控机器人由操纵员远距离操控，而不是按照一系列事先确定的程序做出动作。这种机器人也许被放置在另一个房间或另一个国家，也许在一个与操纵员完全不同的活动范围。与一般开放式手术相比，腹腔镜外科机器人可使医生在患者体内操作，产生较小的创伤，从而大大缩短了愈合的时间。遥控机器人应用的一个有趣的例子是作家玛格丽特·阿特伍德，最近她开始使用一种机器人笔给远处的人在书上签名。与之不同的是另一种机器人——iRobot ConnectR robot，人们可以利用这种机器人与远方的家人和朋友保持联系。医生与患者进行联系还可用到另一种机器人，这种机器人使医生无处不在，从而能够使更多的患者接受医生的监护。

**VIII.**

**Robots**

There are many variations in definitions of what exactly is a robot. Therefore, it is sometimes difficult to compare numbers of robots in different countries. To try to provide a universally acceptable definition, the International Organization for Standardization gives a definition of robot in ISO 8373, which defines a robot as "an automatically controlled, reprogrammable, multipurpose, manipulator programmable in three or more axes, which may be either fixed in place or mobile for use in industrial automation applications."

In spite of the ISO definition, countries, such as the US and Japan have different definitions of robots. Japan, for example, lists very many robots partly because more machines are counted as robots.

There is no one definition of robot which satisfies everyone, and many people have their own. For example, Joseph Engelberger, a pioneer in industrial robotics, once remarked: "I can't define a robot, but I know one when I see one."

Jobs which require speed, accuracy, reliability or endurance can be performed far better by a robot than a human. Hence many jobs in factories which were traditionally performed by people are now robotized. This has led to cheaper mass-produced goods, including automobiles and electronics. Robots have now been working in factories for more than fifty years, ever since the Unimate robot was installed to automatically remove hot metal from a die casting machine. The number of installed robots has grown faster and faster, and today there are more than one million robots in operation worldwide.

## Section 2 Reading for Academic Purposes

**I.**

1. T  2. T  3. F  4. F  5. T  6. T  7. F  8. T  9. F  10. T

**II.**

1. An academic journal is a peer-reviewed periodical in which scholarship relating

to a particular academic discipline is published.

2. The peer-review process is considered critical to establishing a reliable body of research and knowledge.
3. Review articles, also called "reviews of progress", are checks on the research published in journals.
4. Unlike original research articles, review articles tend to be solicited submissions, sometimes planned years in advance.
5. Natural science journals are categorized and ranked in the *Science Citation Index*, and social science journals in the *Social Science Citation Index*.
6. The number of later articles citing articles already published in the journal, the overall number of citations, how quickly articles are cited, and the average "half-life" of articles.
7. Subsidies by universities or professional organizations and advertising fees by advertisers.
8. The Internet has revolutionized the production of, and access to, academic journals, with their contents available online via services subscribed to by academic libraries or even in a way of open access.

## Chapter 3

### Section 1 Reading and Translation

**I.**

1. Dark energy appears to make up the bulk of the universe, and it may also shape the evolution of the universe's inhabitants—stars, galaxies, galaxy clusters.
2. Dark energy, unlike matter, does not clump in some places more than others; by its very nature, it is spread smoothly everywhere. Whatever the location, it has the same density, equivalent to a handful of hydrogen atoms.
3. Scientists think that the expansion rate of galaxies should be slowing down because the inward gravitational attraction exerted by galaxies on one another should have counteracted the outward expansion.
4. This is because the formation and evolution of these systems is partially driven by interactions and mergers between galaxies, which in turn may have been driven strongly by dark energy.
5. The central piece of evidence is the rough coincidence in timing between the end of most galaxy and cluster formation and the onset of the domination of dark energy.

**II.**

1. implications
2. a small asteroid
3. distance; distant; faster
4. slower; accelerating
5. cobweblike; light-years
6. a billion; merging
7. bent out of shape
8. gas clouds; black holes

**III.**

1. has a stranglehold on = has firm control over
   译文：暗能量不仅加快了宇宙膨胀的速度，同时也决定了星系的形状和间距。
2. in the dance of the planets = in the movement of the planets
   译文：太阳系中全部暗能量加在一起也只相当于一颗小行星的质量，因此在行星的运行中起着微不足道的作用。
3. zoom in = carry out further observation
   译文：如果你进一步观察，一直到达星系和星系团的范围，就会发现情况更加复杂。
4. this tug-of-war = competition
   译文：在宇宙中，这两个激烈竞争的对手谁都不可能占有绝对优势。
5. the big bang = the birth of the universe
   译文：根据模型，暗物质在宇宙大爆炸后的一瞬间产生，形成球形块状物，天文学家称之为晕轮。
6. rubbed shoulders = contact each other
   译文：星系相互接触，相互作用，并常常融合在一起。
7. the downsizing of the galaxy population = the decrease in the number of the galaxies
   译文：也许这种结果就是星系数量减少的原因。
8. Neighbors = other galaxies nearby
   译文：因此，与那些低质量星系相比，它们可能会更早地与邻近星系相碰撞。
9. chokes itself off = interrupts itself; stops automatically
   译文：这样一来，恒星的形成自动停止了，因为这些恒星使它们赖以产生的气体升温，结果阻止了新恒星的产生。
10. a zoo of galaxies = a group of various galaxies
    译文：这种星团环境促成了一系列星系的形成。

**IV.**

1. 暗能量似乎占据了宇宙的大部分空间，不仅如此，如果它的存在能够经受时间的考验，那么有可能需要建立新的物理学理论。
2. 对于星系群来说也是如此，星系群是宇宙中最大的联合体，集中了数千个星系，埋藏在巨大的热气云团中，由重力联系在一起。
3. 自那时以来，星系的分布状态就固定了下来，表明星系的分裂与合并不再经常发生。
4. 星系赖以生存的结构被打破，结果，星系合并的速度逐渐减缓。
5. 这些巨大的星系仍可相互融合，但由于缺乏冷气体，因此很少有新的恒星产生。
6. 要做这方面的分析就需要开发新的理论工具，不过今后几年中我们就会掌握分析的方法。
7. 科学家认为，由政府建造一个耗资数十亿美元的永久空间站会有助于预测地球上的自然灾害，进行医学研究，以及收集太阳能并将其送回地面。
8. 因此，寻找宇宙γ线，也就是寻找宇宙射线的来源。

**V.**

1. 月球的玄武岩表层；很多宽度不足1毫米的陨石
2. 在美国工作的年轻的欧洲科学家弗兰茨·哈尔伯格；实验室里小白鼠的白细胞数目
3. 从月球上带回的岩石样品年代不一；46亿年前的高地岩石
4. 应用软件厂商的工作会更容易；提供给用户的应用软件会更好
5. 一根直径10毫米、长86毫米的不锈钢管
6. 美国射电天文学家弗兰克·德雷克
7. 一只蜜蜂靠着翅膀的颤动以及尾部的摇摆这种方式在蜂巢边上跳舞；它刚刚发现的花蜜或花粉的位置
8. 相当于地球直径，被命名为“大黑斑”的逆旋风

**VI.**

1. 早在美国天文学家爱德温·哈勃的时代，观测者就已经了解到，除了距离我们最近的星系，所有的星系都在以很高的速度离我们而去。
2. 通过观察遥远的星系，以及它们在宇宙时间的长河中融合的方式，我和同事们一直在探索检测这些模式的方法。
3. 目前存在的多数恒星都是在宇宙史的前半叶形成的，对此最早使人信服的发现是在20世纪90年代由多个研究组做出的，其中包括由当时多伦多大学西蒙·黎利领导的研究组，由太空望远镜科学研究所皮尔若·马道领导的研究组，以及加州理工学院的查尔斯·史泰德尔领导的研究组。

4. 正如所观察到的那样，其结果是大型星系形成的时间要早于较小的星系。
5. 星系的融合，黑洞的活动，以及恒星的形成都会随着时间而逐渐消失，而且这些活动很可能以某种方式相互关联。

**VII.**

天文学家根据暗物质对可见物体产生的重力效应得知它的存在，并了解到构成暗物质的粒子的情况。相比之下暗能量却仍然是个不解之谜。“暗能量”一词指的是宇宙大部分空荡荡的空间内一定存在着某种“物质”，从而使宇宙的膨胀加速。从这个意义上来说，暗能量是一种类似于电场或磁场的“场”，后两者都是由电磁能量产生的。但是这种比喻只限于此，因为通过携带电磁能量的粒子——光子，我们可以很容易地观察到电磁能量。

有些天文学家认为暗能量与爱因斯坦的宇宙常数是一致的。爱因斯坦之所以在其广义相对论中引入这一常数，是因为他意识到该理论所预测的是一个不断膨胀的宇宙，这与20世纪初他和其他物理学家所证明的宇宙静止的理论正好相反。该常数使膨胀得到平衡，因而使宇宙处于静止状态。此后由于爱德温·哈勃发现宇宙在不断膨胀，爱因斯坦放弃了他的常数理论。后来该理论被认为与量子理论中的真空能量相一致。

就暗能量来说，宇宙常数就像一座水库，储存着能量。随着宇宙的膨胀，其能量也在增加。应用到超新星现象上，就可以将宇宙中一般物质所产生的作用和暗能量所产生的作用区分开来。然而，这种储存的能量虽有需要，但远远超过我们能够观察到的量，从而导致宇宙的膨胀速度大大加快（其速度之快，甚至阻碍了恒星和星系的形成）。物理学家又提出了一种新的物质——第五元素，它像液体一样充满宇宙，并具有负引力质量。然而，通过哈勃太空望远镜获得的数据对宇宙学参数附加了新的约束条件，至少排除了第五元素的简单模型。

**VIII.**

**Dark Matter**

In physics and cosmology, dark matter is hypothetical matter that does not interact with the electromagnetic force, but whose presence can be inferred from gravitational effects on visible matter. According to present observations of structures larger than galaxies, as well as Big Bang cosmology, dark matter and dark energy account for the vast majority of the mass in the observable universe. The observed phenomena which imply the presence of dark matter include the rotational speeds of galaxies, orbital velocities of galaxies in clusters, gravitational lensing of background objects by galaxy clusters such as the Bullet cluster, and the temperature distribution of hot gas in galaxies and clusters of galaxies. Dark matter also plays a central role in structure formation and galaxy evolution, and has measurable effects on the anisotropy of the cosmic microwave background. All these lines of evidence suggest that galaxies, clusters of galaxies, and the universe as a whole contain far more matter than that which interacts with electromagnetic radiation: the remainder is called the "dark matter component".

The dark matter component has much more mass than the "visible" component of the universe. At present, the density of ordinary baryons and radiation in the universe is estimated to be equivalent to about one hydrogen atom per cubic meter of space. Only about 4% of the total energy density in the universe can be seen directly. About 22% is thought to be composed of dark matter. The remaining 74% is thought to consist of dark energy, an even stranger component, distributed diffusely in space. Some hard-to-detect baryonic matter is believed to make a contribution to dark matter but would constitute only a small portion. Determining the nature of this missing mass is one of the most important problems in modern cosmology and particle physics.

## Section 2 Reading for Academic Purposes

**I.**

1. T 2. T 3. F 4. T 5. F 6. T 7. T 8. F 9. T 10. F

**II.**

1. Summary or Abstract, Introduction, Materials and Methods, Results, Discussion, Acknowledgments.
2. It gives a brief background to the topic, describes concisely the major findings of the paper, and relates these findings to the field of study.
3. It describes first the accepted state of knowledge in a specialized field; then it focuses more specifically on a particular aspect, usually describing a finding or set of findings that led directly to the work described in the paper.
4. Its purpose is to describe the materials used in the experiments and the methods by which the experiments were carried out.
5. In some papers, the results are presented without extensive discussion, which is reserved for the following section. In other papers, results are given, and then they are interpreted, perhaps taken together with other findings not in the paper, so as to give the logical basis for later experiments.
6. The data in the paper are interpreted; the findings of the paper are related to other findings in the field; this serves to show how the findings contribute to knowledge,

or correct the errors of previous work; some of the logical arguments are often provided when it is necessary to clarify why later experiments were carried out.

7. Because the data need extensive discussion to allow the reader to follow the train of logic developed in the course of the research.
8. In *Science*, the abstract is self-contained; in *Nature*, the abstract also serves as a brief introduction to the paper.

# Chapter 4

## Section 1 Reading and Translation

**I.**

1. Some 14 billion years ago the cosmos was hotter and denser than the interior of a star, and since then it has been cooling off and thinning out as the fabric of space expands.
2. Entropy in a closed system never decreases.
3. According to Jacob Bekenstein, black holes fit neatly into the second law. Like the hot objects that the second law was originally formulated to describe, black holes emit radiation and have entropy—a lot of it.
4. The existence of dark energy, a form of energy that exists even in empty space and does not appear to dilute away as the universe expands.
5. Because either entropy was increasing as the prior universe approached the crunch—in which case the arrow of time stretches infinitely far into the past—or the entropy was decreasing, in which case an unnatural low—entropy condition occurred in the middle of the universe's history (at the bounce).

**II.**

1. a strong intuition
2. the arrow of time
3. the microstate; the macrostate
4. gravity
5. inflation
6. dark energy
7. the initial configuration
8. fluctuate into existence

**III.**

1. doing observational cosmology = doing something that can reflect cosmological phenomena which can be observed by human beings
   译文：每当你打开一个鸡蛋，你就是在演示可观察到的宇宙学现象。
2. by the luck of the draw = purely by chance
   译文：高熵状态大大多于低熵状态；该系统几乎所有变化都会把它置于高熵状态，这种现象纯属偶然。
3. emptied out to form voids = poured out to form empty space
   译文：粒子稍多的区域形成了恒星和星系，而粒子稍少的区域则形成了空旷的宇宙空间。
4. relate the entropy of a fluid to the behavior of the molecules that constitute it = make connection between the entropy of a fluid and the behavior of the molecules that make up the fluid
   译文：尽管我们能把液体的熵与组成这种液体的分子的行为联系起来，但我们并不知道是什么形成了太空，因此我们不知道哪些重力微观状态与某种特定的宏观状态相对应。
5. a provocative suggestion = a suggestion which challenges popular ideas concerned with the issue
   译文：20世纪70年代，剑桥大学的史蒂芬·霍金证实了如今在耶路撒冷希伯莱大学工作的雅各布·贝肯斯坦提出的说法，那就是黑洞完全符合热力学第二定律，该说法曾引起许多争议。
6. a blunt feature of the universe that escapes explanation = an obvious feature that cannot be systematically explained
   译文：或者我们把深奥的时间不对称性当作宇宙的一个无法解释的、不争的特征，或者我们必须对空间与时间的运行原理进行更仔细的研究。
7. pushes the puzzle back a step = puts another question to the puzzle instead of solving it
   译文：这样只是使这个谜团又增加了一个疑问：膨胀是怎么发生的呢？
8. to stumble by accident into the microstate = to fall into the microstate by chance
   译文：从直觉上说，这么小的区域不会有很多的微观状态，因此，宇宙碰巧进入了与膨胀相对应的微观状态，这一点也不无可能。
9. passed the buck = refused to take responsibility by letting others decide and act for you
   不管是哪种方式，对于为什么在靠近我们称之为大爆炸的地方熵值很小这个问题，我们又一次踢了皮球。

10. pinch off = break off

译文：如果条件正好合适，这小块能量可以发生膨胀，断裂开去，形成自己独有的宇宙——小宇宙。

**IV.**

1. 不要误会，宇宙学家已经在阐释宇宙的构成和发展方面取得了难以置信的成绩。
2. 我们可以清楚地看到时间的不对称性（即由过去指向未来的时间箭头）在我们的日常生活中所起的作用。
3. 根据这种观点，熵值随着时间而增加这一事实就不会出人意料了。
4. 在重力可以忽略不计的情况下（例如一杯咖啡），粒子均匀的分布具有很高的熵值。
5. 的确，当重力发挥作用时，如果我们希望将一定容积内的熵增加到最大值，那么我们知道会得到什么结果：黑洞。
6. 但是我们所了解的是，在一个不断加速的宇宙中，在可以观察到的容积内，熵达到了一个恒定的数值，该数值与其界面面积成正比。
7. 这种膨胀现象对于解释宇宙的许多基本特点很有说服力。
8. 这种能量使宇宙以极高的速率加速膨胀，随后衰变为物质和辐射线，只留下了少量的暗能量，今天这些暗能量再次与我们的研究发生了联系。

**V.**

| | | | | |
|---|---|---|---|---|
| 1. 也是 | 2. 再 | 3. 而且 | 4. 而 | 5. 再 |
| 6. 进而 | 7. 因为 | 8. 而 | 9. 乘 | 10. 又 |

**VI.**

1. 大约140亿年前的宇宙比恒星的内部更热，密度更大，然而从那以后随着宇宙空间的膨胀，宇宙开始冷却下来，密度也在降低。
2. 换句话说，真正的困难不在于如何解释宇宙的熵值明天会比今天高，而是如何解释熵值为什么昨天更低，而前天又低于昨天。
3. 随着宇宙的膨胀、冷却，重力的吸引使上述差异逐渐扩大。
4. 宇宙将处于平衡状态，今后也不再会发生明显的变化。
5. 熵存在于微观状态，然而乍看起来空旷的宇宙空间是没有微观状态的。

**VII.**

熵是物理学中唯一为时间“选择”特定方向的量，有时称作时间箭头。当人在时间上向“前”走，那么根据热力学第二定律，孤立系统的熵值只能增大或保持不变，而不会减少。因此，从某个角度来说，人们将熵的度量看作一种时钟。

人们通常把热力学箭头与宇宙学的时间箭头联系在一起，这是因为归根结底，热力学箭头涉及宇宙早期的临界状态。根据大爆炸理论，宇宙最初温度很高，能量分布均匀，对于重力起重要作用的系统（如宇宙）来说，这是低熵状态（与之对立的是高熵状态，这种状态下全部物质塌缩为黑洞，上述系统最终会发展成为这种状态）。随着宇宙的演变，温度下降，从而使在未来能够发挥作用的能量相对减少。此外，能量的无序状态加强了（最终形成星系和恒星）。这样宇宙本身就有了明确的热力学时间。然而这并没有回答为什么宇宙在早期处于低熵状态的问题。假使宇宙膨胀由于重力的作用而停下来，并逆向发展，其温度将再次升高，但它的熵值会因为无序状态的不断加强而继续增加，最终形成黑洞。

**VIII.**

## The Big Bang

Since its conception, the theory of the Big Bang has been constantly challenged. These challenges have led those who believe in the theory to search for more concrete evidence which would prove them correct. From the point at which this book leaves off, many have tried to go further and several discoveries have been made that paint a more complete picture of the creation of the universe.

Recently, NASA has made some astounding discoveries which lend themselves to the proof of the Big Bang theory. Most importantly, astronomers using the Astro-2 observatory were able to confirm one of the requirements for the foundation of the universe through the Big Bang. In June, 1995, scientists were able to detect primordial helium, such as deuterium, in the far reaches of the universe. These findings are consistent with an important aspect of the Big Bang theory that a mixture of hydrogen and helium was created at the beginning of the universe.

In addition, the Hubble telescope, named after the father of Big Bang theory, has provided certain clues as to what elements were present following creation. Astronomers using Hubble have found the element boron in extremely ancient stars. They postulate that its presence could be either a remnant of energetic events at the birth of galaxies or it could indicate that boron is even older, dating back to the Big Bang itself. If the latter is true, scientists will be forced once again to modify their theory for the birth of the universe and events immediately afterward because, according to the present theory, such a heavy and complex atom could not have existed.

## Section 2 Reading for Academic Purposes

I.

1. F  2. T  3. T  4. F  5. F  6. T  7. F  8. F  9. F  10. T

II.

1. The Abstract.
2. First, it clarifies whether you in fact know enough background to appreciate the paper. Second, it refreshes your memory about the topic. Third, it helps you integrate the new information into your previous knowledge about the topic.
3. The Introduction.
4. Data not shown, unpublished data, preliminary data.
5. Those who are poor writers; those who do not enjoy writing, and do not take the time or effort to ensure that the prose is clear and logical; those who are so familiar with the material that it is difficult to step back and see it from the point of view of a reader not familiar with the topic.
6. First, the logical connections are often left out. Second, papers are often cluttered with a great deal of jargon. Third, the authors often do not provide a clear road-map through the paper.
7. They are relegated to Figure legends or Materials and Methods or clearly identified as side issues, so as not to distract the reader.
8. The authors refer back to previous papers; these refer in turn to previous papers in a long chain.

# Chapter 5

## Section 1 Reading and Translation

I.

1. Because the multi-touch screen interface could improve collaboration without a mouse or keyboard, and it even could allow the commands of multiple hands from multiple people.
2. Jeff Han expects the multi-touch technology to find a home in graphically intense businesses such as energy trading and medical imaging.
3. The multi-touch technology's toughest hurdle is achieving fine-solution fingertip

sensing. The solution required both hardware and software innovations.

4. Microsoft is shipping Surface table computers to four partners in the leisure, retail and entertainment industries, which it believes are most likely to apply the technology.
5. The great strength of multi-touch is letting multiple people work together on a complex activity.

**II.**

1. the multi-touch screen
2. screens
3. the early 1980s
4. the outside of a glass with water
5. software routines; instructions
6. 2006
7. Surface; consumer
8. Mitsubishi

**III.**

1. hit the street = became very popular
   译文：去年，苹果公司的iPhone手机风靡大街小巷，它向公众推出了一款所谓的多点触控式屏幕。
2. vastly outgrown = greatly exceeded
   译文：在iPhone产品推向市场的时候，全世界实验室的多点式触控屏幕已远远超出了用两个手指发出指令的阶段。
3. at the forefront = at the most forward and important position; playing the leading role
   译文：纽约大学计算机学家兼顾问、纽约Perceptive Pixel公司创立人杰夫·韩处于多点触控技术的前沿。
4. unleashes a world of images = makes a variety of images appear
   译文：韩走到电子墙前，用手指一点，墙上就出现了各种各样的图像。
5. toughest hurdles = greatest difficulties; hardest problems
   译文：但是在2000年前后，韩着手攻克最大的技术难题之一：高清晰指尖触摸感应。
6. an insane amount of wiring = an extremely large amount of wiring
   译文：然而，追踪手指的随意运动需要在屏幕背面安装许多线路，而这又会限制屏幕的功能性。
7. the architecture of a computer's operating system = the design of a computer's

operating system

译文：使用者不是利用键盘就是利用鼠标进行输入，这种思想深深扎根于计算机操作系统的设计理念。

8. tear up a lot of plumbing = remove a lot of parts which formed the original system

   译文：由于目前的操作系统（Windows, Macintosh, Linux）过多地依赖那个由鼠标控制的单一光标，因此韩说："我们不得不除去系统的许多组成部分，以创立一个全新的多点触控制图系统。"

9. rolling out = introducing, disclosing

   译文：软件巨头微软公司现在推出了一款更小的多点式触控计算机——"平面"，并准备把这种类型的硬件品牌命名为"平面计算机"。

10. untether us from the ubiquitous mouse = liberate us from the mouse that is present everywhere

    译文：很快这种多点式触控界面将会帮助我们从无处不在的鼠标使用中解放出来。

**IV.**

1. 通过指尖的移动可以使屏幕上的图像移动，如果把两个手指放在图像的边缘，将它们分开或收拢，就可以使图像扩大或缩小。
2. 微软研究所的主要研究员比尔·巴克斯顿说，研究多点触控人机界面的基础性工作早在20世纪80年代初期就开始了。
3. 最后韩发明了一种由纯丙烯酸树脂制成的长方形薄屏，起到类似光波导向器的作用，其关键部件是一个传导光波的管子。
4. 较难解决的问题是设计出一种常用软件，这种软件要能够追踪手指的移动，并将其转变为使屏幕上的图像发生变化的指令。
5. 在上述工作中，韩发现还有另一种方法达到指压传感的目的，即将薄薄的一层聚合物覆盖在丙烯酸树脂屏的正面，并在其表面制作出微小的隆起。
6. 如果将无线数码相机置于触摸式电脑上，便可以利用微软公司的其他软件将相机中的图像信息上传到电脑而不需要连线。
7. 两位研究人员在试制了多达85个产品原型后，终于研制出台式触摸屏，该触摸屏具有丙烯酸透明台面，台下的地板上安装着投影器。
8. 技术开发人员也许对一家叫做Circle Twelve的新公司生产的DiamondTouch触摸屏感兴趣，该公司位于马萨诸塞州的弗雷明汉。

**V.**

1. 对生物学的历史发展乃至现代科学
2. 在这方面，空气污染最能说明问题。

3. 如今有一种普遍观点；转而注意地区性问题（例如酸雨），最后，在未来的某个时候，我们应当提出全球温室气体问题
4. 必须继续努力开展计算分析，发展计算方法
5. 在这方面存在着巨大的潜力
6. 把计算器放进支票簿内；让只有砖块大小的立体扬声器播放巨大的声响
7. 正在拟定计划，要把一台机器人送入太空，去寻找并修理在轨道上飞行的卫星
8. 生殖产生新的后代，使种族得以延续

**VI.**

1. 只有通过硬件和软件这两方面的革新才能解决这种问题。
2. 这样我们即可以利用这个显示屏输出图像，也可以利用它输入该图像上的触摸信息。
3. 这一工程始于2001年，当时微软公司硬件开发部的斯蒂维・贝思切和微软研究院的安迪・威尔逊着手研发桌面人机互动屏，这种显示屏可以辨认放在上面的某些物体。
4. 三菱研制的桌面触摸屏是经过特殊设计的，使用者可以按照自己设想的应用方式编制软件，科研人员和一般消费者已在使用的这种触摸屏已达数十个之多。
5. 多点触控显示屏具有突出的优越性，可以让多人同时进行某项复杂的活动。

**VII.**

那些彻底改变我们日常生活的技术往往不知不觉进入我们的生活，直到突然有一天我们很难想象如果没有它们（例如即时信息和微波炉）世界将会是什么样子。不久我们还将看到其他重要技术的出现，我们可以预见，这些技术的应用将引起巨大的社会效应。未来技术的一个突出的例子是无线射频识别芯片（简称RFID）。

RFID芯片是比米粒还小的硅片，存有各种信息——从零售价格、烹饪指南，到你的全部病历，一应俱全。还有一种较大的设备，称作RFID“阅读器”，它不必与RFID芯片直接连接就可以从芯片中下载信息，并将其输入任何电子设备（如现金收款机、视频屏幕、家用电器，甚至直接进入互联网）。目前RFID技术已用于桥梁与隧道的自动收款业务，方法是发给司机一个装有RFID芯片的小塑料盒子，这样他就可以通过收费关卡而无需停车。收款亭中的RFID阅读器会感知芯片上的信息，于是路费就会自动由司机的账户中扣除。

RFID技术将首先大规模应用将于零售业。在下周举行的一个重要的工业会议上，预计沃尔玛将敦促它的供应商采用FRID技术，20年前，这个零售业巨头就是这样抢先使用条形码的。实际上有些生产商已经开始应用这一技术，例如吉列公司最近订购了5亿个FRID芯片，准备用来对每一套剃须刀进行跟踪。

VIII.

**Multi-touch Technology**

Multi-touch is a human-computer interaction technique and the hardware devices that implement it, which allow users to compute without conventional input devices such as mouse and keyboard. Multi-touch consists of a touch screen or touchpad, as well as software that recognizes multiple simultaneous touch points, as opposed to the standard touchscreen, which recognizes only one touch point.

Multi-touch technology dates back to 1982, when the University of Toronto developed the first finger pressure multi-touch display. The same year, Bell Labs published what is believed to be the first paper discussing touch-screen based interfaces. Various companies expanded upon the development of this technology at the beginning of the 21st century. In 2007, Apple unveiled the iPhone and Microsoft debuted surface computing. The iPhone in particular has spawned a wave of interest in multi-touch computing, since it permits greatly increased user interaction on a small scale. More robust and customizable multi-touch solutions are beginning to become available. For example, the wall displays and tables developed by Perceptive Pixel can accommodate up to 20 fingers.

The use of multi-touch technology is expected to rapidly become common place. For example, touch screen telephones are expected to increase from 200,000 shipped in 2006 to 21 million in 2012. In fact, developers of the technology have suggested a variety of ways that multi-touch can be used.

## Section 2 Reading for Academic Purposes

I.

1. F　2. F　3. T　4. T　5. F　6. T　7. T　8. F　9. F　10. T

II.

1. Descriptive research, comparative research and analytical research.
2. It goes from general to the specific, eventually framing a question or set of questions.
3. Studying the abstract of the paper.
4. The logical connection between the data and the interpretation is not sound; there might be other interpretations that might be consistent with the data.
5. Developing your ability to evaluate the quality of the evidence.
6. To understand thoroughly the methods used in the experiments.

7. The increasing availability of journals on the web.
8. If the controls are missing, it is harder to be confident that the results really show what is happening in the experiment.

# Chapter 6

## Section 1 Reading and Translation

**I.**

1. The book takes the position that the logical structure, while important, is insufficient even to begin to account for what is really going on in mathematical practice, much less to account for the enormously successful applications of mathematics to almost all fields of human thought.
2. The book puts forward a new vision of what mathematics is all about. It concerns itself not only with the culture of mathematics in its own right, but also with the place of mathematics in the larger scientific and general culture.
3. It is dangerous because it ignores one of the most basic aspects of human nature—in mathematics or elsewhere—our aesthetic dimension, our originality and ability to innovate.
4. Mathematics is a way of approaching the world that is absolutely unique. It cannot be reduced to some other subject that is more elementary in the way that it is claimed that chemistry can be reduced to physics. Other subjects may use mathematics, may even be expressed in a totally mathematical form, but mathematics has no other subject that stands in relation to it in the way that it stands in relation to other subjects.
5. The mathematics that is discussed is there for two reasons: first, because it is intrinsically interesting, and second, because it contributes to the discussion of the nature of mathematics in general.

**II.**

1. the precision of its ideas and its systematic use of the most stringent logical criteria
2. The phenomenon of ambiguity
3. a rigid, deductive structure
4. the “light of reason”

5. ambiguity
6. the nature of the natural world
7. the truth of mathematics; its contingency
8. *this mathematization of the world*

**III.**

1. the ultimate in rational expression = the most fundamental way to express rationality
   译文：其中一种解释把数学看作理性表述的终极形式；事实上，“理性之光”这个说法可以用来指代数学。
2. one dimension of a larger picture = one aspect of a larger area
   译文：该书使人们看到数学的另一面，即逻辑性只是大范围中的一个方面。
3. in the birth of the extraordinary leaps of creativity = in the new and unusual progress of the original ideas and practices
   译文：我要做的就是说明数学如何超越这两种相反的观点：使人们看到数学既有逻辑性又有模糊性这一现象，这种现象既存在于及其复杂的宏大演绎系统的发展中，也存在于非凡的创新成就的产生过程中，这些创新成就不仅已经改变了这个世界，也改变了我们对世界的认识。
4. the true descendents of the Greek mathematicians and philosophers = those who hold the same ideas as the Greek mathematicians and philosophers
   译文：在对于理性的看法上面，我们是希腊数学家和哲学家真正的继承者。
5. plumb its depths = understand it thoroughly and deeply
   译文：要了解数学，并进行深入研究，就必须重新审视自己对模糊性、甚至是对悖论的看法（对“模糊性”的定义见第一章）。
6. the larger implications of what is being said = more matters related to what is being discussed in the book
   译文：虽然我会尽力在数学范围内进行探讨，却不会忽略所讨论话题的进一步的延伸。
7. hard-wired into our brains = strongly planted into our mind
   译文：有人可能会提出，以数学的方式看待这个世界的倾向已经建立在我们的发展结构中，植入到我们的大脑里，也许就暗含在我们基因的DNA结构中。
8. viewed globally = considered in a comprehensive way
   译文：从全面的角度来看，数学的本质向我们揭示的有关人类、人类的思考方式、人类所创造的文化的本质是怎样的？
9. mathematics boasts genius in abundance = it is a privilege that there is a great deal of genius in the field of mathematics research

译文：通过天才的行为，把不可能转化为可能——这就是天才行为的定义，而数学则是以盛产天才而著称的。

10. is the language of much of science = is used to express the ideas of many branches of science

    首先，数学是许多科学领域的通用语言。

**IV.**

1. 模糊现象是本书对数学的描述中最重要的部分。
2. 数千年以来，数学理论告诉人们的似乎是有关自然界本质的深奥的东西，某些只能用数学才能表达的东西。
3. 其他学科也许会用到数学，甚至需要通过纯粹的数学形式进行表达，但是就数学与其他学科的关系来说，其他任何学科都不能替代数学的地位。
4. 如果读者发现某个专题晦涩难懂，我的建议是跳过这个专题，继续读下去。
5. 本书之所以讨论这样一个数学问题出于两个原因：首先是因为它本身很有意思，其次是因为它可以丰富对数学的一般性质的讨论。
6. 直觉告诉人们数学反映真理，这种直觉是正确的，但并不是按照人们通常理解的方式。
7. 我们今天的文明是建筑在数学基础之上的。
8. 医学、政治、社会政策越来越多地使用数学和统计学的语言来表达。

**V.**

1. 找到了新的计算方法
2. 通常用功率曲线图
3. 动植物种类的范围可能发生区域性的变化
4. 数控能够用数学方法将确定的曲线转换为产品
5. 与对照组无区别
6. 其正式名称为M998系列
7. 且对当今科学思维的影响最为深刻的学科
8. 这种积累的主要原因是矿物燃料和氯氟烃的使用，以及大片森林的砍伐和各种农业活动

**VI.**

1. 相反，本书的观点是，逻辑结构固然重要，但是要说明数学研究到底在做什么这个问题，逻辑结构远远不够。
2. 这个较大的范围包括了以前的数学描述忽略掉的一些因素，即跨逻辑的因素（在逻辑以外，但不是非逻辑性的）。

3. 因此，（本书）讨论的内容包括诸如模糊性、矛盾和悖论等方面的问题，出人意料的是，这些问题在数学研究中也起着不可替代的作用。
4. 从本质上看，模糊性与创造性是相互关联的。
5. 人们有一种误解，认为数学真理只能用刻板、演绎性的结构来表达。

**VII.**

卡尔·弗里德里希·高斯把数学叫做“科学之王”。在拉丁文的*Regina Scientiarum*以及德文的*Königin der Wissenschaften*中，“科学” 这个词的意思是（某个领域的）知识。的确，在英语中这也是它的原意，无疑，从这个意义上来说数学是一门科学。后来才将科学的意义限定在“自然”科学这个专业范畴之内。如果我们把科学看作只与自然界有关，那么数学（至少是理论数学）就不是一门科学。阿尔伯特·爱因斯坦曾指出：“如果数学定律与现实有关，那么这些定律就是不确定的；如果这些定律是确定的，那么它们就不会与现实有关。”

许多哲学家认为数学不能通过实验的方法伪证，因此根据卡尔·波普的定义，数学不是一门科学。然而，20世纪30年代数学逻辑领域的重要研究表明，数学不能简单地归于逻辑学，卡尔·波普得出的结论是：“像物理学和生物学理论一样，大多数数学理论都是基于假设-演绎方法：因此理论数学与自然科学（其假设相当于推理）的关系实际上要比以前（甚至是不久前）显得紧密得多。”其他思想家（尤其是伊姆里·拉卡托斯）将证伪论应用于数学本身。

**VIII.**

**The World as Mirror of Mathematics**

The fundamental answer to the critiques of mathematics was given by Galileo in his *Assayer* of 1623, when he wrote that the universe “is written in the language of mathematics.” Galileo was expressing the widely held notion among practitioners that mathematics, far from being devoid of all subject matter as claimed by its critics, had the entire natural world as its object. But while most agreed that mathematics was closely integrated with the physical world, the precise nature of their relationship remained a matter of intense dispute.

One leading approach accepted the classical view of mathematics as a rigorous deductive science of number and magnitude. The universal laws of mathematics, in this view, were the fundamental laws that governed material reality. Thus when one is investigating mathematical and geometrical relationships, one is in fact investigating the basic structure of matter.

The chief promoter of this approach was René Descartes, who viewed mathematics as a fundamental rational law laid down by God for his creation. Once

God, the divine architect, had set in motion his perfectly rational universe, it would henceforth operate forever in accordance with mathematical principles. Mathematical investigations are accordingly studies of the divine plan for the natural world, and the world is the direct expression of abstract mathematical principles.

## Section 2 Reading for Academic Purposes

**I.**

1. F 2. T 3. F 4. F 5. T 6. T 7. F 8. T 9. F 10. F

**II.**

1. via electronic transmission
2. by subscription, or by pay-per-view access
3. HTML and PDF formats
4. ASCII text
5. the publisher or its agent
6. (the content of journal) articles
7. ephemeral, "masthead" information
8. archives, publishers, and scholars

# Chapter 7

## Section 1 Reading and Translation

**I.**

1. Studies of Earth's climate history involve any material that contains a record of past climate, for example, deep-ocean cores collected from sea-going research vessels, ice cores drilled by fossil-fuel machine power in the Antarctic or Greenland ice sheets or by hand or solar power in mountain glaciers; soft-sediment cores hand-driven into lake muds; hand-augered drills that extract thin wood cores from trees; coral samples drilled from tropical reefs.
2. The four great revolutions are: in the 1700s James Hutton concluded that Earth is an ancient planet with a long history of gradually accumulated changes produced mainly by processes working at very slow rates; in 1859 Charles Darwin published his theory of natural selection; in 1912 Alfred Wegener proposed the concept of

continental drift, which eventually led to the theory of plate tectonics; and the major advances in this field began in the late 1900s, continue today, and seem destined to go on for decades.

3. Basalt is the solid rock that is formed by the skeletal remains found in ancient lake sediments sandwiched between two layers of lava, and the basalt layers can be dated by the radioactive decay of key types of minerals enclosed within.
4. These research fields have much in common with the field of crime solving because both of them concentrate on the past.
5. If any hypothesis survives years of challenges and can explain a large amount of old and new evidence, it may become recognized as a theory.

**II.**

1. geologists; geochemists; meteorologists; glaciologists; ecologists; biological oceanographers; climatologists
2. deep-ocean cores; ice cores; soft-sediment cores; wood cores from trees; coral samples
3. several billions of
4. slow natural selection
5. marine geophysical; 100 million
6. the late 1700s; the late 1800s; early 1900s
7. brain size; stone tools for cutting, crushing, and digging; control of fire
8. the teeth and bones

**III.**

1. "gentleman" geologists and geographers = geologists and geographers of high social class
   译文：半个世纪前，做这项工作的只有数十人，其中大多数来自西欧和美国东部，具有一定的社会地位，是受过高等教育或自学成才的地质学家和地理学家。
2. enlarged by new insights = enriched by deeper understanding
   译文：虽然达尔文理论的基本框架得到普遍认可，但仍不断受到挑战，并通过人们更深入的了解而得到充实。
3. intellectual boundaries = limit of the ability to reason and acquire knowledge
   译文：该领域也大大超出了半个世纪前人们对它的认识范围。
4. stumbling upon = finding unexpectedly or by chance
   译文：原来长达100万年的缺口现在从总体上看已不到原来的十分之一，这些缺口是由少数人类学家和他们的助手填补起来的，他们坚持不懈地在非洲勘察露出地表的岩层，偶然会发现已成为化石的生物遗骸。

5. sandwiched = located (between two layers)
   译文：假设这些遗骸是在古代湖泊的沉积物中发现的，它们被夹在两层熔岩之间，这些熔岩很久以前就已变成了坚硬的岩石（即玄武岩）。
6. gave way to = were replaced by
   译文：处于原始人类和类人猿（南猿，即“南方猿人”）之间的动物生活在450万年到250万年前，大约在那个时期，这种动物被可以勉强看作人类但又不完全是人类的动物（人属，意为“人”）所取代。
7. has come into much sharper focus = has become clearer
   译文：通过上述以及其他一些探索，过去1.2万年间人类历史的发展历程清晰了许多。
8. zero in on = focus their attention on
   译文：根据全部证据以及犯罪嫌疑人的一贯手法，侦探逐渐将注意力集中到罪犯本人。
9. are not immune from = cannot avoid
   译文：然而即使再著名的范例也会不断受到检验。
10. pieces the story together = discovers the fact from separate pieces of evidence
   译文：在《神探可伦坡》接下来的一集集中，佛克逐渐将各种线索汇集起来，弄清了真相。

**IV.**

1. 我可以肯定地说，气候学领域使用首字母缩略词的机构数目已超过我开始从事这方面研究时的人数。
2. 如果把焦点再缩小一些，在数十年或数百年间，气候的变化与大规模火山爆发以及太阳强度的微小变化密切相关。
3. 有些科学家把当前气候史研究的结果看作地球科学四大革命中最新的一个，尽管对气候的了解就像此前大多数革命一样是在逐渐加深。
4. 到10万年前，或者把时间再向前推一些，完全意义上的现代人就已经存在了。
5. 今天该领域涵盖了诸如历史生态学和环境地质学等学科，或与这些学科建立了关系，这些学科所研究的人类历史活动远离城市地区，大大拓宽了在地球上的覆盖面。
6. 上文提到的地球轨道的改变是按照每数万年一个周期有规律地发生的。
7. 在气候学领域，对于某观察结果的解释（如曾经有冰层覆盖的地方，如今冰层已不复存在，或在沙漠中可以看到的古老河床）最终往往为那些看似有理的论点提供了一些支持。
8. 1859年，达尔文发表了物竞天择的理论，其中有一部分是基于他此前发表的论著——在那部著作中他指出，在时间的长河中，生物的出现和消失是按照不断变化但又十分明确的顺序进行的，我们有保存完好的化石记录（约6亿年）说明这一点。

**V.**

1. 温度计受热后
2. 因为氯气是一种有毒的气体
3. 如果各种污染物、热气流和气候变化以及紫外线辐射增强等多重压力一起出现的话
4. 如果器壁垂直
5. 自从用电来照明、制冷、灌溉以及通信以来
6. 这些器官在供氧减少时可能受到损害
7. 这使电力的发展成为可能
8. 当采用二进位制时，数字只用0和1这两个数位来表示
9. 如果一只盘子里装满了不同种类的水果
10. 如果软件事先把患者出现异常结果的逻辑可能性告诉他们

**VI.**

1. 这些生物死亡之后，遗体腐烂，经过数百万年在海底形成矿泥层。由于地壳运动的挤压和由此产生的热量的共同作用，这些矿泥层变成了石油。
2. 今天，电子学领域的科研活动与技术开发工作仍在紧张地进行，因此，可以预期电子将在许多方面得以应用。
3. 长期以来研究人员就一直认为遗传可能在某些病历（即使不是所有病历）中起一定的作用，而艾梅西家族提供了一个检验这种假说的理想环境。
4. 压力传感器一般都是精密仪器，如果使用得当，可以得到相当于设计压力千分之一或更高的准确度。
5. 温度降到4℃之前，水越冷越收缩。在4℃以下，如温度继续下降，水就会膨胀。

**VII.**

长期以来，欧盟一直处于努力应对气候变化的前沿，并在制定两项有关该问题的重要协议中发挥了关键作用，这两项协议即1992年的《联合国气候变化框架公约》和1997年通过的《京都议定书》。

自20世纪90年代初，欧盟一直在采取认真措施解决其自身的温室气体排放问题。2000年，欧盟委员会发起了“欧洲气候变化计划”(ECCP)。由于该计划，欧盟出台了一系列新政策和新措施，其中包括具有开拓性的“欧盟排放交易体系”，该体系已成为欧盟在降低成本的同时努力减少温室气体排放的基础，同时通过立法形式应对含氟温

室气体的排放问题。

监测数据和预测表明，2002年欧盟批准《京都议定书》时的15个成员国将达到议定书提出的减少温室气体排放的目标，该目标要求2008–2012年排放量比1990年减少8%。

然而《京都议定书》只是第一步，其目标只设定到2012年。在《联合国气候变化框架公约》指导下，正在展开国际谈判，旨在就2012年后如何采取行动应对气候变化问题达成全球协议。

2007年1月，作为气候变化与能源政策的一个组成部分，欧盟委员会在其通讯《将全球气候变化限制在2摄氏度：展望2020年及此后的前景》中提出了一系列建议和方案，目的是促成一个目标远大的全球协议。欧盟领导人于2007年3月表示支持这项建议，他们承诺欧盟将在2020年把温室气体排放量减至1990年的30%，条件是根据全球协议，其他发达国家也要努力使温室气体排放量减少到同样的水平。

**VIII.**

## Human Influences on Climate Change

Anthropogenic factors are human activities that change the environment and influence climate. In some cases the chain of causality is direct and unambiguous (e.g., by the effects of irrigation on temperature and humidity), while in others it is less clear. Various hypotheses for human-induced climate change have been debated for many years, though it is important to note that the scientific debate has moved on from scepticism, as there is scientific consensus on climate change that human activity is beyond reasonable doubt as the main explanation for the current rapid changes in the world's climate.

### Fossil fuels

Beginning with the industrial revolution in the 1880s and accelerating ever since, the human consumption of fossil fuels has elevated $CO_2$ levels from a concentration of 280 ppm to 387 ppm today. These increasing concentrations are projected to reach a range of 535 to 983 ppm by the end of the 21st century. It is known that carbon dioxide levels are substantially higher now than at any time in the last 750,000 years. Along with rising methane levels, these changes are anticipated to cause an increase of 1.4–5.6 °C between 1990 and 2100.

### Land use

Prior to widespread fossil fuel use, humanity's largest effect on local climate is likely to have resulted from land use. Irrigation, deforestation, and agriculture fundamentally change the environment. For example, they change the amount of water going into and out of a given location. They also may change the local albedo by influencing the ground cover and altering the amount of sunlight that is absorbed. For example, there is evidence to suggest that the climate of Greece and other Mediterranean countries was permanently changed by widespread deforestation between 700 BC and 1 AD (the wood being used for shipbuilding, construction and fuel), with the result that the modern climate in the region is significantly hotter and drier, and the species of trees that were used for shipbuilding in the ancient world can no longer be found in the area.

In modern times, a 2007 study found that the average temperature of California has risen about 2 degrees over the past 50 years, with a much higher increase in urban areas. The change was attributed mostly to extensive human development of the landscape.

## Section 2 Reading for Academic Purposes

**I.**

1. T 2. F 3. F 4. T 5. T 6. F 7. T 8. F 9. F 10. F

**II.**

1. In a 1988 report to the Corporation for National Research Initiatives
2. In 1994
3. An online collection of information must be managed by and made accessible to a community of users
4. Physical collections and digital collections
5. To conform to the goals of open access
6. Institutional, truly free, and corporate repositories
7. Containing primary sources of information rather than the secondary sources found in a library; having their contents organized in groups rather than individual items; having unique contents
8. Distributed searching and searching previously harvested metadata

## Chapter 8

### Section 1 Reading and Translation

**I.**

1. U.S. may end its dependence on foreign oil and cut down greenhouse gas emissions.
2. The great limiting factor of solar power is that it generates little electricity when skies are cloudy and none at night.
3. This is because the existing system of alternating-current power lines is not strong enough to carry power from these centers to consumers everywhere and would lose too much energy over long distances.
4. By 2020, about 84 GW of photovoltaics and concentrated solar power plants would be built. Besides, the DC transmission system would be laid.
5. Because the use of solar power needn't involve energy consumption in extracting and processing fossil fuels and no more energy will be wasted in burning the fossil fuels and controlling their emissions.

**II.**

1. Photovoltaic panels; solar heating troughs
2. Plug-in hybrids
3. 14 percent
4. concentrated solar power
5. Spain
6. finding ways to boost the temperature of heat exchanger fluids
7. $420 billion
8. the suitable Southwest land

**III.**

1. looms large = comes near or nearer; comes into view
   译文：由于中国、印度和其他国家对矿物燃料的需求激增，未来的能源争夺战已近在眼前。
2. off the chart = immeasurably great or large
   译文：太阳能潜力无穷。
3. putting a major brake on = causing a great lessening or alleviation of
   译文：到2050年，美国二氧化碳排放量将比2005年降低62%，从而大幅缓解全球变暖现象。

4. opening the way for = making sth. very possible or feasible

   译文：近几年来，光电池及组件的生产成本大幅降低，为大规模运用开辟了道路。

5. environmentally sensitive areas = areas where the environment may be influenced by sth.

   译文：科罗拉多州戈登市国家再生能源实验室的研究显示，美国西南部的土地供过于求，无需用到环境敏感地区、人口密集区或崎岖地带。

6. came online = came into operation

   译文：内华达州64兆瓦的新发电厂于2007年3月开始运作。

7. dot the landscape = spread over the whole country

   译文：目前煤、油、天然气和核能发电厂遍布全美，通常都比较接近需要电力的地区。

8. is not robust enough = does not have enough ability

   译文：而现有交流电网络的承载力还不足以将这些发电站发出的电传送给各地的客户，而且在长途传输过程中将会损失大量能源。

9. rights-of-way = the strip of land over which facilities such as railroads, highways, or power lines are built

   译文：这个系统将通过州际高速公路沿线现有的公用事业用地进行扩展，这样可以将用地和管理方面的障碍减到最小。

10. self-sustained growth = development without depending on external factors

    译文：最重要的是对主要市场采取的激励措施必须持续到2020年，以使其以后能够依靠自身力量继续发展。

**IV.**

1.

（1）美国需要制定一个大胆的计划，促使人们不再使用矿物燃料。

（2）光电池发电厂不存在环境污染问题（也不耗水），因此可以最大程度地消除人们对环境的忧虑。

（3）当然，太阳能发电的一个突出局限性是阴天时几乎无法发电，而夜晚则根本不能发电。

（4）此外，由于交流电供电系统的发电能力不够，导致加利福尼亚以及其他地区明显缺电。

（5）这套系统依靠水的自然循环而不是用水泵进行冷却。

（6）计算机将来是否能够按照自主意识行事或创造仍是一个悬而未决的问题。

2.

（1）冰的密度比水小，因此能浮在水面上。

（2）在压力的作用下，任何材料或多或少都会变形。

（3）由于主轴和轴承刚性良好，液体轴承能够永久保持其精确度。

（4）直到成本较低的只读存储器（ROS）出现之后，微程序设计才用得多起来。

**V.**

1. 没有这四种力的作用
2. 昆虫改变其正常生理行为，对杀虫剂不再敏感
3. 光一秒钟的行程
4. 实验室的仪表全都
5. 电的用途
6. 符合下述两项标准的软件将受到规定的制约
7. 毒性很强的尼古丁
8. 使它不致对心脏和人体其他系统造成损害

**VI.**

1. 随着组件的改良，屋顶上安装的太阳光电池在成本上将更具竞争性，因为它可以减少白天的电力需求，这一点同样值得关注。
2. 按照规划，这些电厂需要16小时来储存电力，以提供24小时的电能。
3. 善意的科学家、工程师、经济学家和政治家曾提出各种各样的措施，哪怕只对减少矿物燃料的使用和气体排放稍有帮助。
4. 在只以天然气为燃料的情况下，涡轮机仅燃烧40%的天然气，如果采用更好的热回收技术，这一数字有可能降到30%。
5. 为了储存能量，管道通到装满盐水的大型高效保温槽。

**VII.**

潮汐和太阳一样，会永远与我们相伴，至少从11世纪人们就开始利用潮涨潮落过程中蕴含的能量。在过去的几个世纪中，人们利用水推动碾磨机来加工粮食，然而直到1966年才开始大规模利用潮汐发电，那一年世界上第一座潮汐发电厂在法国开始全面运营。

这种发电厂需要大面积的潮汐，潮水穿过狭窄的海湾或河流进出，通过修建堤坝可以将这些海湾或河流截断。涨潮时，海湾或河流中的水位就会升高，在退潮之前堤坝将海湾或河流关闭。退潮时，堤坝外面的水位降到堤内水位以下，此时闸门打开，储存的水向较低位置流动，就会推动涡轮机发电。

潮水有升有降，但是人们需要的是连续不断的电力，为此，利用水的落差带动水泵，将水存储在蓄水池内，以备潮汐周期之间用。

北美第一座潮汐发电厂建在新斯科舍海滨的安纳波利斯盆地，那里最长达8.7米的海潮将水送入安纳波利斯河，当潮水再次涌出进入番迪湾时，水流推动涡流发电机，

每年可发电5,000千瓦。如果该发电厂获得成功，它的经验可以用于在番迪湾海口处建立规模更大的发电厂，那里的海潮长度为15米。然而，由于这种发电厂只能建在有限的几个地域，因此不可能推广。

**VIII.**

**Photovoltaic Cells**

A photovoltaic cell (or solar cell) is a device that converts light into direct current using the photoelectric effect. The first photovoltaic cell was constructed in the 1880s. Although the prototype selenium cells converted less than 1% of incident light into electricity, researchers recognized the importance of this discovery.

The earliest significant application of photovoltaic cells was as a back-up power source to the *Vanguard I* satellite, which allowed it to continue transmitting for over a year after its chemical battery was exhausted. The successful operation of photovoltaic cells on this mission was duplicated in many other satellites, and by the late 1960s PV had become the established source of power for them. Photovoltaics went on to play an essential part in the success of early commercial satellites and remain vital to the telecommunications infrastructure today.

The 1973 oil crisis stimulated a rapid rise in the production of PV during the 1970s and early 1980s. Since the mid-1990s, leadership in the PV sector has shifted from the US to Japan and Germany. Between 1992 and 1994 Japan increased R&D funding, and introduced a subsidy program to encourage the installation of residential PV systems. As a result, PV installations in the country climbed from 31.2 MW in 1994 to 318 MW in 1999, and worldwide production growth increased to 30% in the late 1990s. Since Germany became the leading PV market worldwide, installed PV capacity has risen from 100 MW in 2000 to approximately 4,150 MW. Spain became the third largest PV market in 2004, while France, Italy, South Korea and the US have seen rapid growth recently due to various incentive programs and local market conditions.

## Section 2 Reading for Academic Purposes

**I.**

1. T　2. F　3. T　4. T　5. T　6. F　7. T　8. F　9. T　10. F

**II.**

1. Traditional libraries are limited by storage space; digital libraries have the potential to store much more information, simply because digital information requires very

little physical space to contain it.

2. Providing users with improvements in electronic and audio book technology as well as presenting new forms of communication such as wikis and blogs.
3. Increased accessibility to users.
4. The same resources can be used simultaneously by a number of institutions and patrons.
5. Purpose of use, nature of the work, market impact, and amount or substantiality used.
6. Because they digitize out-of-copyright works and make them freely available to the public, or license content and distribute it on a commercial basis.
7. Influencing the creative process that results from the visual resources.
8. In spite of the current intellectual property rights issues, digitization will receive a growing acceptance.

# Chapter 9

## Section 1 Reading and Translation

**I.**

1. It has two objectives: to inhibit undesirable interactions and to promote desirable ones, so that a stable and productive bond forms between protein partners.
2. Because it is a multisubunit ring, which, in cooperation with HSP70, can disassemble damaged proteins or undesirable protein aggregates or can even cause a fully folded protein to unfold.
3. Under emergency conditions, such as extreme heat or cold, oxygen deprivation, dehydration or starvation, a cell would be struggling just to survive. Critical proteins might be degraded by the harsh environment, even as the cell would try to churn out replacements. In these circumstances, heat shock proteins would mitigate the stress by rescuing essential proteins, dismantling and recycling damaged ones, and generally keeping cell operations running as smoothly as possible.
4. When antigen-presenting cells encounter an HSP-peptide complex, they internalize it through the CD91 doorway and present the HSP-chaperoned peptides to the T cells, which can then multiply and fight off the cancer or pathogen.
5. He noticed that immunization with very high doses of HSPs did not elicit immunity

but rather caused suppression of immune responses.

**II.**

1. the heat shock loci
2. higher organisms
3. correct shape
4. amino acid sequences
5. an immune response
6. tumor-immunizing activity
7. the antigen-chaperoning property
8. in cancer patients; chemotherapy

**III.**

1. these “heat shocked” flies = the fruit flies in an incubator where the temperature was turned up

   译文：那时费鲁乔·里托萨还是个年轻的遗传学者，他研究了这些经过“热休克”处理的果蝇的细胞，注意到它们的染色体上不连续的位置出现了膨胀现象。
2. to act as “chaperones” for other proteins = to act as a guarding escort for other proteins, to prevent undesirable interactions and to promote desirable ones, so that a stable and productive bond forms between protein partners

   译文：要了解这些用途多样的蛋白质如何在治疗上得到应用，可以先看看它们如何用多种方式来执行核心任务，那就是为其他蛋白质做“陪护”。
3. a wide range of “client” proteins = a great variety of proteins that HSPs serve or protect

   译文：与此不同的是，热休克蛋白质与多种“客户”蛋白相联系，因而得以执行一系列令人眼花缭乱的工作。
4. grab them by the “elbows” to help them along = lead a variety of different peptides out of critical consequences

   译文：与那些笼子状陪护不同，大多数的热休克蛋白没有把它们的酶作用物包裹起来，而是抓住它们的“臂肘”带领它们前进。
5. the “antigenic fingerprint” = the typical antigenic characteristics

   译文：因此，与热休克蛋白相联系的肽代表了它们的来源细胞或组织的“抗原指纹”。
6. chemically silenced = chemically suppressed

   译文：热休克蛋白对肽的陪伴作用对肽最终注入主组织相容性复合体I分子非常关键；如果热休克蛋白的化学作用遭到抑制，主组织相容性复合体I分子依然没有肽，因而不能被T细胞所识别。

7. friend and foe antigens = antigens that will do good or harm to the cells
   译文：但是热休克蛋白——肽复合体在T细胞对敌友抗原的识别方面还有另外一个关键作用，即通过它们与被称为抗原呈递细胞的各种免疫细胞发生交互作用。
8. home in on = to trace down
   译文：它们把遇到的一切都呈递给T细胞，而T细胞则予以追踪，并试图摧毁癌细胞或者被感染的细胞。
9. the recurrence-free survival time = a period when the patient can live without the disease reoccurring any more
   译文：在针对肾癌的试验中，这种疫苗把一些试验组里的病人的无复发生存时间延长了一年半有余。
10. to drive their malignancy = to promote their fatal and uncontrollable growth
   译文：与此类似，热休克蛋白被认为会对不断积累的突变起到缓冲作用，这种突变本应使癌细胞渐渐丧失再生能力，但却似乎进一步促进了癌细胞的恶性扩散。

**IV.**

1. 在水中快速冷却热钢可以产生高强度钢材，其强度之高在某些情况下可用于切削较软的钢。
2. 其动作极快，快得能使这些光点汇成电视演播室中的活动场面。
3. 太阳上的温度高得没有任何东西能以固态存在。
4. 可是，由于存在着许多光纤传输模式，因此可以假定噪声已被平均分配，小到可以忽略不计。
5. 反射器彼此靠得很近，因此反射波重叠，形成单一连续的反射信号链。
6. 这些计算相当复杂，一个计算人员要花几年的时间才能计算出来。
7. 对波导管电路元件进行布局设计十分困难，因此精确的计算几乎是不可能的。
8. 该产品所含杂质极少，用普通方法已经不能测定。

**V.**

**Making up the missing words**

groundwater地下水；heating供暖系统；jawless fish无颚纲鱼形动物；letter press活字印刷；liquid hydrogen液态氢；nerve gas神经毒气；periodic table元素周期表；rapid-transit system城市高速铁路交通系统；red giant红巨星；satellite dish圆盘式卫星电视天线；self-fertilization自体受精；solar cell太阳能电池；surrounded-sound headphone环绕立体声耳机；trial and error试错法

**Reversing the word order**

lead oxide氧化铅；light-sensitive感光的；playback回放，重放；rainfall降雨量；

ROM (Read Only Memory) 只读存储器；sand-blast喷沙；sodium chloride氯化钠

**Free translation**

fresh water淡水；general anaesthetic全身麻醉；good conductor良导体；gunpowder 黑色炸药，有烟火药；half-life（放射性）半衰期；hay fever花粉热；high explosive 烈性炸药；identical twin同卵双生；mouthpart（昆虫等的）口器；pacemaker（心脏）起搏器；pig-iron生铁；redwood红杉，红树；seed bank（保存濒于灭绝的植物品种的）种子库；space shuttle航天飞机；starfish海星；voice box喉；warhead弹头；windpipe气管；bush baby丛猴，夜猴

**VI.**

1. ……但是如果缺少三磷酸腺苷，HSP70上的一种像盖子一样的结构就会盖在结合肽上，并将较大的蛋白质链固定住。
2. 在紧急情况下（如极热或极冷、缺氧、脱水或营养缺乏），细胞只能为了生存而挣扎。
3. 存在于肿瘤和正常组织中的gp96分子在氨基酸序列上完全一致，因此由肿瘤产生的gp96并非肿瘤所特有的。
4. 我和一位在西奈山医学院我的实验室工作的博士后将HSP70从肿瘤中分离出来，以检测它是否也能产生肿瘤免疫。
5. 当SP70或HSP90由肿瘤或受到病毒或结核菌感染的细胞中取出时，二者都带有来自只存在于肿瘤抗原、病毒抗原和结核抗原的缩氨酸，几乎无一例外。

**VII.**

美国国家人类基因组研究所主任说，美国科学家正在考虑为肿瘤的异常基因特性建立一套完整的记录，为战胜这种致命疾病做出新的努力。该研究所计划至少为12,500个肿瘤标本建立DNA序列，为50种最常见的肿瘤分别制作250张基因序列图，以将这些图与健康细胞的基因序列图进行对比。人类基因研究所主任弗朗西斯·科林斯在一份说明中指出："随着2003年4月'人类基因工程'的完成，现在辨认每一类肿瘤的基因总谱已经成为可能。"他说："这就是我们提出'人类癌症基因组项目'这样一个大胆的新建议的目的。这种记录将给科学家提供一个强大的新武器，对癌症的各种主要形式进行预防、诊断和治疗。"

虽然该项目"还处于策划阶段"，科林斯领导下的科学家都希望能够与国家肿瘤研究所合作，探索该项目的实施方法。附属于哈佛大学和麻省理工学院的布罗德研究所主任埃里克·兰德告诉《泰晤士报》记者：了解癌细胞的缺陷"就能帮助我们找到肿瘤的致命弱点。"二月，兰德和诺贝尔奖得主、佛瑞德哈钦森癌症研究中心（位于华盛顿州的西雅图）主任利兰·哈特维尔将他们起草的癌症基因组研究项目交给了国家肿

瘤研究所的咨询委员会。

**VIII.**

**Heat Shock Proteins**

Heat shock proteins (HSPs) are a group of proteins that are present in all cells in all life forms. They are induced when a cell undergoes various types of environmental stresses like heat, cold and oxygen deprivation.

Heat shock proteins are also present in cells under perfectly normal conditions. They act like "chaperones", making sure that the cell's proteins are in the right shape and in the right place at the right time. For example, HSPs help new or distorted proteins fold into shape, which is essential for their function. They also shuttle proteins from one compartment to another inside the cell, and transport old proteins to "garbage disposals" inside the cell. Heat shock proteins are also believed to play a role in the presentation of pieces of proteins (or peptides) on the cell surface to help the immune system recognize diseased cells.

For decades it has been known that animals can be 'vaccinated' against cancer. This is how it works: Tumor cells can be weakened and injected like a vaccine into a mouse. Afterwards, if these same tumor cells, at full strength, are injected into the mouse, the mouse will reject the tumor cells and cancer will not develop. However, if a mouse has not been vaccinated in this manner, the tumor cells will "take root" and the mouse will develop cancer.

Although it was clear that animals could be vaccinated against cancer, for a long time it was not known how it worked. Then about 25 years ago, a graduate student named Pramod Srivastava began a series of experiments. He took tumor cells, broke them open, and separated the different parts of the cells into fractions. He then used each of the fractions as "vaccines" to see which fraction protected the mice from developing cancer. After many experiments, he found that the element responsible for protecting the mice was heat shock proteins.

## Section 2 Reading for Academic Purposes

**I.**

1. F 2. T 3. F 4. T 5. T 6. F 7. T 8. T 9. F 10. T

**II.**

1. The systematic scientific investigation of properties and phenomena and their

relationships.

2. To develop and employ mathematical models, theories and hypotheses pertaining to natural phenomena.
3. Because it provides the fundamental connection between empirical observation and mathematical expression of quantitative relationships.
4. Large amounts of qualitative work have usually been prerequisite to fruitful quantification in the physical sciences.
5. Statistics.
6. Correlation does not imply causation.
7. The field of study concerned with the theory and technique for measuring social and psychological attributes and phenomena.
8. To use eclectic approaches.

## Chapter 10

### Section 1 Reading and Translation

I.

1. The scientists wanted to find the Higgs particle. The LHC could help us refine such questions as why gravity is so much weaker than the other forces of nature and that could reveal what the unknown dark matter that fills the universe is, etc. The collider provides the greatest leap in capability of any instrument in the history of particle physics.
2. What distinguishes these two categories is a property akin to electric charge, called color. This name is metaphorical; it has nothing to do with ordinary colors. Quarks have color, and leptons do not.
3. The guiding principle of the Standard Model is that its equations are symmetrical. Just as a sphere looks the same whatever your viewing angle is, the equations remain unchanged even when you change the perspective from which they are defined. Moreover, they remain unchanged even when the perspective shifts by different amounts at different points in space and time.
4. The Standard Model inverts Louis Sullivan's architectural dictum: Louis Sullivan believes "form follows function", while the Standard Model insists on "function follows form".
5. Although the laws of electromagnetism themselves are symmetrical, the behavior of electromagnetism within the superconducting material is not. A photon gains

mass within a superconductor, thereby limiting the intrusion of magnetic fields into the material.

**II.**

1. electromagnetism; the weak interactions
2. quarks; leptons; electromagnetism; interactions
3. consolidation
4. the Standard Model
5. equations
6. superconductivity
7. quantum effects
8. thought experiment

**III.**

1. the marquee attraction = having public appeal, or very attractive to the public
   译文：希格斯粒子作为现代物质理论中唯一尚未被发现的物质，对公众具有强大的吸引力。
2. echo through = exert influence and produce response
   译文：我们不知道它将发现什么，但是我们做出的新发现和遇到的新困惑必将改变粒子物理学的面貌，同时引起邻近学科的共鸣。
3. set us on the road to answering them = provide us with the ways to answer them
   译文：大型强子对撞机将帮助我们对这些问题进行提炼，找到解答问题的途径。
4. fell into place = began to make sense in relationship to each other
   译文：在令人振奋的20世纪七八十年代，各种具有里程碑意义的试验发现参与创新性的讨论，并产生了很多理论想法，标准模型的主要原则才逐渐开始清晰。
5. brewing = growing in force; developing
   译文：也许我们只要回顾一下，就可以看到那场革命处于酝酿当中。
6. immune to = not affected by
   译文：轻子（其中最被人们所熟知的是电子）不受强力的影响。
7. fall under the rubric of = be categorized into a certain type
   译文：另外两种力量——电磁和弱核力——被归类为“弱电力”，它们建立在一个不同的对称性之上。
8. detected the work of such an unseen hand = discovered the effects made by such an unseen factor
   译文：位于欧洲粒子物理研究所的大型正负电子对撞机（该机以前安装在隧道里，目前在大型强子对撞机中使用）发现一只看不见的手中在起作用。
9. the end of the story = the final result of the study

译文：然而，人们对这些研究成果的期待值虽然很高，但这还不是研究的最终结果。

10. lifting the electroweak veil = uncovering and revealing the nature of the electroweak
    译文：揭开弱电力的神秘面纱很有可能更加清楚地揭示这些问题的面貌，改变我们对这些问题的思维方式，并启发未来研究的本质。

**IV.**

1. 标准模型虽然在实验中得到了越来越多的支持，但也有越来越多的现象是它未能包括的，新的理论和思想扩大了我们的眼界，使我们看到了更加丰富、更加宽广的世界。
2. 这一灵感来源于一个看上去毫无关联的现象：超导性，即在低温状态下某些材料可以传输电流，而不会遇到任何阻力。
3. 构成夸克和轻子的要素通过玻色子相互作用，这一现象彻底改变了我们有关物质的概念，并表明这样一种可能性：把极高的能量附加在粒子上，强相互作用、弱相互作用和电磁相互作用将汇于一体。
4. 这种现象是不定原则的另一个结果，表明诸如希格斯玻色子这一类粒子虽然存在的时间太短，无法直接观察到，但已足够在粒子运动过程中留下细微的痕迹。
5. 如果该标准模型一直到1015 GeV（在该水平强相互作用和电弱相互作用似乎合并在一起）仍然有效，那么能量极高的粒子就会对希格斯玻色子产生影响，并给予它相对较高的质量。
6. 假如自然界处于对称状态，粒子和超对称伙伴的质量就会完全相同，而它们对希格斯玻色子的影响将会相互抵消。
7. 如果该粒子与弱力发生相互作用，大爆炸就会按所需要的数目产生粒子，条件是其质量处于大约100 GeV和1 TeV之间。
8. 塞西尔·鲍威尔在高山上将高敏感度胶片感光乳剂暴露于宇宙射线，从而发现了π介子粒子（汤川秀树曾于1935年论述核力时提出过π介子的想法），因而获得1950年的诺贝尔物理学奖。

**V.**

1. 蕴含着深深的沮丧
2. 起着缓冲作用
3. 随时可以动用的；功能
4. 完全凭仪表操纵
5. 早期阶段
6. 比比皆是
7. 空想理论家

8. 突飞猛进

**VI.**

1. 许多粒子物理学家把过去的15年看作是成果巩固时期，与最初几十年的蓬勃发展形成了鲜明的对比。
2. 希格斯玻色子对其他粒子的影响我们看不到，其他粒子对希格斯玻色子的影响也是如此。
3. 上述粒子具有一系列的能量，它们产生的最终影响取决于标准模型何时被更深刻的理论所取代。
4. 大型强子对撞机取代了我们目前使用的对撞机——费米国家加速器实验室的一万亿电子伏加速器对撞机，这样我们的研究设备大大推动了研究目标的实现，并与研究目标形成了很好的配合。
5. 虽然开始时购买机器人的费用很高，但由于在某些工作岗位上机器人干得既快又好，提高了生产率并降低了生产成本，因此不要多久本钱就会赚回来。
6. 很久以来，科学家一直承认“欺骗因素”的存在，许多科学家都倾向于通过使数据转向有利于自己的方向以得到所需结果。
7. 如果不设法消除感染，而患者又不实行严格的口腔卫生，哪怕是第一流的手术也不能阻止病情日益恶化。
8. 虽然这种飞机还只处于设计阶段，有关这种飞机的若干事实已被披露出来。

**VII.**

全球粒子物理学家对于粒子物理学在近期与中期最重要的研究目标看法一致，其中心目标是，通过某些独特的方式发现和了解标准模型以外的物理现象。根据实验，我们有充分理由期待新的物理现象的出现，包括暗物质和中微子质量。然而，最重要的是，也许还存在着人们尚未预料的令人惊叹的现象，因而给予我们了解自然的最佳机会。

发现新的物理现象的努力主要集中在对撞机的实验方面。一个相对来说比较近期的目标是在2008年完成对大型强子对撞机的研究，进而研究希格斯玻色子、超对称粒子以及其他新的物理现象。中期目标则是建造“国际直线对撞机”(International Linear Collider)，该对撞机将对新发现的粒子特性进行更加精确的测量，是对大型强子对撞机的补充。国际线性对撞机的技术决策已于2004年做出，但尚未就装机地点达成一致。

此外，正在进行的还有重要的非对撞机实验，以发现和了解标准模型以外的物理现象。其中一项重要的实验是确定中微子质量，因为这些质量有可能来自与重粒子混合的中微子。除此之外，尽管不使用对撞机也许不可能确定暗物质的准确特性，但宇宙观察提供了许多暗物质的有效制约因素。

## VIII.

### Particle Physics

Particle physics is a branch of physics that studies the elementary constituents of matter and radiation, and the interactions between them. It is also called high energy physics, because many elementary particles do not occur under normal circumstances in nature, but can be created and detected during energetic collisions of other particles, as is done in particle accelerators.

Modern particle physics research is focused on subatomic particles, which include atomic constituents such as electrons, protons, and neutrons, particles produced by radiative and scattering processes, such as photons, neutrinos, and muons, as well as a wide range of exotic particles.

Strictly speaking, the term *particle* is a misnomer because the dynamics of particle physics are governed by quantum mechanics. As such, they exhibit wave-particle duality, displaying particle-like behavior under certain experimental conditions and wave—like behavior in others. Following the convention of particle physicists, "elementary particles" refer to objects such as electrons and photons, with the understanding that these "particles" display wave-like properties as well.

All the particles and their interactions observed to date can almost be described entirely by a quantum field theory called the Standard Model. The Standard Model has 40 species of elementary particles, which can combine to form composite particles, accounting for the hundreds of other species of particles discovered since the 1960s. The Standard Model has been found to agree with almost all the experimental tests conducted to date. However, most particle physicists believe that it is an incomplete description of nature, and that a more fundamental theory awaits discovery. In recent years, measurements of neutrino mass have provided the first experimental deviations from the Standard Model.

## Section 2 Reading for Academic Purposes

**I.**

1. T  2. F  3. T  4. F  5. T  6. F  7. T  8. F  9. T  10. T

**II.**

1. An in-depth understanding of human behavior and the reasons that control human behavior.

2. Largely quantitative research is conclusive and qualitative research is exploratory.
3. The fields included women's studies, disability studies, education studies, social work studies, information studies, management studies, nursing service studies, human service studies, psychology, communication studies, and other.
4. Looking at events, collecting data, analyzing information and reporting the results.
5. Education (in particular educational evaluation).
6. Structures of consciousness as experienced from one's own point of view.
7. Ontology, epistemology, logic and ethics.
8. Generation of theory from data in the process of conducting research.

# Chapter 11

## Section 1 Reading and Translation

**I.**

1. The Constellation program intends to build a space transportation system that can not only bring humans to the moon and back but also resupply the International Space Station (ISS) and eventually place people on the planet Mars.
2. The *Orion* system is mainly made up of the rocket launchers, crew and service modules, upper stages and landing systems.
3. The service module contains power generation and storage systems, radiators that expel surplus heat into space, all necessary fluids and a science equipment bay, the avionics system, and the environmental control and life-support subsystems.
4. After a four-day trip outbound, the crew enters into lunar orbit, having dumped the Earth departure stage along the way. The four astronauts climb into the lander, which consists of the descent stage and the ascent stage, leaving the crew capsule and service module to wait for them in orbit. Then they touch down on the surface of the moon.
5. By means of a skip reentry and a "land landing" mode.

**II.**

1. tried-and-true technical principles and know-how
2. 4; 3
3. a powerful escape rocket

4. a liquid-oxygen-hydrogen-fueled J-2X engine
5. umbrella-shaped solar arrays
6. the crew module; 10
7. lunar sortie missions; lunar outpost missions
8. thermal protection system of a returning vehicle

**III.**

1. the resemblance is only skin-deep = the similarity is only limited to the surface
   译文："猎户座"飞船初期的一般功能有很多与"阿波罗"飞船相同，乘员舱的形状也类似，但其实相同的部分只有外表。
2. widening safety margins = improving the safety level
   译文：工程师在不断提高飞船的安全性。
3. stands poised to blast off = stands quietly, waiting for the take-off
   译文：高达110米的发射塔耸立在美国佛罗里达肯尼迪航天中心的盐碱滩上，两级式战神5号（Ares V）货运火箭静静地矗立着，等待发射升空。
4. anticipating imminent liftoff = expecting the approaching rise of the spacecraft from the launching site
   译文：同一天，4名登月宇航员在肯尼迪发射中心的另一座发射平台上待命，在距离地面98米的"猎户座"乘员舱中等待升空。
5. envelop both to shield them from = cover both modules to protect them from
   译文：保护整流罩包裹着乘员舱和服务舱，避免它们在升空时遭到强烈气流及其他严酷环境的冲击。
6. capping the tall stack = lying on top of the tall rocket
   译文：位于火箭顶端的是逃逸塔，如果火箭发射失败，它能带着宇航员脱离危险。
7. impart a thrust = supply a propelling force
   译文：相反，"猎户座"飞船的"发射中断系统"（LAS）可以在几秒钟内提供相当于自身加上分离的剩员仓总质量15倍的推力。
8. damping out any contact forces = weakening any forces in the touching places
   译文：当力反馈和机电组件"感受"到负载，就能自动捕获航天器的接合环，主动减弱所有的触点压力。
9. is lined with = is covered with
   译文：乘员舱的外部覆盖有热防护系统，不仅能保护居住舱免受返回地球大气层时高温灼烧的影响，还与坚硬的防撞击保护层一起，保护乘员舱的外壁免受微流星体和其他碎片的高速撞击。
10. compounds the challenge = increases the difficulty (in developing the thermal

protection system)

译文：此外，“猎户座”乘员舱比“阿波罗”指挥舱更大，这也增加了热防护系统的制造难度。

**IV.**

1. *English-Chinese Translation*

（1）飞行任务的设计人员估计，发射中断系统与“猎户座”飞船先进的导向和控制系统可以使宇航员安全返回地球，安全率可达所需的99.9%。

（2）由于“猎户座”飞船返回地球大气层时的速度（速度级别达到每秒11公里）比航天飞机从近地轨道下降时的速度快41%，其热负荷要比后者大数倍。

（3）由于“猎户座”飞船热防护罩的面积是“星尘”号飞船热防护罩的40倍，因此需要分块制作，使得这一工作更加复杂。

（4）科学家认为，明年的扩充工作将使这个系统的速度达到每秒处理10亿条信息，为目前速度的4倍。

（5）有些实例证明，与通用处理机相比，使用微程序设计的针对具体问题的处理机，其成本最多只有通用处理机的二十分之一。

2. *Chinese-English Translation*

(1) If the air system pumped in just 3 percent carbon dioxide, the breathing rate of the astronauts would *double*, and they would have trouble hearing.

(2) There are solar kitchens working with paraboloid mirrors which make it possible to increase the concentration of solar rays *1,000 times*.

(3) Nearly 790,000 Americans die each year of heart and artery diseases——almost *twice as many as* die of cancer and about *15 times as many as* are killed in automobile accidents.

(4) Traditional jet engines burn hydrocarbon fuel made from oil, but these engines become too hot at *about three times the speed of sound.*

(5) Wind-tunnel time savings of a *factor of five* have been achieved with the use of this type of integrated test technique in performance evaluation of a cruise missile configuration.

**V.**

1. a. 研究表明灵长目动物与恐龙曾同时在地球上生存
2. b. 研究发现烟雾污染造成更多的人死亡
3. a. 科学家已发现细胞的“质量监控”机制
4. b. 救护车中的虚拟医生

5. b. 电话疗法——抗忧郁的好助手

**VI.**

(1) 火星研究的明天——美国宇航局未来10年的红色星球探测计划
(2) 新的食品结构敦促人们多吃全谷类食品，加强锻炼
(3) 大脑：老年性失聪的原因所在
(4) 科学家把细菌变为计算机
(5) 研究指出全球变暖将持续100年
(6) 超级计算机将助核爆炸一臂之力
(7) 泡泡中的生命——数学家解释昆虫如何水下呼吸
(8) 消化系统专家采用新的冷冻疗法杀灭食道癌
(9) 纳米粒子+光=肿瘤细胞的死亡
(10) 新型材料制成的微波炊具加热快，能耗少
(11) 规划智能机器人的未来
(12) 与体育的敌人——服用兴奋剂和操纵基因——赛跑，科学家领先一步

**VII.**

美国国家航空航天局正在为今后十年的火星探测进行规划，他们听取了各方的意见，包括局外的学术界和局内的研究小组，以及白宫。如果一切都能按计划进行，通过日益强大的宇宙飞行器对这个神秘世界进行探测，火星就会敞开它的大门，将星球表面和内部的秘密展现给我们。目前人们面对的挑战是如何应对预计将要蜂拥而至的新发现，从而将最新研究成果充分运用到宇宙探索中。

行星与月球探测委员会是（美国）国家研究委员会的一个研究部门，该部门最近提出了一个方案，建议利用大批能够将标本送回地球的机器人飞行器，彻底查清火星的历史，揭示该星球上过去或现在是否存在生命。

根据行星与月球探测委员会的研究，2011年将首次对这个红色星球进行机器人采集标本并送回地球的飞行任务，以后这样的飞行任务有可能达到10次，这种大规模自动搜索采集标本并送回地球的工作也许要延续30至40年，甚至长达一个世纪。

实际上，几十年前美国国家航空航天局就计划通过机器人飞行器将火星土壤和岩石标本送回地球，但这个任务十分艰巨，其中最大的障碍是经费问题。这几年航空航天局以及各工业团体对该项目做出了详细的规划，欧洲宇航局的工程师和法国、俄国的独立团体也做了这方面的工作。谁能够掌握在火星上着陆、采集标本并安全地将火星标本送回地球的技术（所有工作都通过机器人硬件来做），谁就能在太空时代的激烈竞争中立于不败。

**VIII.**

## Space Stations

After the geophysical exploration of the moon via the *Apollo* program was completed, the United States continued human space exploration with *Skylab*, an earth-orbiting space station that served as workshop and living quarters for three astronauts. The main capsule was launched by a booster; the crews arrived later in an *Apollo*-type craft that docked to the main capsule. *Skylab* had an operational lifetime of eight months, during which three three-astronaut crews remained in the space station for periods of about one month, two months, and three months. The first crew reached *Skylab* in May, 1972.

*Skylab*'s scientific mission alternated between predominantly solar astrophysical research and study of the earth's natural resources; in addition, the crews evaluated their response to prolonged conditions of weightlessness. The solar observatory contained eight high-resolution telescopes, each designed to study a different part of the spectrum (e.g., visible, ultraviolet, X-ray, or infrared light). Particular attention was given to the study of solar flares. The earth applications, which involved remote sensing of natural resources, relied on visible and infrared light in a technique called multispectral scanning. The data collected helped scientists to forecast crop and timber yields, locate potentially productive land, detect insect infestation, map deserts, measure snow and ice cover, locate mineral deposits, trace marine and wildlife migrations, and detect the dispersal patterns of air and water pollution. In addition, radar studies yielded information about the surface roughness and electrical properties of the sea on a global basis. *Skylab* fell out of orbit in July, 1979; despite diligent efforts, several large pieces of debris fell on land.

After that time the only continuing presence of humans in earth orbit were the Russian space stations, in which cosmonauts worked for periods ranging to more than 14 months. In addition to conducting remote sensing and gathering medical data, cosmonauts used their microgravity environment to produce electronic and medical artifacts impossible to create on earth. In preparation for the International Space Station (ISS)—a cooperative program of the United States, Russia, Japan, Canada, Brazil, and the ESA—astronauts and cosmonauts from Afghanistan, Austria, Britain, Bulgaria, France, Germany, Japan, Kazakhstan, Syria, and the United States worked alongside their Russian counterparts.

## Section 2 Reading for Academic Purposes

**I.**

1. F  2. T  3. T  4. F  5. T  6. F  7. F  8. T  9. T  10. T

**II.**

1. Computer hardware and software, chemistry, the aerospace industry, robotics, finance, consumer electronics, and biotechnology.
2. Information developers, documentation specialist, documentation engineer, or technical content developers.
3. To explain complex ideas to technical and nontechnical audiences.
4. Technical writers, programmers, product or project managers, or other technical staff.
5. A coverpage, a titlepage and copyright page, a preface, a contents page, a guide on how to use at least the main functions of the system, a troubleshooting section, an FAQ (frequently asked questions), where to find further help and contact details and a glossary and an index for larger documents.
6. A description of the location and operation of all controls, a schedule and descriptions of maintenance required, both by the owner and by a mechanic, and specifications such as oil and fuel capacity and part numbers of light bulbs used.
7. A booklet that instructs the player on how to play the game, gives descriptions of the controls and their effects, and shows a general outline of the concepts and goals of the game.
8. The trend in recent years is towards smaller manuals (sometimes just a single instruction sheet or even electronic manuals).

# Chapter 12

## Section 1 Reading and Translation

**I.**

1. The practice of turning the soil before planting buries crop residues, animal manure and troublesome weeds and also aerates and warms the soil. But clearing and disturbing the soil in this way can also leave it vulnerable to erosion by wind and water. Tillage is a root cause of agricultural land degradation—one of the most

serious environmental problems worldwide—which poses a threat to food production and rural livelihoods, particularly in poor and densely populated areas of the developing world.

2. In the transition from hunting and gathering to raising crops, our Neolithic predecessors planted garden plots near their dwellings and foraged for other foods in the wild. Some performed the earliest version of no-till by punching holes in the land with a stick, dropping seeds in each divot and then covering it with soil. Others scratched the ground with a stick, an incipient form of tillage, to place seeds under the surface.
3. The specially designed seeders and chemical herbicides are two of the main technologies that have at last enabled growers to effectively practice no-till on a commercial scale.
4. The no-till approach not only benefits soil protection, but also fosters the diversity of soil flora and fauna by providing soil organism, such as earthworms. Wildlife, too, gains from no-till because standing crop residues and inevitable harvest losses of grain provide cover and food for upland game birds and other species. Furthermore, reducing tillage increases soil carbon sequestration, compared with conventional moldboard plowing; In addition, no-till can offer economic advantages to farmers.
5. The common perception is that no-till approach incurs a greater risk of crop failure or lower net returns than conventional agriculture. Although farmers accept that agriculture is not a fail-safe profession, they will hesitate to adopt a new farming practice if the risk of failure is greater than in conventional practice.

**II.**

1. 100; 25 to 75
2. 10,000
3. seeders; germinate
4. 1931; 1939; topsoil
5. half an inch; three inches
6. 43
7. 50 to 80; 30 to 50; 50
8. biological; physical; chemical

**III.**

1. taking a toll = causing loss or damage

译文：帕洛斯的山区地带几十年来都用那种方式耕种，但是耕作法使帕洛斯遭受了损失，当地闻名遐迩的肥沃土壤正在以惊人的速度流失。

2. trade-offs = balancing of various factors in order to achieve the best combination; compromises

   译文：和其他新系统一样，免耕法也面临着各种挑战和利弊的权衡。

3. the explosion of pioneer agriculture = the rapid and sudden development of new methods in farming

   译文：犁具设计方面的不断改进推动了19世纪中期农业耕作新方法的迅猛发展。

4. undergo profound scrutiny = receive close and thorough examination

   译文：然而，人们即将对犁耕法进行深入的剖析和研究。

5. buffer them against stresses = protect them by reducing stresses

   译文：随着土壤有机质的增加，这些条件促使土壤形成更加稳定的内部结构，进一步提高土壤的全面培育能力，并缓解耕作和环境危害所带来的压力。

6. tailored to their particular needs = adapted to meet their specific demands

   译文：在南美洲，经过大学农业函授教育工作者和当地农业社区的共同努力，根据农民的不同需求，开发了可行的免耕种植体系，因此人们相对较快地接受了免耕法。

7. a fail-safe profession = a profession that is not going to fail

   译文：尽管农民们都知道农业不是万无一失的行业，但是如果新耕作方法比传统方法失败的风险更大的话，他们对采纳新方法会持犹豫态度。

8. keep weeds at bay = keep weeds under control and prevent them from growing

   译文：在不翻耕土壤的情况下，农民们更加依赖除草剂来控制杂草。

9. a cure-all = a remedy that can solve all the problems

   译文：但是免耕法并不是解决问题的万灵药；这样的事物在农业生产中根本就不存在。

10. has followed a gradual, cautious path = has been developed step by step with great care

    译文：他一步步谨慎地实施免耕法，从而把减产和净利润亏损的风险降到最低程度。

**IV.**

1. 从农业的角度来看，减少温室气体的一个主要方法是将碳存储在土壤中，通过这种手段，农作物在光合作用的过程中可以消耗大气中的二氧化碳，与此同时，作物收割后的遗留物和残根等转变为有机物，其中58%是碳。
2. 免耕法带来的某些变化可能几年、甚至是几十年后才能显现出来，因此农民应当对新的、有时是未曾预料到的情况保持警觉，提高适应能力。

3. 如果认为免耕法的风险大于传统技术的观点在从事农业的人们当中越来越普遍，银行有可能拒绝给采用免耕法的农民贷款。
4. 无疑，改善农民、大学、农业综合企业和政府部门之间的信息交流将对克服这些困难起到很大作用。
5. 在最初的4－6年间，由于土壤表面越积越多的有机物阻碍营养物（包括氮）的扩散，……因此如采用免耕法，就需要增加氮肥的使用以满足某些农作物对营养的需求。
6. 除了成本问题以外，更多地依赖农用化学品还会对目标以外的物种造成负面影响，或污染空气、水和土壤。
7. 然而以中西部的玉米产区带为例，为了生产乙醇，人们当前只注重玉米的种植，这种作法实际上是在鼓励单一栽培，……而且有可能使在该地区实施免耕法更加困难。
8. 专家们还在对农田里种植燃料作物的利弊争论不休，但是，如果一旦决定继续种植生物燃料作物，那么就要考虑采用免耕法配以轮作的方式，以进行可持续生产。

**V.**

1. 即带有能量的光粒子
2. 仅为原容量的1%
3. 至少目前如此
4. 由肾脏产生的一种调节血压的激素
5. 又称质数
6. 大约是可见光子能量的10万倍
7. 亦称等离子区
8. 即流量值

**VI.**

1. 翻耕是导致农业用地退化（全球性最严重的环境问题之一）的根本原因，这种情况威胁到粮食生产和农民的生计，特别是在发展中国家的贫困和人口稠密地区。
2. 但是由于我们现在对地球的生产力有了了解，我们已经意识到仅生产充足的粮食是不够的，粮食生产还必须具有可持续性。
3. 第二次世界大战后，随着除草剂（如2,4-D、莠去津、百草枯）的发展，福克纳激进的主张显得越来越有说服力，而对现代免耕法的认真研究开始于20世纪60年代。
4. 然而，2004年（这是有数据可查的最近的年份）全世界采用免耕法的面积仅为2.36亿英亩，还不到全球农业用地的7%。

5. “轮作制”就是这样一种技术，即按季节顺序在同一空间种植各种不同的农作物，这种技术通过切断杂草、害虫和疾病的循环周期（这种循环周期发生在某单一品种连续不断生长的情况下），帮助人们加强免耕法对害虫和杂草的控制。

**VII.**

“沙暴”（亦称“肮脏的三十年代”）指的是1930—1936年（某些地区延续到1940年）由于严重的干旱，加之几十年大规模的耕作以及未采用轮作制或其他防止土壤流失的措施，而发生的严重的沙尘暴。这一时期的沙尘暴使美国和加拿大草原地区的生态和农业遭受了严重的损失。

“沙暴”是一场人为的灾难，它发生在（美国）大平原，起因是在未开垦的土地上实行表土层的深耕作业，从而破坏了天然草场。一般情况下，草可以固定土壤、保持水分，即使在干旱和大风季节也是这样。在20世纪30年代的干旱时期，由于失去了草的覆盖，土壤变干，进而变成粉尘，像滚滚黑云一样腾空而起，向东、南方向移动。有时沙尘将天空遮住，一直到达位于东海岸的城市，如纽约和华盛顿。大部分尘土最终落入大西洋。“沙暴”使1亿英亩（40万平方公里）受灾，主要集中在德克萨斯、俄克拉荷马、新墨西哥、科罗拉多和堪萨斯几个州。

“沙暴”被人们称作“黑色风暴”和“黑浪”，因为当时的能见度降低到几英尺(约1米)。这是一场由滥用土地和连年干旱造成的生态与人类的灾难，数百万英亩的农田一片荒芜，几十万人被迫离开家园，干旱土地的退化破坏了人们的文化遗产和生计。来自“沙暴”受灾地区的数十万家庭（因为其中很多人来自俄克拉荷马州，因此常被称作“俄州佬”）迁移到加利福尼亚和其他州，可是他们发现那里的生活状况也不过如此。许多人因为没有土地，只得在不同的农场采摘水果或其他作物，得到的只是能够勉强维持生命的工资。

**VIII.**

**Soil Conservation**

Soil conservation is a set of management strategies for prevention of soil being eroded from the earth's surface or becoming chemically altered by overuse, salinization, acidification, or other chemical soil contamination. Many scientific disciplines are involved in these pursuits, including agronomy, hydrology, soil science, meteorology, microbiology, and environmental chemistry.

Decisions regarding appropriate crop rotation, cover crops, and planted windbreaks are central to the ability of surface soils to retain their integrity, both with respect to erosive forces and chemical change from nutrient depletion. Crop rotation is simply the conventional alternation of crops on a given field, so that nutrient depletion is avoided from repetitive chemical uptake/deposition of single crop growth.

Cover crops serve the function of protecting the soil from erosion, weed establishment or excess evapotranspiration; however, they may also serve vital soil chemistry functions. For example, legumes can be ploughed under to augment soil nitrates, and other plants have the ability to metabolize soil contaminants or alter adverse pH. Some of these same precepts are applicable to urban landscaping, especially with respect to ground-cover selection for erosion control and weed suppression.

Windbreaks are created by planting sufficiently dense rows or stands of trees at the windward exposure of an agricultural field subject to wind erosion. Evergreen species are preferred to achieve year-round protection; however, as long as foliage is present in the seasons of bare soil surfaces, the effect of deciduous trees may also be adequate. Trees, shrubs and groundcovers are also effective perimeter treatment for soil erosion prevention, by insuring any surface flows are impeded.

## Section 2 Reading for Academic Purposes

**I.**

1. F  2. F  3. T  4. T  5. T  6. F  7. T  8. T  9. F  10. F

**II.**

1. Its subject matter, its scope, its method of organization, and its method of production.
2. Usually people with an academic degree and people who are well-educated and well-informed.
3. To make space for modern topics, valuable material of historic use regularly had to be discarded.
4. The Interpedia proposal on Usenet in 1993 outlined an Internet-based online encyclopedia to which anyone could submit content and that could be freely accessible.
5. The English Wikipedia.
6. Disk-based publications have the advantage of being cheaply produced and easily portable.
7. Because new information can be presented almost immediately rather than waiting for the next release of a static format.

8. Many printed encyclopedias traditionally published annual supplemental volumes ("yearbooks") to update events between editions, as a partial solution to the problem of staying up-to-date.